THE LIFE REGION

Our planet urgently needs strategies for assuring sustainable development. However, it is not sufficient to limit radical thinking to environmental protection and resource management. It is necessary to focus at the same time on the socio-cultural symptoms of crisis which affect modern society. What is required now are holistic models of development which, within the same context, pay attention to the social, cultural and environmental issues of the contemporary global crisis.

The Life Region launches a strategy for sustainable development, setting out from a socio-ecological position and developing a model for a socially and culturally supportive community, or 'Life Region'. Special emphasis is placed on the situation of the provincial and peripheral regions of Europe and the world, and the introduction of self-reliant civic strategies in national and international politics. Three aspects of regional policy are looked at in detail: spatial planning, cultural policy and social development, and regional economics and regional science are also touched upon.

Presenting the ideas of a number of distinguished European researchers and debaters, *The Life Region* will be a major contribution to the emerging debate on social and cultural strategies for sustainable global development.

Per Råberg is Director of the Department of Cultural Ecology and Associate Professor in the Faculty of Arts at Umeå University.

Routledge Studies in Development and Society

1 **Searching for Security**
Women's Responses to Economic Tranformations
Edited by Isa Baud and Ines Smyth

2 **The Life Region**
Edited by Per Råberg

THE LIFE REGION

The Social and Cultural Ecology of Sustainable Development

Edited by Per Råberg

ROUTLEDGE

London and New York

First published 1997
by Routledge
11 New Fetter Lane, London EC4P 4EE

Simultaneously published in the USA and Canada
by Routledge
29 West 35th Street, New York, NY 10001

The book is illustrated by three artists from the County of Västernorrland:
Kirstie Ekelund, Stöde
"*Thoughts of the Mountain*", 6 pencil drawings (also cover)
Hans Marklund, Härnösand
"*Window Vision*", a series of 6 photographs
Margareta Klingberg, Junsele
"*Angelica Silvestris*", a series of 5 photographs

Ilustrations on pages 141, 143, 145 by
Abelardo Gonzalez and Maria Udriot

Printed and bound in Great Britain by
TJ International Ltd, Padstow, Cornwall

British Library Cataloguing in Publication Data
A catalogue record for this book is available from the British Library

Library of Congress Cataloging in Publication Data
The life region : the social and cultural ecology of sustainable development / edited by Per Råberg.
p. cm.
Includes bibliographical references and index.
1. Sustainable development. 2. Sustainable development—Sweden. 3. Social ecology. 4. Social ecology—Sweden. 5. Social planning. 6. Social planning—Sweden. I. Råberg, Per G., 1934– .
HC79.E5L535 1996
363.7—dc20 96–32983
CIP

ISBN 0–415–15905–9

Contents

Preface

The globalization of human activities is a pervasive theme in the changes which are currently impacting on society. It is a process whose effects have influenced development for several centuries but which today has assumed a particularly distinct character. Day by day the economic and technological power net is being spun more tightly around the planet and the life world of its inhabitants. A global communication structure is taking shape based on the new advances in information technology. Political representatives of the nation states, who had long since felt left behind, are carving out international careers and establishing supranational organizations: NATO, OECD, NAFTA and the EU, are typical examples deriving from different parts of the contemporary world.

Globalization appears as an evolutionary force which carries everyone and everything along with it. The development of such an epic power does not come about without convulsions, since in most cases it involves a painful departure from century-old traditions. It is characterized by a certain ruthlessness and many are the innocent victims of its destructive behaviour. By virtue of its brutal force, the process may even pose a threat to existential and environmental values which are of paramount importance to the survival of mankind. This is, in actual fact, what is now happening everywhere and at an increasing tempo. The ecological crisis of the environment has, up till now, been the most debated example. This has inspired a discussion of strategies for a sustainable global future. Several world conferences have been held on the subject over the last decades.

Globalization, however, has given rise to another crisis which so far has not attracted comparable attention. We are subjected also to a social crisis, the effects of which are beginning to strike at the core of our personal and existential awareness. This crisis is a direct consequence of the enormous increase in scale which has lately occurred in all community structures, and the ever more complex hierarchies which have evolved in our society. The human consequence of this is personal alienation and social disintegration throughout society and, especially, at its basic civil level. All over the world the fragile webs of close social relations are now at different stages of disintegration. Individuals and groups are separated from each other, family ties are irrevocably broken, generations and ethnic groupings turn against each other in an endless destructive spiral.

This social disintegration also encompasses a collapse of socity's system of values. Fundamental ideas such as freedom, human rights and democracy, lose their significance and assume an extraneous, rhetorical content. From the frustration which derives from the loss of social identity human desperation grows, and out of the same predicament grow implacable hatred and aggression. Mindless violence permeates modern society, and its manifestations are in no way limited to the street or the private sector. They infiltrate both the political structures and the agencies that exercise power. They give rise to regional strife and create relentless ethnic conflicts.

How is society to deal with the contemporary crisis confronting our civilization? As hinted at already, there is a debate concerning sustainable global development. To far-sighted observers, however, it is quite clear that the social crisis will also have to be addressed if the worth-while objectives are to be attained. *Sustainable social development* in fact is an essential prerequisite for retaining balance within the global ecological environment.

In recent years a number of UN agencies within the social and humanitarian field have argued - with growing commitment - in favour of a broader social approach. The World Health Organization (WHO) is among those which are to the fore in this debate. In Agenda 21 from the Rio Conference too there is a strong emphasis on the ideological connections which exist between ecological and social issues. The World Summit for Social Development in Copenhagen, especially through the UNDP and UNRISD contributions (March 1995), at the moment of writing is the latest event in a series of remarkable UN social initiatives.

This book is a contribution to this new and more wide-ranging debate concerning strategies for a sustainable global future. The book is the result of a regional R&D project which has been conducted by the University Group for Humanist Futurology at Umeå in Sweden. The book presents - in the form of a model - a proposal for a socially and ecologically viable society. The model sketch is presented under the title *the ecological life-region.* It is distinct from the current geopolitical concepts by virtue of the aspirations vested in the provincial regions as the fertile soil for a progressive development. Further, it is based on an epistemology which inludes humanitarian and existential needs within the scientific study - we describe this as a *socio-ecological* approach. It is our sincere hope that the socio-ecological prototypes will be able to serve as an inspiration for a radicalized social policy discussion in municipalities and regions in many parts of the world community.

In the actual book the model of the life-region is thematized in a number of essays by researchers and debaters from the University Group's international network. The participants have been chosen as belonging to a critical humanist tradition in social debate. For several decades this tradition has appeared under various banners such as "grass-roots commitment", "participatory planning", "self-reliance", "ecological public health", "endogenous development", "bottom-up politics", "communitarianism", "civil society", "regionalism" etc.

In addition to the expert papers, the book includes a number of reports from regional study groups. The material has been sifted over the course of several years of project work, in collaboration with authorities and interested parties in the County of Västernorrland (located in northern Sweden). The contributions were first presented at three R&D conferences held by the University Group in the region in 1992 and 1993.

On behalf of Umeå University I would like to thank all those who have helped to make this book possible. These consist primarily of the specially invited social scientists and members of the regional study groups, whose names will appear later in the book.

Here, however, I would like to address the municipal organizers who made the actual arrangement of the various conferences possible. First and foremost should be mentioned the Planning Committee of the Municipality of Härnösand, the Cultural Amenities Committee of Örnsköldsvik and the Social Services Committee of the same municipality, who acted as hosts to the three conferences, and who also made substantial financial contributions to the project. Special mention must be made of Hans Thunell, Town Architect in Härnösand, Birgitta Bergfeldt, Director of Cultural Amenities in Örnsköldsvik and Stig Erncstål, Director of Social Services in Örnsköldsvik, and their staffs, who set up the conferences in an efficient and personal manner.

Financial contributions to the project were also received from several funds and donor organizations, both regional and national. I wish to place on record our gratitude to the Swedish Council for Building Research, the Swedish National Council for Cultural Affairs, the Swedish National Rural Area Development Authority, the Västernorrland Research Council, Västernorrland County Council, the Expert Committee on Regional Development (ERU) and the Västernorrland County Administrative Board. However, the basic financial contribution was granted by UNIKUM - the Contact Centre for Umeå University. The centre and its staff (Leif Sollén and Elisabeth Nyberg) also showed considerable interest in the realization of the entire project.

Researchers at many universities and experts from the county have generously supplied us with the necessary facts for our analyses and for the regional study groups which took part in the project. Furthermore, several eminent architects have made major contributions by visualizing the ideas put forward. They are Abelardo Gonzalez (Malmö), Professor Jan Henriksson (Stockholm), Krister Wiberg (Lund), and Arne Wistedt, Per-Eddy Bjuggstam and Kjell Mattsson (all of Örnsköldsvik).

Finally, the Cultural Work Agency of Umeå must be mentioned; its courtesy has made it possible to engage the three artists from Västernorrland Kirstie Ekelund, Margareta Klingberg and Hans Marklund, to illustrate the book.

Per Råberg

Main Contributors

Karl-Eric Axenström
Former Principal Secretary of the Örnsköldsvik Municipality Secretariat.

Michel Bassand
Professor of sociology, Department of Architecture, École Polytechnique Fédérale de Lausanne. Head of the Council of Europe project "Culture and Regions" 1983-1991.

Barbro Björk
Bachelor of Arts. Head Curator at the Sundsvall Museum.

Jean Fischer
Master of Arts, Culture Sociologist. Lecturer at the Institute of Culture Sociology, University of Copenhagen.

Jan Henriksson
Architect SAR, Professor at the Royal Institute of Technology, Department of Architecture, Stockholm.

Inger Lilliequist
Bachelor of Arts. County Curator, County of Västernorrland.

Gunilla Lundahl
Bachelor of Arts. Architecture critic, Stockholm.

Torsten Malmberg
Professor, former head of the Department of Human Ecology, University of Lund.

Sidsel Mörck
Writer, active in the public debate on cultural issues, Oslo.

Per Råberg
Associate Professor, Department of Museology, University of Umeå. Director of the University Group for Humanist Futurology. In charge of the project "The Ecological Life-Region". (Also author of sections I, IV, V and chapters starting at p. 120, 227, 250, 259, 270.)

Walter B. Stöhr
Professor and former Director of the Interdisciplinary Institute for Urban and Regional Studies. University of Economics and Business Administration, Vienna.

Arne Söderström
Curator, Örnsköldsvik Museum.

Arne Wistedt
Architect SAR, ARKING AB, Örnsköldsvik/Umeå.

Harry Ördell
Former Director of Culture Office, Västernorrland County Council, Härnösand.

I THE PERSPECTIVE

"Thoughts of the Mountain". Pencil drawing by Kirstie Ekelund

Introduction

What is a community? What is meant by a life of human credibility? We are today in a predicament where these simple questions are no longer obviously pathetic but are an expression of a growing despair that many people experience in their daily life. We raise the questions because the welfare visions on which we have set our hopes seem to have lost their substance and to lack a distinct human meaning. Deep instincts within us demand an answer and refuse to accept the manipulation of terms which is notoriously conducted by the leading powers of the times. We can all clearly perceive how the ceaseless exploitation of the human and environmental capital going on has embarked our society on an extremely threatening course of events. Many people are therefore anxiously searching for new models for thinking and living, which may infuse a more dignified and, at the same time, a more joyful content into society and human life.

That, of course, is also a task for the human and social sciences, indeed, for the whole scientific community. However, this is more easily said than done, as modern science considers itself to be bound by other obligations and is firmly occupied with its current duties. The reluctance of the human and social sciences is, in fact, part of the contemporary social dilemma. Fortunately, the prevailing openness of the academic community has made it possible, in spite of dominating trends, to set up platforms for radical initiatives in a number of places. In several universities all around the world alternative experiments are in progress, forming humanitarian oases in the traditional academic milieu. The contacts between these centres are slowly being cemented and transformed into solid networks. Various theoretical approaches in the social field begin to give rise to community oriented projects.

One of these alternative courses, which we are going to consider more closely in the following, is the *socio-ecological* approach. This has close links with the university in northern Sweden where the actual book was conceptualized and published.

This book contains a series of essays which attempt, from various angles, to give concrete form to a socio-ecological vision. Hereby we mean a model of society which is founded on solid and uncompromising knowledge of man's basic existential needs as they interact with the living environment, and which uses this knowledge skilfully to promote long-term social development. In modern parlance this may be called a civil or communitarian perspective, and its social implementation endogenous (self-generating) or self-reliant development. Consistently applied, the socio-ecological principle may lead to regional models of social organization. A vision containing these features clashes in several respects with currently prevailing political doctrines. It has in fact been formulated for the purpose of alleviating the socio-cultural crisis, which threatens the core of post-industrial society, and which these doctrines have not succeeded in overcoming. Unlike these doctrines, the socio-ecological alternative pays much attention to the detailed soft data available in the humanist and social spheres of knowledge. Nevertheless it claims to represent a rational and scientific attitude to the material.

The broad humanitarian perspective presented by the socio-ecological approach is supported today by a widespread net of eminent intellectual radicals. The Umeå Group had the opportunity to bring a few of these people together on the occasion of a research and development project, which was conducted in cooperation with municipal and civil agents of a region in its vicinity, the County of Västernorrland. This book is the joint result of this initiative. It contains a selection of essays based on lectures and study reports delivered at a series of R&D conferences held in the county

The task of this introductory section is to give a presentation in summary of the scientific contributions and their authors. Also, we shall give a brief account of the R&D project which provided the organizational frame for the conferences. But before tackling this task we must try to sketch at least the outlines of the socio-cultural crisis which has provoked our social-scientific concern. To enable the reader to follow the reasoning, it is necessary to explain how this crisis has arisen and to uncover its historical roots. The new socio-ecological vision has also to be placed in the larger paradigmatic context of which it is a part.

A convenient model to shed light on these partly complicated matters is to explain the socio-cultural crisis as a consequence of the process of *globalization* which our society is currently undergoing, and which looks back on a history of several centuries. In sharp contrast to the expansionist globalization paradigm, we are now perceiving the contours of an emerging *globalism,* which there is good reason to declare a consolidation paradigm. The socio-ecological vision is introduced as an outline for a new social theory consistent with the paradigm of globalism. Its is demonstrated that its theoretical program has its roots in the spectrum of modern ecological thought, which increasingly stands out as a substantial part of the philosophy of the coming era.

A Global Paradigm

THE PROCESS OF GLOBALIZATION

Human society in our time is undergoing a process of epic change. The complex and baffling processes involved can be described in many ways, but perhaps they are best summed up by the rather vague and glamorous term "globalization".

Across the whole spectrum of human activity man's intentions and strategies today are expanding into astounding, previously unexpected, dimensions. The ends and consequences of this powerful process certainly remain beyond the scope of observance for most of us. But as the process inevitably includes a series of strong feedback mechanisms we are everywhere in our daily surroundings exposed to its effects. Even on invisible, structural planes our lives are deeply influenced. Wherever we live on the globe we are inexorably involved in networks whose functions have the entire planet for their arena and whose aim is to establish closer contact with its outermost limits.

The activating forces which have set this epic process going occupy a central position everywhere in community life. The very forefront of the public scene has long been dominated by the boundless media world, particularly its expanding electronic sector. The international commodities trade and its vast range of offerings from all countries and continents is another example. Many of us look with eager expectations to the seemingly endless opportunities for human exploitation now opened up in every direction and at all levels. Others instead feel a growing fear in face of the unforeseen consequences that expansionist development brings with it. The fact remains that every single day our working situation and economic conditions change as a result of interventions on the part of remote invisible hands. The whole of our awareness and our view of society are thus poised for change. Actively or reluctantly we are participating in the transformatory course of global events.

Heading the great transformation - and particularly far beyond most people's range of insight - is economic development, with transnational enterprises as the top agents. Year by year, their empire-building is joining the world together in an increasingly close-knit web of corporate dependencies. At the head of world economic development and as its main generator are the high-tech industrial regions of the USA, East Asia and Western Europe. Slowly but distinctly the centre of gravity of the productive world economy seems to be moving from the Atlantic to the Pacific basin. The overarching stratum of the planetary hierarchy, however, is not the productive economy but the global money market. Its huge transactions are conducted with incomprehensible rapidity in an electronic 24-hour market from the global stock exchanges in Tokyo, London and New York plus the 25 or so financial subcentres through which the entire world economy is controlled today.

This, if anything, is the proper home of the economists' famous invisible hand, and its logical goal is to create one single, all-embracing, financial world system.

The trend towards globalization we are experiencing today may give the impression of being uniquely related to our era. But closer consideration reveals that it is by no means new. As every historian knows it merely represents the latest phase in an escalating geopolitical transformation process which has been going on since the 16th century, or indeed, in rudimentary forms, for much longer. This is a long-term historical process which emanates from European territory and starts with the circumnavigations of the Spanish and Portuguese Condottieri as from about 1500. By way of these circumnavigations large parts of the previously "unknown" world were joined to the European continent in a primitive trading network.

Before long, however, increasingly sophisticated commercial routines developed. Then, during the second half of the 18th century, the Condottieri and trading network phase was superseded by the first flush of imperialism. As a result of dramatic events large parts of the southern and eastern continents were, as producers of raw materials, incorporated into the national power structures of the European states. This Euro-centric stage was superseded, after the First World War, by the phase of grand internationalism. From now on the further consolidation of global - mainly economic - relations was confirmed on the basis of the new nation states and their territorial power. The various geopolitical stages of the historical process recorded here reflect quite clearly a development towards gradually more sophisticated and efficient systems of human interaction on the global scale.

In this process, the current economic restructuring phase - with all its dramatic events - evidently represents a further logical step towards rational large-scale interaction. Its main idea is that planetary communication will work better still if market forces are allowed to mould the global structure entirely on their own terms. The consequence of this radical economic philosophy is a concentration of productive inputs to selected urban nodes which are linked directly to global distributive and financial networks. Because of its territorial limits, the nation state is regarded as far too unwieldy a colossus for the lofty visions of the modern network builders. Therefore its role is now open to debate in wide circles of the economic avant-garde. The question now posed by many political players is whether the nation state can be re-used and to what end. Liberal conservatism's advocacy for reinventing the strong state is the last bid in the sequence of policy trends.

THE DRIVING FORCES

Economic forces have - as stated above - played a decisive part in the centuries-long process of globalization. But no part of this dynamic course of event would have been possible without the technological progress made during the epoch; in fact the development cannot be understood unless the contribution of technology is acknowledged.

We are faced here with a multiple process in which several sectors have constantly inspired each other to make new radical advances. The ever-growing domain for advanced human technology can be followed through the centuries from the design of the Portuguese caravels and classical navigation instruments, all the way to the modern microchips and fibre-optic cable networks, and in the field of energy from the watermill via the steam engine to contemporary nuclear power. Within a short time this development has brought about a dramatic transformation of systems of transport and communication, and its supreme intellectual achievement is the advanced information technology of today. Not only systems of production but also the conditions of trade and the money market have been radically transformed by the new information culture. Flexibility and speed have become the watchwords for both economic and social politics. Knowledge intensity and academic competence are the new objects of capital investment. The aura of the new technology is so pervasive that we consider ourselves to be living in a post-industrial phase of development, in an ever expanding information society.

But although the machinery of the globalization process has continually changed, we can clearly discern how stable the basic motivating forces behind global change have remained throughout the process. These basic motivating forces comprise an extremely dynamic alloy of economics, technology and science, governed by one single imperative: *development through expansion and growth*. Expansion and growth is a scheme of success which we already find in the European environment of early capitalism, represented e.g. by the banking house of the Fugger family. But in exactly the same fashion it leaves its imprint on the thinking habits of present-day transnational giants such as Unilever and Toyota. The main difference is that, in the meantime, the mentality of expansionism has come to characterize the whole of society. In governing circles today it has the character of a super-ideology, an unchallenged universal creed.

THE EMERGING GLOBAL CONSCIENCE

But a new and quite unforeseen problem looms up on the horizon of historical development. Within a short time it has thrust itself into the foreground of our awareness. At the same time as globalization is celebrating its greatest triumphs ever, we are becoming aware that expansionism has, finally, come up against its limits. In its exertion of power it has reached a stage where it is beginning to devastate its own foundation of resources and human capital.

The pollution of soil and water and the laying waste of the planet's ecological base, both consequences of modern industrial development, have grown into veritable threats to the survival of our species and to the whole of our biosphere. Alarming symptoms of crisis have also existed for a long time all over the world in the form of open warfare and even regional genocide. Immense arsenals of nuclear, chemical and biological weapons cover the face of our planet. The population explosion of the Third World, world starvation and widespread poverty are all, basically, the effects of modern expansionism. The notorious violation of human rights through imprisonment, torture and liquidation in many states and, in many more, the undermining of democratic liberty, illustrate the way in which victorious expansionism, in its blind eagerness to act, increasingly lapses into purely self-destructive behaviour.

Many observers now maintain that the destructive manifestations of the last phase of globalization exceed the limits of what our planet's ecosystem can support. But it is likewise evident that the symptoms are beginning to exceed the limit of what we are humanly and mentally capable of tolerating. The many well-known crisis reports of the sixties and onwards have had the effect of arousing *a world conscience*. This new conscience first awakened among the rank and file of citizens' movements and intellectuals but, after some delay, has also begun to penetrate the political and public levels of society.

Perhaps the most important manifestation of the new world conscience has been a number of inquiries conducted by various UN agencies, the best-known being the Palme Report (1980), the Brandt report (1980 and 1983) and, more recently, the study by the Brundtland Commission (1987). The global conference on the environment held in Rio de Janeiro in the summer of 1992, in its field one of the biggest UN undertakings, came as a forceful sequel to the Brundtland Report. The commitment to environmental matters has long served as a unifying and inspiring forum for the awakening world conscience. A debate concerning sustainable global development has been initiated and has built up actual momentum within the last few decades.

However, by means of a series of initiatives in recent years, particularly by the humanitarian UN agencies, a new orientation has already occurred. The main message of this new tendency is that *the social dimension of the world crisis* is brought to the fore in the sustainability debate. A new wing of radical critics point, with increasing emphasis, at the socio-cultural crisis in the midst of our society which is an alarming aspect of the modern and post-modern development.They claim that this social crisis may even represent the very nucleus of the contemporary dilemma. The essence of this insidious crisis, according to critics, is the social disintegration in progress all over the human community. Everywhere, these days, we can observe how the primary web of communal relations between individuals, groups, generations and friends, is being broken down or just eroded away, without being replaced by other, equivalent, structures. The whole of the vital sphere of life which is often described as *the civil society,* and which has encompassed all through history the basic life processes, is about to be wiped out before our very eyes.

This is a process which involves an actual diminution of the quality of life for most humans and which gives rise to traumatic reactions among widening groups, not least among the young. It creates a deep alienation before society which, in turn, gives rise to compensatory behaviour. A large part of the neurotic generation of power which our culture exerts, and which we try to regard as normal behaviour, has very probably come about from this breakdown of the basic social value system. Also the frightening growth of violence mentality throughout contemporary society can probably be attributed to the ongoing social disintegration. In this case social disintegration reaches a level which certainly must be characterized as pathologic.

The emergence of a broadened crisis awareness has opened up for radical social perspectives in the debate on global development issues. A true humanitarian devotion for the first time permeates the futurological discourse. Innovative ideas directed at a more profound social development are now sought in widening quarters. This new social philosophy has recently been displayed at a number of UN conferences in several countries. The World Health Organization conference "Environment and Health" (Sundsvall, Sweden, 1991), the World Conference on Human Rights in Vienna in 1993, the Population Conference in Cairo in 1994, the World Summit for Social Development (Copenhagen 1995) and the Fourth Women's Conference in Peking, also in 1995, belong to the top echelon of events which deserve mentioning here.

The view is rapidly spreading that the avoidance of an ecological environmental crisis presupposes the crisis of the social structure of the world community first being brought under control and remedied.

THE SHIFT OF PARADIGM

One can rightly claim that a humanitarian awareness of crisis is now establishing itself in the global development debate. But this new broader awareness seems not, as yet, in any way to have shaken the convictions held by the political establishment. The contemporary economic regression promptly gave rise to deep cuts in the social welfare system, but at the same time huge sums were calmly spent on new weapons systems. In the face of the establishment's unwillingness to acknowledge first the environmental and now the social world crisis, we are awakened to the fact that we are facing a set of problems of extraordinary proportions, in fact *a dilemma of paradigm.* The symptoms confronting us no longer concern minor deviations from the good norm, they are signs of a deep system fault, of insufficiencies in the basic social structure which surrounds the establishment as well as all of us in our professional and civilian roles. It is, at the deepest level, a crisis of the social self-understanding of modern man.

With this, our criticism regarding various social problems is dramatically widened to a critique of the whole present-day societal system, including its history and established institutions. In the new light we realize that the alarming symptoms of social crisis cannot be remedied by way of the standard methodology of the ruling paradigm. We also begin to understand, on still closer reflection, that the celebrated expansionist development model in no way represents an eternal truth, but is intimately connected with a unique phase in our history, namely that of globalization. For a long time the techniques of the globalization project have provided a sovereign instrument for realizing the dynamic goals of this gigantic undertaking.

However, it is also clear that the evolutionary task of this project comes to a natural end with the achievement of the global limit. Its strategies then lose their previous undisputed relevance. Applied with customary vigour they now become anarchic and finally assume the shape of a threat to the human community they have served so long and so effectively. This is, by all appearances, the dramatic turning point which we have finally reached today.

The only logical conclusion to such reflections is that more relevant strategies for handling the contemporary crisis must be sought. With great probability this has to be done in radically new areas. The field is open to inventory research within the existing schools of thought, and for new planning philosophies to be staged. We anticipate a seminal debate to take place, where different radical wings will compete for public interest and political influence. Still, seriously committed people will probably agree on one important issue: the search for a new strategy of civilization should primarily be undertaken in the visionary realm of *the global paradigm* which will, hopefully, sooner or later supersede the globalization project and its obsolete forms of dominance.

The demanding question which then turns up is: Which are the new carrying issues to be developed for a global paradigm, and where do we find the golden seeds and signs which shall inspire the endeavour? In actual fact, the formulation of a new paradigm of globalism is already taking place in people's minds all over the world. The new paradigm is being revealed in the moral attitudes of many communitarian groups, in the social criticism of marginalized individuals and in the aesthetic protest of radical artists and writers in unexpected places. New attitudes to life and society are growing out of personal revolt against the ecological evils of expansionism and a new civic awareness of the human and social damage caused by the prevailing system. Broad population groups in many countries of the world are showing growing unease in front of the paths that mainstream development is taking these days.

As yet the visionary synthesis is lacking and the grasp of goals and means remains rather diffuse. But leading philosophers and social thinkers of the time (such as Fritjof Capra, Johan Galtung, Jürgen Habermas, Robert Jungk and Erich Jantsch) by way of their critical scrutiny have generated new attitudes towards world development, which are successively closing the gap between the utopian dream and a viable planetarian future. The established powers' painful perplexity in front of the alarming symptoms of crisis slowly opens the ears and eyes even within political fringe movements for new radical issues in social thought.

Even at this early stage we can claim, with great plausibility, that the break-through of a global consciousness will entail a radical change of basic political attitude. In fact, the prerequisites for such a change have been created by the ageing globalization paradigm itself. In the global awareness that is taking shape with new generations entering the stage the unity and coherence of the planet are no longer remote goals which we have to struggle to achieve, but a simple fact and a logical starting point for action. This leads to the appearance of totally new ideals for development and the elevation of new idols. Holistic thinking, the shaping of alliances and the achievement of systemic balance become natural ideals of an innate globalism, rather than aggression, arrogant exercise of power and territorial violence. A growing need is felt for a transparent and open civic culture to replace the technocratic labyrinth, and for accounts to be settled with the tiresome manipulations and deceptions of contemporary political pragmatism. In the image of a coming global civilization a conservationist (but not conservative) attitude to environmental resources and social capital spontaneously replaces shortsighted exploitation and waste.

But the transition to a global paradigm will also involve a step from quantitative to qualitative thinking in human matters, from economism to social humanism and - hopefully - from a technological to an organic view in the sciences and in politics.

The Ecological Imperative

THE ECO-MOVEMENT

While conjuring up concrete visions of the alluring global paradigm, we must keep one thing in mind: without ideological renewal, there will be no meaningful breakthrough. The big question just now is where we are to find the philosophical foundations for a coming global civilization. Several competing lines of thought open up in various directions. Still, there seems to be a growing conviction among intellectuals that a significant part of the ideological material can be scooped from the sources of *the modern ecological movement* and its attendant philosophy.

No doubt the contemporary ecological movement's programmatic commitment to the environmental questions and the survival of the Earth as a biotic habitat is congenial with the panorama of a new global paradigm. It is a vision which adopts a position in favour of organic life and the biological diversity of the planet in contrast to the ruling mechanistic paradigm's self-assertion, and which - at the same time - constitutes a clear ideological marking. This is an ideological marking which does not content itself with political convictions, but which seeks legitimacy in new knowledge and new methods which derive from the field of modern science, and primarily from the life sciences and the science of ecology. At the same time it takes issue with the Cartesian paradigm on its home ground: the rational discourse. All this gives the modern ecological tendency the shape of a presumptive alternative to the Cartesian cosmology.

In the same breath as we state this there is a reservation which must be attached. It is the sight of the contemporary ecological flow as a reformatory movement which gives rise to the objections. In its practical application the eco-movement has, first and foremost, addressed environmental matters and the never-ceasing stream of alarming symptoms at the local and global level. By way of its critical commitment a unique rescue effort has, indeed, been made within this vast field. But environmentalists have failed in taking the bull by the horns by examining the political causative context lying behind. Even though a socio-critical attitude has always existed, they have hesitated to start a paradigmatic discussion. This is due to the fact that radical criticism is not widened to a holistic analysis where the social crisis is also brought to light, and, above all: ecologists have not clearly acknowledged the necessity of drawing up a counter-image, an alternative vision of society which includes a strategy for human development and for the social organization of the community.

The alternative movement, the green political parties and - in their wake

- the universities' human ecologists, no doubt have launched a wealth of environmental issues over the last couple of decades, sometimes even with social content. An ideo-historical rhetoric has occasionally served as an overlay. But none of them has managed hitherto to live up to the great expectations aroused by the rhetoric and by the requirements which must logically be imposed upon an alternative vision. Both the holistic attitude and the creative prominence are lacking in the effort.

That is why we now experience the ecological movement being so easily absorbed by the dominating paradigm and assimilated by the social institutions which it originally criticized. The movement has become conformist, academic and politically mainstream. This change is frequently described as a victory for the ecological programme, but it actually involves a loss of the paradigmatic idea and a defeat for its development potential. It is, instead, a victory for the technocratic system élite which, yet again, has shown its ability to assimilate a foreign matter, without altering its determined course by an iota.

DEEP ECOLOGY

The ecological idealism which is required in order to realize the new society of the global paradigm will have to assume a far more radical attitude. It will have to depart from a *deep ecological view* which is able to liberate the unredeemed creative potential within the ecological project.

Closer investigation reveals that a deep ecological awareness has existed in the form of a parallel flow to the environmental trend throughout the modern history of eco-idealism. One of its earliest protagonists was the Norwegian Arne Næss who, in his book "Ecology, Community and Lifestyle" (Norwegian original 1974) makes the distinction between *deep* and *shallow* ecology. Other leading exponents of the last three decades' persistent alternative civilization debate are Theodore Roszak, Gregory Bateson, E.F. Schumacher, Hazel Henderson and Henryk Skolimowski. As background figures to the deep ecology we include names from three centuries of philosophic discourse. They range from Herbert Marcuse via Max Weber back to Bahruch Spinoza in the seventeenth century. The latter represented, in his time, the ecological philosophy's front against the pioneers of mechanistic rationality, Bacon and Descartes.

Within the deep ecology movement the contemporary environmental crisis affords the impulse to consider also the crisis of co-existence which prevails between modern man and the other biological life forms of the planet. New life is breathed into the self-evident fact, long suppressed by our cultural norms, that man is himself a biological being, who constitutes a part of the ecological web of life which covers the planet. This realization provides the motive for a confrontation with the anthropocentric dominance perspective which our civilization has established in its arrogance, as well as for an attempt to restore the upset balance between the ecosystem of man and the other species.

A terminal point in this chain of reflections is the undertaking of a review of man's own society on the basis of ecological principles which are valid for the other planetary life systems. This is also the core of an ecological basic view, as the idea is being applied by the deep ecology.

At the scientific level the deep ecology movement does not content itself with referring to the successful biosciences' methodology and academic status. By means of the life sciences' perspective the whole human and social sphere of scientific knowledge is brought up for discussion and criticism is directed at its established empiricism and value concepts. Instead, the organic and natural interactive processes are placed programmatically in the foreground, in an attempt to overcome the scientific monopoly of positivist reductionism. Spurred on by critical ambition, scientific radicalism tries to dismantle the walls between the disciplines and the faculties, as well as towards society and the ecological environment, and to permit life-giving oxygen to flow with full vigour into the scientific community. In this way the search for knowledge can assume, it is believed, at the same time a new character and a new, more committed, direction.

This new mentality also leaves its imprint in the realm of the theory of knowledge. Classical empiricism's linear rationality is questioned and a holistic methodology, based on a process-orientated rationality, is introduced. It is an epistemology which is well in keeping with the experiences of the new life sciences, but which is also supported by the new conception of reality flourishing in modern physics and chemistry.

By means of these radically extended perspectives the deep ecologists lay the foundation for a new socio-philosophical view on the same wavelength as the global paradigm. But this theoretical base subsequently forms the point of departure for an even bolder intellectual experiment. It gives rise to speculations about an updated world picture, a cosmology which is able to replace the worn-out Newtonian cosmology and its technological metaphor. In the Utopia of the coming global era the universal totality is seen as the natural anchorage of human thought. The cosmos assumes the form of an ecological megasystem with strata and facets which perpetually interact, and where the interaction between the smallest and the largest unit occurs with the speed of thought. The planetary ecosystem, like the human, takes the form of logical microspheres in this cosmic ecological panorama.

A decisive position in the deep ecological cosmology is occupied by the idea of a living consciousness, with the task of serving as a regulatory factor in the eco-cosmos' dynamic processes. But this consciousness - if it does exist - is no deterministic force, but a creative one which actively contributes towards the universe's development and transformations. Man's consciousness, too, is a facet of the ecological consciousness, and the spiritual and cultural valuations, our subjective experiences of beauty and empathetic concerns for the world, constitute a decisive expression of it. At the heart of the eco-cosmological scenario there is a spiritual hypothesis which, in traditional terms, we would characterize as humanist, but which might more correctly be described as a sign of cosmo-ecological empathy.

THE SOCIO-ECOLOGICAL APPROACH

This eco-humanist mentality forms, to continue the argument, an important ground for *the socio-ecological direction* which we also find in the deep ecological model's cloak, and which constitutes a particularly interesting contribution by its reference back to the crucial civilization debate and the question of man's future. Murray Bookchin is the best known exponent of this direction, and his activity as a leading figure dates back as far as the 70's. Where Bookchin and the social ecologists are concerned, the focus is upon the transformation of the socio-cultural community core in accordance with an ecological consciousness. The reason for this priority is both one of principle and strategy: this is the route which must be followed to restore the threatened global environmental balance and also, perhaps, the rescue of the species from a collapse of the system.

The political debate has, for a couple of decades, featured several other "social" theory moves identifying themselves as ecological. Eco-socialism, eco-anarchism, eco-communalism and eco-feminism are among the better known. In spite of the fact that discussions reiterate, for the most part, the theses of the well-known original movements, they are important as a symptom of the time. Robyn Eckersley has recently described these trends and their relevance to new eco-political thinking ("Environmentalism and Political Theory. Toward an Ecocentric Approach", 1992). However, we are still waiting for the different fractions to start cooperating in order to broaden the front against the prevailing humanitarian apathy.

But how are we going to handle the new humanitarian ideas in order to make them a useful resource for a coming global civilization? In which way can the deep-ecological awareness and the socio-ecological approach contribute to the construction of a planetary paradigm with a social signature? In order to place the efforts needed in a realistic perspective we have to use provocative terms: what is needed is nothing less than a revolution. But by this should not be understood an ostentatious political revolution like the Russian and before it the French revolution. What we are conssidering is a revolution of a more discrete but also much more pervasive kind, such as the technological revolution which has taken the power over human society right across all political and ideological borders. The technological revolution is, in esssence, a revolution of intellectual knowledge, and its position today marks the definite frontline of human thought and thus of the species' evolution.

The amalgamation of the social and the ecological perspectives create, in a single stroke, the preconditions for a new intellectual revolution which may take rational consciousness over and past the technological position and which, at the same time, might give it a friendly, human face. When we start regarding social functions as ecological processes and the whole society as an ecological system, the doors open wide for a rational search for knowledge in largely new areas of human living and experience. This revolution of rational thought is also conducted with the aid of radically new methods, more in the holistic style of the biosciences. As within these, the final target is to augment our systemic knowledge by uncovering the universal laws functioning behind the unique manifestations of life, in the actual case within the human system of life. From such an effort a totally new base of human knowledge may be obtained, which may eventually be capable of changing the future course of human events.

The bulk of human knowledge which may, in this way, be uncovered should then be used for designing alternative models and futures strategies, i.e. strategies which aim at sustainable social and environmental development at the local and global levels. An alternative social theory has to be articulated, which can be used to restrain possible tendencies towards one-eyed economic or technological planning, but which, strictly speaking, incorporates both perspectives as parts of a comprehensive theory of society. The prerequisites for this to happen is, of course, that the radical social theory holds such quality that it can match technological rationality and, in actual fact, surpass the latter.

What must be specially underlined is the following: It is at the very frontline of intellectual research and knowledge that the battle for a change of system, in favour of a human dimension in politics and civil life, has to be fought in order for lasting success to be achieved.

A Theory of Social Ecology

THE UMEÅ ENDEAVOUR

Which, then, are the prospects within the field of science for an intellectual effort along the radical lines sketched in the previous chapter? The answer is that work is already in progress in various niches in an academic environment which, by and large, remains alien to the newcomers. I will refer in the following to one single example, namely the R&D work which has been accomplished by the University Group for Humanist Futurology at Umeå for a couple of years. I will do this in some detail because the thematic design of the actual book, although containing the contributions of many independent authors, makes reference on several points to the Umeå studies.

The Group for Humanist Futurology (UGHF) is a network of humanist and social researchers with its residential seat in the university town of Umeå. For nearly ten years the network has conducted theoretical and applied studies which involve issues of humanitarian development originating from a new paradigmatic view. The Group has applied a scientific methodology which we have described as *socio-ecological*. The objective of studies has been consistent: to discuss, and subsequently concretize, models for a balanced social and ecological development of the actual and future human community.

The preliminary results to which the Group can point so far comprise a comprehensive proposal for a socially sustainable society; it has been named *the ecological life-region*. The intention is step by step to calibrate this theoretical model to the extent where it is applicable to practical community planning at relevant levels - including the inter-regional and cosmopolitan levels. It is hoped that the actual model will form, by means of its alignment towards quality of life and humanitarian needs, a competitive alternative to the current geopolitical schemes of development and planning.

We shall clarify the picture of the Umeå researchers' socio- ecological theory by hinting at some of the strategic ideas which have guided the project. The epistemological point of departure has been our belief in the ability of rational scientific thought to formulate a sensible norm of action for human and social development. One logical first stage of the study has been to establish a theoretical platform by conducting a rational criticism of the pseudorationalism permeating the contemporary society. Unfortunately the socio-ecological attempt, even at this early point, ends up in a problematic position, first and foremost towards the theory-of-science tradition at many academic institutions. In contrast with the fashionable Cartesian epistemology, with its software human and social science appendix, socio-ecology of the Umeå kind pleads a holistic rationality which enjoys the great advantage of also including the human and social sciences in the rational search for universal knowledge. Human qualitative needs should also (in reasonable proportions) be treated as hard data, which it is both legitimate and necessary to maintain in social politics alongside the claims of economists and modern technology.

The socio-ecological effort attempts, in short, to afford scientific legitimacy to fundamental existential and humanitarian needs in the community, with the final objective of better asserting these needs at the overall political level. But in order for such an objective to be realized, it is necessary to build up a solid fund of knowledge through an empirical-scientific charting of man's basic nature. The sketching of a *human identity model* has therefore long been at the centre of the Umeå college's research efforts.

A guiding principle in this was to study man as an ecological being, i.e. as a product of the organism's interaction with its living environment in an evolutionary chain of events. During the evolutionary process, possession is taken successively of more and more dimensions of the surrounding world, and new levels in the species' ecological consciousness are thereby established. Archaic behavioral archetypes are gradually overlaid by more advanced strata. This is the leading idea.

But for a social vision to achieve a place in the contemporary political debate, this procedure is not enough. It is also necessary to transpose the knowledge obtained into *models of the ecological habitat* which constitutes the natural element for interacting man. To the ecological scientist (with routine knowledge of the species' endogenous and exogenous interaction) this is a rather self-evident requirement. Another major part of the Umeå group's socio-ecological commitment has thus been to try and outline the frames of an ecological habitat, based on the evolutionary schemes but adapted to the level of modern man. - The fact that this has to occur in tandem with a struggle with deep-rooted human-scientific dogmas, which deny the possibility of general social solutions, constitutes a hardening part of the task.

The habitat model, which is the outcome of the college's tentative research, shows a regional community of about the same format as the existing counties of many countries. It reveals the same mixed composition of medium sized towns and rural territories as do many of the provincial regions. Socio-culturally it combines the cosmopolitan aspirations of the metropolis with the proximity and closeness to earth of the rural community, while - at the same time - avoiding the well-known negative effects of both variants. Through its moderate complication and scale the ideal human habitat represents a harmonic bioecological life system which is capable of retaining control over its main system functions.

But this dual ecological habitat also encompasses a post- regional level. In contrast with the majority of the other species' ecosystems, man's eco-society develops rationally in the interregional macro-space. The cosmopolitan eco-society is organized according to network principles; it assumes the shape of a web of life-regions which, on an equal footing, cooperate around joint economic and public projects. This coordination also presupposes the introduction of large-scale organizational structures in planning. But an important point is that these are not politically superior to the basic webs and do not thereby represent a threat to the freedom of the life-region. The ecological society is politically organized as an inverted hierarchy, according to bottom-up principles.

In this way the socio-ecological grass-roots vision finally makes contact even with the global mega level and sets up sensible institutions to deal with the inevitable coordination problems which arise here. At this level the social and ecological environment commitment are finally united, and evolve joint strategies directed at a sustainable future of the planet.

A RADICAL REGIONALISM

The Umeå researchers' ideas concerning the life-region are logically connected to the discussion surrounding the renaissance of the regions, which constitutes an important element of the 90's geopolitical debate. Throughout the world the role of the nation state as the main political arena is today being questioned. A flight in open or covert forms is taking place, either towards higher and more abstract arenas or towards levels which are local and closer to man.

A common feature of most regionalist efforts, which also unite them with the Umeå project, is the struggle to construct a more efficient productive environment. But the concept of regionalism today hardly covers any entitative flow. It contains such conflicting trends as the technological society's hegemonic metropolitan regions, as well as peripheral regions which are the focus for popular mobilization. To discuss these heterogenous schemes under the same heading is bound to create confusion.

The regionalism which the Umeå college represents, departing from an ecological view, is closer to the latter variant, assuming that this is supported by an ideological will and is not merely conditioned by short-term policy decisions. In connection with this kind of social regionalism, there has also been a socio-scientific discussion which seeks criteria for a more radical democratic community development. Among the arguments which are examined are ethnic affiliations, confinement to localities, and local historical or cultural identification. These are all arguments which the life-region approach is pleased to accept, since they are conceptually close to it. On the other hand it may be claimed that the life-region involves a distinct theoretical deepening of the discussion and therefore, in the long run, contains powerful arguments for lasting policy reforms.

The Umeå model's ideas belong, however, within a far wider ideological framework than that of the contemporary regional debate. When the life-region looks for kindred notions, it turns as much to initiatives in quite different areas of competence. These include the communitarian tradition, both in its modern and its classic 18th century version. Here are to be found popular mobilization movements, grass-roots and alternative thinkers, which have long since represented a civil democratic initiative throughout the world. The movement for collective habitation and its research complement, the folk cultural tradition, the initiative for IT in participatory development, for humanitarian care and for an ecological public health, are other examples of the expanding flora.

All these represent currents of our time which are trying to reconstruct a social and humanitarian alternative which has been more or less lost to sight in post-modern society. Many of them derive their inspiration from the living sphere which is usually referred to as the civil society. One may perhaps claim that the reformulation of the civil society's functions in front of the challenges of a new era is the major joint task of the radical humanitarian efforts. - This also includes the life-region's project.

SUMMING UP

One objective of the Umeå Group's work has been to transpose a number of system-ecological principles into guidelines pertaining to community organization. The study has led to a proposal for basic criteria for a socio-ecological planning practice. The following six postulations may be quoted:

• The functional organization of the ecosociety is based on an *endogenous perspective.* This means that planning is effected from inside the living civil community and serves the interests of a local or regional population. Self-reliant safeguarding of the steering mechanisms of the vital processes is a basic law in most living systems, and does not exclude awareness of the demands for coexistence with other species or population groups in the ecological space.

It should be noted, however, that the vital core of the control system rests within the single organism and depends on its ecological consciousness. All this is bioscientific basics whis remains appplicable also to the advanced level of human existence.

• The organization of the eco-community takes place according to *interactionist principles.* Ecological life is realized, with man as everywhere in the biocommunity, through the dynamic interplay of a local population with its surrounding environment, that is between a living-centre and a resource periphery. In the tension between the inner and outer poles socio-ecological existence unfolds in its rich and improvised diversity.

The structural patterns which arise at the same time mirror the species' ecological constitution, generated in the course of a lengthy evolution.

• A third organizing factor is *the holistic view.* It is a prerequisite for a properly functioning life-system that the individual can maintain an overview of his elementary species and resource environment. This principle is realized at the human level through collective efforts to arrange the community functions within a unitary spatial frame. In this way a community on a human scale arises which remains visible to every single citizen and is accessible to human identification. Such a transparent living system can rightfully be termed an ecological habitat of man.

• The ecological society, in this way, takes the shape of *a qualitative system.* Its aim is the satisfaction of man's existential needs by means of aesthetic, empathetic and sensory experience. It is the achievement of the qualitative goals which makes life meaningful and valuable in the ecological communty. But the fulfilment of man's sensory needs has a double objective: it also affords the motivation for his continued interaction with the local and regional resource environment.

• A final significant point is *the dimensional restrictions* to the organization of man's ecosystem. The most basic dimensional restriction is set by man's physical habitus, which poses very precise limits on functional dynamics. The second restriction is dictated by the spatial dimension, as realized under the conditions of planetary gravitation and local topography. The third restriction relates to the temporal dimension regarded as a sphere of ecological interaction.

All these dimensional restrictions raise specific demands on the social organization in order to be suitably explored and combined. One consequence of their impact is the division of the ecological community plan into a micro and a macro level. The microplan comprises the local collective's permanent habitat. The macroplan covers the territory of the encompassing world, which man can personally only experience occasionally and in fragment.

The division into micro and macro levels most profoundly reflects a leap in the living systems' consciousness evolution from an elementary sensory (spatial) to an advanced cognitive (temporal) stage. Both aspects represent indispensable parts of man's eco-identity and must be considered in community organization. The crux is at the same time to mark out both sub-identities and then to combine them within a single model-synthesis, which reflects the basic constitutive unity of man.

The ecological life-region is a proposal for a solution to this evolutionary dilemma.

The Sustainable Region

A REGIONAL R&D PROJECT

The purpose of the discourse in the foregoing pages has been to elucidate the ideological and historical background to the themes treated in this book. Let us now turn to the more hands-on issues. The initial objective of the book is to account for the outcome of a series of R&D conferences on regional development, regarded from an endogenous (self-generative) and socio-ecological viewpoint. The latter implies, as mentioned earlier, that regional development is conducted on the region's own premises across the whole spectrum of social needs, and with the citizens' quality of life as the final target. In the actual bok this is only a limited discussion within the paradigmatic transformation which is on the horizon.

The book is the result of a devoted effort by a research network within the proximal action field of its home university. The initiator of the project is - as has been mentioned already - the University Group for Humanist Futurology at Umeå. The project was conducted together with public agencies in a northern Swedish county, and may formally be classified as a R&D (research and development) project. Parallel with a strong wish for solid local grounding there has been a persistent striving to transsfer data to a universal level of understanding and knowledge. In this chapter we are presenting a summary of the conference contributions, thematically arranged. But first we shall give a short presentation of the project proceedings.

Ideas regarding development are best formulated at a certain distance from the sphere of influence of daily politics. But, in order to avoid ending up in league with Utopians, before to long they must be confronted with the reality they try to capture. This is the reason why the Umeå researchers were eager at an early stage to test their theoretical theses in a real life situation and at the same time to expose themselves to criticism, primarily from the local actors in the field. Considering the comprehensive scope of the socio-ecological model the experimental situation had to comprise quite a large and complex field of study, preferably a whole regional community with its institutions, citizens and territory. The project planners at Umeå were happy to find a suitable object for study in their immediate vicinity: the County of Västernorrland is its administrative title today. Contact was made with municipal representatives in the region and with leading county authorities. In connection with the various studies civil and grass-roots groups and several centres of competence in the county were also consulted.

The realization of the project had to take place in several stages focusing on selected subjects. Three part-projects were subsequently constructed around the themes of *territorial planning*, *cultural policy* and *social development*. Regional study groups were also set up in order to analyse the themes, and a dialogue was established with the Umeå researchers. In order to inspire the regional participants, special interest was taken in the practical applicability within the county. It has to be stressed, however, that the project groups' intervention in the county had nothing of the character of a professional mobilization effort. It remained clearly and purely an experiment in the application of theoretical models in order to test their viability. The study results were compiled in short papers for later presentation.

The culminating event of the Umeå project were a couple of R&D-conferences on the chosen topics. The conferences were held in two of the particpating municipalities in the county, and were realized with the assistance of their authorities. The following three conferences were held during the period of 1992-93:

- The Region as Living Space (Härnösand, April 92);
- Culture and the Living Region (Örnsköldsvik, June 92);
- Social Identity and Life-Region (Örnsköldsvik, June 93);

To these conferences a number of eminent experts from the Umeå group's Swedish and international contact network were invited. Their expertise covered a broad range of humanitarian competence in the field. The lectures given at the conferences were of special significance for elucidating the endogenous development theme, and were met with considerable interest by the regional audience. The formulation of a common policy declaration however never came about (and was never intended). Nor did any of the speakers declare that they identified themselves with the Umeå project's special vision. This is said here in order to underline the intellectual openness which characterized the discussions. At the conferences the regional study groups' reports were also debated in special sessions.

A selection of these papers and studies is presented in this book, which thus basically constitutes a project report. The contributions are complemented by an editorial postscript, in which the economic and scientific aspects of the socio-ecological approach have also been thematized. In spite of the somewhat fragmented disposition of the book, which was a result of unforeseeable circumstances, it is easy to see what unites the many contributions: it is the committed attempt to re-establish, in political practice, humanitarian forms of existence, since these constitute an indispensable - but today seriously threatened - capital for a civilized society. In this way the following is also a contribution towards the movement for sustainable social development and towards a strategy to meet the looming global paradigm.

REGION AND TERRITORY

Turning now to a brief presentation of the individual contributions, we shall start with the territorial planning theme. Holistic planning is a key topic of endogenous developmental thinking. For this reason the territorial community plan occupies a central position in the discussions on regional organization. Spatial planning is the sovereign implement for uniting disparate functions into a coherent whole in the region; at the same time, through its physical form, the plan makes this wholeness visible. This is why spatial planning has been allotted a prominent position in the university project, and it also accounts for the theme being treated first in the introductory section of this book.

As a prelude to discussing endogenous regional planning, it seemed appropriate (considering the location of the actual project) to take a panorama of the work of the European integration and its regional effects in the various parts of the continent. An up-to-date overview of regionalist tendencies in contemporary Europe was also of interest, and Professor Walter B. Stöhr of Wirtschaftsuniversität, Vienna, was invited to supply this. All through the 1980s, W.S. has been a leading world representative of an endogenous planning approach rooted in civic interests, and he is one of those who have launched the model.

In his contribution (entitled *Regional Development in the New Europe*) W.S. shows how the creation of a common European market has given rise to ethnic and social movements, not only on the peripheries of Europe but in several of the central EU countries. One vital reason for this is the one-sided economic focus of the integration process. The social opposition often has a clear regional profile and has given rise to regional separatist movements. In this way, unexpectedly, socio-cultural and regionalist attitudes have become determinants of the discussion concerning the integrated Europe of the future.

In his article W.S. underlines the importance of introducing endogenous and socio-territorial strategies to channel the socio-cultural unease created by economic integration. Strategies of this kind are of great value in regions, especially peripheral ones, under the threat of crisis. He also points to the great synergistic effects which regional holistic thinking can produce even on the economic plane. He particularly refers to the importance of introducing intra-regional feedback mechanisms for the preservation of a viable regional structure. Through these mechanisms, the threat of structural disintegration frequently posed to the region by external large-scale start-ups can be more easily averted.

W.S.'s observations concerning the European situation add up to a plaidoyer for the regional territory as the point of intersection where the dynamic economic and social forces of present-day Europe can meet in a fruitful manner. The analyses reveal how an endogenous and socio-territorial approach forces itself upon us in the process of social transformation, regardless of the grandiose views which politicians and state administrators may possibly see before them. In actual fact, the more the social deep dimension is neglected, the more provocative forms the reactions apparently assume.

"Territoriality" has long been a misunderstood and disreputable term in social and planning research. In a discussion of endogenous regional development, however, territoriality has to be credited with decisive importance. As stated earlier, a holistic view of individual and collective existence is fundamental to endogenous thinking. The territorial plane forms the unifying holistic frame in which the elementary living processes are held together.

Professor Torsten Malmberg, human ecologist and ethologist from Lund, is a leading researcher into the territoriality of man and the other species. In his essay *Region, Territory and Identity* he describes modern ethological and ecological territorial research and outlines its importance for a radical study of society. Malmberg begins by making clear the central position of territorial behaviour in human nature. That position is founded on the role played in life processes by the interaction between man and his environment. Territorial behaviour is deeply rooted in man's neural system, and its dynamics is to a great extent of an instinctive character. T.M. hints that we must assume this behaviour to have taken shape in a process which possesses evolutionary proportions.

Man's relation to the world around him is profoundly influenced by territorial behaviour and by the functions indissolubly linked with it. Simple perception analysis is not enough to explain its manifestations. T.M. mentions various types of spatial comprehension in man and describes the sense of place as the most basic spatial category. One elementary instinct in man, as in many other species, is the bid to identify with a selected point in space, with a habitat. Emotive reactions such as local patriotism and home sickness originate presumably from this instinct.

But man also makes specifiable demands on the socio-territorial organization of the habitat, demands which should be taken into account in careful community planning for human needs. T.M. quotes Lewis Mumford, who has castigated modern urban planning for its indifference to people's qualitative demands on the living environment; that indifference is documented, for example, by the indiscriminate herding together of crowds of people which we find in many metropolitan cities. Carrying on from Mumford, he points to the great merits of the provincial region as a human habitat. Västernorrland, according to T.M., is one example of a region having good qualities as "behavioural territory".

The primary field of action for endogenous planning is the proximal life-space, what we have designated the local community. And its genuine creators are not abstract planning agencies but the local collective. The ideal self-sufficient community takes shape through collective action and is manifested most clearly through its holistic architectural form. This is one reason why the long succession of local and civic community experiments in the past few decades is so interesting to study in the present context. They represent a kind of endogenous thinking which every community should apply if it intends to introduce human ecological planning in earnest.

At the same time they represent the living flame of the endogenous planning process, a flame which will also have to be kept burning in future, perhaps more large-scale phases. In the following paper in this book, entitled *The Forces of the Social Life-Space,* social and architectural critic Gunilla Lundahl provides an overview of recent examples of civic initiatives in the planning of the local environment. She herself has taken part in several mobilization projects and research assignments, and can quote from a broad personal experience.

Most of the examples to which she refers come from the Nordic countries. They range all the way from alternative living experiments in city centres to rural eco-villages and collective experiments in suburbia. Many of them are characterized by the aim of applying ecological principles both to technical construction (solar collectors, air circulation) and to functional organization (greenhouses, plantings). Above all, though, G.L. describes the social dimension as the mainspring of the alternative experiments. The great majority have come about as a result of civic initiative and devote much importance to community functions within a larger unit than that of the primary group. Interesting experiments have also been conducted by a Nordic women's research group with the aim, in area planning, of identifying an intermediate level with special tasks to perform in the life of the local community.

G.L. does not rest content with describing the radical alternatives together with their ecological, social and also architectural qualities. She broadens her perspective so as to include an analysis of the contemporary ecological crisis of society and shows how it emanates from a Cartesian mechanistic way of thinking. This, she underlines, is inextricably bound up with a male-dominated culture. That culture, however, is in the process of dissolution and is being superseded by an ecological paradigm in which organic thinking and concern for the natural life processes occupy the forefront. The feminist culture is an important factor in the realization of the new vision. In the campaign for the reorganization of the intimate living sphere, the radical vision of the feminist movement is manifested in community practice. Here, G.L. writes, we have the social roots of sustainable development.

Can the endogenous vision of society be transformed into a hands-on theory of territorial planning? This would certainly facilitate the realization of endogenous community plans under civic auspices. Per Råberg discusses this intriguing question in his article *The Ecological Life-Region*. It begins with a critical appraisal of contemporary planning and shows that the predominant theories today tend to neglect the spatial factors in community organization. The low environmental quality which modern urban planning has been exhibiting for a long time emanates partly from this widespread professional indifference, but also from the fact that the major planning decisions are made at national and international level, where ignorance of the human living dimension is more or less the rule. This omission with the leading powers is one important cause of the social disintegration which is plaguing contemporary society, says P.R.

Looking for proposals which can support the building up of an endogenous planning theory with social signature, P.R. highlights a number of inspiring approaches of historic and contemporary origin. The examples are taken from regional research (John Friedmann), phenomenological philosophy (Husserl, Heidegger), the theory of architecture (Norberg-Schulz) and bioscience (especially ecology and ethology). He concludes by presenting a theoretical model launched by the Umeå University Group and based on a socio-ecological view of man and his habitat. Both, he maintains, have been fashioned through a dynamic process in a number of evolutionary stages. The name which he gives to the model is *the ecological life-region*.

The primary task of an endogenous planning methodology, according to P.R., is to organize the existential space into a comprehensive element for ecological man. This, first of all, requires a sound knowledge of man's ecological needs. In constructing the community plan, for example, it must be remembered that ecological man possesses two disparate action horizons: the local and the global. Logically, the alternative community model should contain a local and a global functional level. On the inter-regional scale, the human action space then assumes the shape of an intricate network linking life-regions near and far. The importance of the ecological habitat model lies in its overcoming the polarized development fix of contemporary society, which has long caused great structural damage to the social and natural environment. The alternative can be described as an effort to create a planning theory for sustainable development.

In addition, a planning experiment was undertaken in order to concretize the idea of a socio-ecological model of planning. It was performed jointly with a group of architects headed by architect Jan Henriksson, a Professor at the Royal Institute of Technology in Stockholm (*Life-Region Västernorrland)*. The experiment was intended to outline an ecological plan of renewal for the County of Västernorrland and, in architectural form, to illustrate a number of important points of theory. It provided an opportunity for assessing the realism and social viability of the Umeå model.

Following on-the-spot studies and consultations with regional expertise, a suite of sketches was prepared for the local and global functional levels of the region. The planning experiment above all set out to test the possibility of treating the regional habitat in its totality as a field of architectural design, and thus to try and expand the limits of the sensorial living-space. The point of this would be to facilitate the individual's identification with the regional habitat and at the same time to infuse into the palpable regional space a true cosmopolitan character.

REGION AND CULTURE

The second part of this book deals with the aesthetic culture as a factor of regional development. Aesthetic culture also occupies a prominent position in the model of the endogenous society. The socio-ecological theory inverts our conventional notions of the human needs, with their absolute priority for economic strategies. Ecological logic tells us that the design for the qualitative objectives of society (the cultural plan of society) should precede the quantitative economic strategies for the realization of these objectives (the economic plan of society).

Culture as a regional motive force has been a fashionable topic of discussion for some years now. Often, that discussion is narrowly connected with the technological media and culture is often defined as museum and concert hall culture of a traditional kind. Less frequently the discussion takes as its starting point the terms of cultural creativity at professional and popular level. There is hardly ever any discussion of the role of culture as a motivating force for an endogenous - green - development or as a factor for creating quality of life in an alternative future society.

One of the exceptions to this rule is Michel Bassand, a leading European cultural researcher and Professor of Sociology at the Lausanne School of Architecture. For eight years he was in charge of the multinational Council of Europe Project “Culture et Regions”, the final conference of which took place in Lyon in the autumn of 1991. In his introductory lecture at the cultural conference in Örnsköldsvik, which opens the sequence of essays (*Culture and Regional Identity*), Bassand describes his experiences from the Council of Europe project.

M.B.'s research into the cultural dimension is rooted in a democratic and social perspective. He shows how modern development has led to a strong polarization of the regional structure of countries into urban centres and rural, often marginalized, peripheries. He is above all interested in the question of how the threatened regions are to escape further decline and eventually, by mobilizing their resources, can achieve a sustainable future.

At the centre of Bassand's argument is the question of regional identity. Consolidating a sense of regional identity is the primary task in order to create viable regions. Regional identity has many different faces and can be promoted by a variety of means, but M.B. points especially to the great inspiratory importance of aesthetic culture. He criticizes the narrow folklore approach sometimes characterizing the agents of regionalism, and pleads for the participation of cosmopolitan avant-garde culture in a progressive regionalism. On the other hand, regional identity also has a historical dimension which is best safeguarded by keeping the cultural heritage alive in museums, festivals and occasional exhibitions.

M.B. also highlights the regional environment as an identity-creating factor. He attaches great importance to architecture - both contemporary and historical - because of the unique way in which it embodies and visualizes collective awareness and expresses altruistic, not to say pantheistic elements of regional identity. M.B. gives an example of how, by means of unconventional cultural inputs, professional agents today can actively contribute towards restoring regional self-esteem to a region stigmatized by development (the Nord/Pas de Calais region).

Professional aesthetic culture in pictorial art, music, the theatre etc. is normally regarded as an urban phenomenon having its absolute centre in a few of the world's metropolitan cities. In the Umeå project we have asked ourselves to what extent the endogenous region can be turned into a creative arena for world art and what a metamorphosis of this kind would mean both for regional development and for the development of art itself. Cultural sociologist Jean Fischer, Copenhagen, is one of relatively few cultural debaters in the Nordic area who combine a genuine interest in the aesthetic avant-garde culture with commitment to the popular and ecological alternative movement. His partly unconventional view of the latest developments in aesthetic culture is highly relevant to a progressive discussion of the position of professional art in a regional culture with ecological inflections.

Fischer opens his article, *The Avant-Garde - A Door to the Cosmos,* by challenging the conventional image, presented by modern art history, of the breakthrough of modernism in the early 20th century and its climax during the years following 1950. He attempts to show how the 1950s modernism, viewed in a wider perspective, merely fulfils and concludes a development which began in the late 18th century and whose fundamental idea was the isolation and individualization of aesthetics from religion, dance and social life. An art culture appears which is created by genuine individualists for individual experiential consumption at the deeper levels of the spirit. This modernism was closely connected with the individualism and rapid dynamics of modern industrial society. It developed an absurd aesthetics whose extreme demands for renewal eventually led to the decay of the movement.

The truly radical breakthrough in 20th century development of art, J.F. maintains, actually came about around 1960. It was closely linked to the appearance of a succession of new aesthetic departures, such as process art and arte povera, installations, land art and site works, body art and sound meditation, religiosity and new spirituality. The radically new and common feature of these developments is their tendency to integrate the arts and transcend the boundaries between them, and also the boundaries between art culture and social life. In this way aesthetic culture recovers its integrated function in the community.

Another radical innovation is the tendency to recover concrete space and slow time, after the rapid time paradigm of the industrial epoch. This also means a recovery of the collective and bodily in the form of rituals and meditation, after the abstraction and cerebralization of modernism. Fourthly, there are also elements here of a new view of nature with human-ecological implications, abolishing and obliterating the modernist boundary between man and biospherical space. At the bottom of the new scheme, however, is a transcendental dimension which among other things has been strongly inspired by oriental philosophy and religion. Through this transcending of boundaries the new art also abolishes the modernist boundary between human perception and cosmic identification.

The antithesis of professional world art is local popular culture. In the endogenous perspective, both are indispensable elements of regional life and possess the same degree of generality. They do not represent high and low variants on the same scale; instead they are complementary entities. In our age, creative popular culture has been replaced by a consumer popular culture created by media manufacturers on the global scene. How can a living popular culture for the 21st century be created in competition with the media, and what role can media culture play in such a process?

The writer Sidsel Mörck has for a long time taken part as a radical debater and cultural critic in the Norwegian debate on media. In her contribution to this book *(Mass Culture Meets Man.)* she launches an attack on the victorious media world with a critical commitment which differs from the frequently conformist media research of today. In a society of increasing social fragmentation, mass media, and audiovisual media not least, have come to occupy a central position in meeting the need for social and cultural communication. Despite the tremendous expansion of our horizons of awareness, this in itself is a dubious process, confirming as it does the alienation of the individual from social and local reality.

But the dubious traits are augmented by the lack of social responsibility with which the media world plays its role in the niche it has created for itself. Through its monopoly status, arbitrary power interests are given the opportunity of invading our private lives and manipulating us. On the other hand, the media people have not really been capable of filling the technological apparatus with a content which measures up to average intellectual and cultural demands. The audiovisual media in particular distribute, as a matter of routine, a substandard output which, through its powerful impact, has a stupefying and passivating influence. In the hands of the powers of development, S.M. maintains, mass media are in several ways a socially disintegrative and anti-democratic force.

Can the media nevertheless play a positive part in the birth of a modern regional popular culture? S.M. has no doubt that they can, and she refers to the as yet under-utilized capacity represented by local radio broadcasting and regional television channels under civic control. With a media culture subject to democratic influence, output can be guided in the direction of higher quality and better information value and can become an activating implement of regional communication.

But the creation of a future popular culture with or to one side of the media will also demand a vigorous mobilization of individual powers of resistance and civic commitment. The first of these can be achieved by fostering independent, less deferential attitudes on the part of individuals, which in turn calls for a thoroughgoing reform of the entire school system. The second can be achieved by reconstructing the local and regional society, so that human networks can once again be made to function. From this a new local culture can arise.

Per Råberg returns in the second part of the book with an article sketching guidelines for an endogenous cultural policy for the region (*The Culture of Nature*). By way of introduction he states that contemporary cultural policy is of limited inspirational value to such an endeavour. Its attitude is that of the technocratic administration and its cultural perspective is normally conservative. Cultural policy practise consists mainly of distributing economic subsidies to interested parties in the established cultural branches. In this respect, there is no decisive difference between the liberal cultural policy of today and its predecessor, the social democratic policy of the early post world-war years. One important notion is that modern cultural policy has no suggestion for how we should tackle the ongoing aesthetic deconstruction of society, which is an alarming aspect of the contemporary crisis. In fact, it seems not even to have noticed the dilemma.

A cultural policy for the sustainable life region must take as its point of departure a more profound socio-cultural awareness and have a clear creative vision in mind, claims P.R. Its goal should be the promotion of basic cultural values in citizens' civil life but also in overall politics. In order to realize this objective the indispensable import of aesthetic experience for the general quality of life in the community has to be understood by those in charge. The foundations for such an understanding can be found in socio-ecological thinking of the kind developed by the University Group for Humanist Futurology.

P.R. goes on to concretize the vision of the Umeå group by pointing out the main domains of an *ecological cultural policy*. These firstly consist of the non-controversial fields of *local culture*, which is seen as a genuinely civic concern, and *cosmopolitan culture,* which is realized mainly by professional artists. But in addition, ecological culture incorporates two huge aesthetic domains which have, up till recently, remained outside the responsibility of cultural politics. The first is the *cultivated environment,* which is mainly upheld by the public institutions and architectonic infrastructure of the permanent human habitat. The second is the *natural aesthetics,* which in modern society is mostly located to a perpheral zone of man's living space. Both are indispensable spheres of aesthetic life which are largely neglected in our age. In an ecological cultural policy they are placed instead at the center of interest, because of their crucial impact on the quality of life and the public welfare of the whole community.

This is a statement which, in one single stroke, changes our view of the values of the provincial regions. We can easily see that many of them possess, latently, aesthetic qualities which place them far ahead of most of the celebrated metropolises of our day. One way for citizens to improve their quality of life is, therefore, to choose the provincial region as a habitat and to contribute to the development of the aesthetic resources hidden therein. In many cases, this is both a cheaper and more effective way of raising living standards than making the economic and metropolitan detour.

One important part of the Umeå Group's Västernorrland project was the establishing of an active interchange with groups of regional agents. During the discussions within these groups, the theoretical ideas of the researchers could be scrutinized and practical alternatives sketched for cultural mobilization measures in the region. One element of the preparations for the cultural symposium involved setting up regional study groups and devising programmes for their activities. The following three themes were chosen as objects for closer consideration by the groups:

- Local culture (co-ordinators Karl-Eric Axenström, Arne Söderström, Arne Wistedt).
- World culture (co-ordinator Barbro Björk).
- Environmental culture (co-ordinator Inger Lilliequist).

During the symposium, the three groups presented their conclusions at a session with a special county profile. This was introduced by Harry Ördell, Västernorrland County Council Director of Cultural Affairs, with a historical retrospect of cultural developments in the county. In this way the theoretical contributions were usefully thrown into relief against past and present regional realities. The various contributions are reproduced here, forming the concluding vignette of the cultural debate in the book.

REGION AND COMMUNITY

The third of the Umeå Futurology Group's R&D conferences dealt with the social dimension as a regional development factor. This, too, was a theme of paramount intrest in a project which aimed at bringing out endogenous and humanist perspectives in community planning. The subjects of the conference and the main participants are listed in the conference programme in the appendix of the book. All the specially invited speakers are well-known people in their respective areas of expertise: Ilona Kickbusch (from the WHO in Geneva) is a leading proponent of the organization's socio-ecological health programme, Alf Ronnby (from Mid-Sweden University, Östersund) is one of Scandinavias foremost humanists in social research, and Lars Qvortrup (from the University of Odense) is an outstanding European expert on information technology for local and participatory use.

In connection with the actual conference a series of studies were carried out in collaboration with working groups in the region. They were reported on in a special session, which contrasted with the outside expert contributions in emphasizing the "regional voice". Since the conference these studies have undergone some revision. In particular a number of specially consulted architects have brought concrete form to the ideas. In the English version of the conference report we have limited the account to the reports of regional study groups in order to avoid tedious repetition. In relevant cases, the comments of the external experts have been acknowledged in the studies.

Per Råberg's paper, *Social Life and the Living Region,* forms an introduction to the series of group studies. The essay is a contribution to the theoretical discussion on socio-ecological issues which forms the consistent theme of this book. It serves at the same time as a uniting link with the other sections.

P.R. points out the discrepancy which exists between rhetoric and reality in modern politics, both municipal and national. Although social welfare and care have long been high-priority themes in public rhetoric, there has been in reality a dismantling of social life-values in the community. The dissolution of vital social networks has actually gone so far that veritable sociopathological reaction can now be recorded in many places, using sociomedical diagnostics. The metropolitan community is especially exposed. The sociopathology of modern society comprises a whole spectrum of symptoms, extending from individual and group level all the way to the public and political levels. The destructive effects of social ignorance are today so widespread that they constitute a threat to the common welfare objectives and to a sound ecological way of living.

Understanding of the socio-cultural crisis and its consequences has long been very limited. In recent years, however, the social dilemma has attracted increasing notice in circles working with themes of global development, P.R.

points out. The Rio Conference in 1992 was a turning point by way of its attempt to couple the social dimension with the environmental and ecological concerns about the future. More and more people now begin to regard a sound social environment as a necessary prerequisite for the global environmental crisis to be averted. A series of social and humanitarian UN conferences in recent years signal a realignment in global futurological thinking.

But for the contemporary drift into a social crisis to be restrained, a radical overhaul of social policy is needed. It is necessary for the fragmented policy bodies to unite in an inter-social policy, which is engaged in creating efficient alternatives to the economic models of today. In order to to give an offensive social policy wind in its sails, also better scientific grounding is needed than now available. There is an acute need for a devoted socio-ecological research which tries out models for a sustainable social development. P.R. points to the R&D work which has been done over a number of years by the Humanist Futurology Group in Umeå following these aims. A proposal for indicators of viable social development, based on the Umeå model, is presented. P.R. describes the world's provincial regions as a nursery for supportive social models, on account of the ideal preconditions for social interaction on a human scale which are found here.

The reports of study groups then begin with the paper *Grass-Roots Mobilization in the New Age.* The reason for taking the grass-roots theme first is the expectations held by ecologists with regard to the ability of the primary life networks to contribute to sustainable social structures in the regions. It is the same basic idea as all the world's alternative grass-roots have championed for over twenty years in polemics with the political and sociological establishment. Emphasizing the grass-roots level and creating respect for its generative dynamics is a major task in the reconstruction of the disintegrating civil community and the promotion of endogenous social values.

The study group draws attention to the tasks to be tackled by the grass-roots in working on behalf of the regions. A strategy for regional grass-roots mobilization in the 90s needs to consider the altered role of the movement by comparison with the pioneering days of the 60s and 70s. Whereas the movement of those years was highly an urban phenomenon and closely connected with the alternative youth culture, the progressive efforts of the 90s are often made in the countryside. They form a part of mobilization strategies which agencies in many crisis-ridden provinces are trying out in order to keep their native communities alive.

In the green movement of the 90s, ideals such as neighbourhood democracy, self-reliance and quality of life are as much prized as ever they were. But the crowd now gathering under these banners is a more heterogeneous one than formerly: pensioners' and youth groups, entrepreneurs and co-operatives, social workers and civil servants. In its modern form grass-roots idealism has taken the step from alternative activism to progressive social action. The dilemmas of the regions in the 90s also present the grass-roots with new tasks, which extend far beyond the internal interests of the local groups.

Västernorrland actually possesses a grass-roots movement through which important social development goals in the region may be realized. This is the Västernorrland rural community council "Sambygd" (Community Coop), which was formed in the winter of 1993. "Sambygd" is an association of a large number of development groups. Its purpose is to co-ordinate local projects and to pursue a consistent policy on their behalf in dealings with the regional authorities. The network also forms part of a nation-wide communitarian movement, and has already gained admission to bodies such as the Popular Movements Council (Svenska Folkrörelserådet) and the government's Regional Policy Council. This is an initiative which is now being applauded at central political level. But it may be unwise to set too many hopes merely on the expression of official goodwill. It is evident that grass roots idealism in crisis-threatened regions can also be used as a cosmetic feature by a regional policy which has neither the resources nor a real vision. Within the movement there should be an even bigger effort to build up inner social cogency.

However, grass-roots idealism alone will not be enough to defend social values or to introduce a bottom-up approach to regional policy-making. In order to realize a qualitative social vision, the public tier of society must also be involved. This is a theme which runs through the next two group reports, both of them addressed to the administrative echelon in the local and regional community. A mobilization of the public social units should be undertaken, and first of all this must tackle two questions of attitude: the first concerns implementing a bottom-up perspective in public sector work and the second supplementing administrative work with creativity. How social service agencies can help to strengthen local and regional identity is one of the main issues for the reformed social sector.

The leading area for identity-strengthening efforts, in the opinion of the Umeå Group, is Education because of the crucial role of the school in upholding the normative values of society. Joining regional experts in discussing the design of an educational curriculum which can strengthen local and regional identity appeared a highly relevant study project. The essay *The Eco-School in Härnösand* reports the results of a discussion which took place on this subject with educationists in Härnösand municipality.

To begin with, the group points out that the contemporary school gives a one-sided - cognitive - view of reality, which fixes the child's attention on spheres of consciousness far beyond the local community and its elemental values of life. The whole of the child's basic sensory-aesthetic and practical-instrumental endowment is in fact highly ignored by official education policy.

Today's schooling represents a lop-sided rationalist indoctrination of the children which fatally hinders their normal personality development. All this has finally to be paid for, and it is ultimately society that suffers.

Against this background, the group asserts that there is a need for supplementation of the present schooling with a curriculum which affirms the primary sensory and communicative levels of young people's consciousness. These levels are intimately connected with local existence and its currently scorned ideals. But a teaching aimed at reviving the archetypes of local life may also become an instrument for strengthening regional identity and articulating the overall qualities of regional life. Referring to well-known developmental psychology theories, the group outlines a socio-ecological curriculum for the compulsory school and the upper secondary school. Together with an architect they then devise an ecological schooling centre intended to form an autonomous supplement to a traditional school environment in the region.

A very different, but equally important, creative task of the ecological public sector is the organization of the social living environment. In its social study the project group was given a convenient opportunity to test its theories in environmental planning. A study circle was formed with architects and planning experts from Örnsköldsvik. The object of study was the Gullänget district in the municipality. The experiment resulted in a socio-ecological renewal plan for the district. The draft is reported under the title of *Gullänget Town-Neighbourhood. An Experiment in Socio-ecological Planning.*

An initial observation in this study is that modern society lacks thoroughgoing social environment planning. Elements of social planning are found in many residential areas, but normally the collective environment takes shape as a purely cumulative structure, where the family dwelling is the standard module. Only a selection of supplementary services here enliven the social aridity. However, thanks to the communitarian tradition, there has been in periods some influx of more fertile social ideals. In the town planning of the 70s and 80s, in particular, community ideals experienced a renaissance, and many residential areas of high environmental standard were built. An additional beneficial contribution also comes from the environmental ecology trend in recent years.

The Gullänget Group introduces a socio-ecological renewal plan which to some extent draws on the community tradition, but which also contains a few new ideas. The keynote is the return of community functions to the residential area and town district. Self-care and self-administration are new social passwords and become a crucial source of inspiration in zonal planning. In the socio-ecological zone plan the primary planning module is not the housing unit, but the district seen as a whole. The ecological planner transforms the town district into a qualitative social life sphere in the municipality. Meeting places and social functions are given a prominent position and architecturally, too, become a main theme. But the social district plan is also a part of a comprehensive socio-ecological plan for the local community. Cultural and productive life zones surround the social zone and form, together with the latter, the functional cycle of the municipality.

The Gullänget Group coins the term "town neighbourhood" to indicate the social sphere of the self-administering commune. To local politicians the idea should be interesting in view of the current crisis in the public sector. It represents a transfer of both responsibility and practical activities to the citizens, which entails substantial savings to the public sector. From the point of view of civil welfare the situation has quite different advantages. The restoration of social functions to the citizens means a welcome vitalization of collective life and a raising of the overall quality of living in the community.

When social mobilization is discussed, thoughts turn almost automatically to primary community networks and their reinforcement. But the social life sphere also contains a cosmopolitan dimension which demands attention at municipal and county level. This is a thesis which is forcefully argued by the Umeå Group, in this case with reference to the cognitive imprinting of human nature. For the average citizen the need for cosmopolitan social communication is very inadequately met today. It is catered for mainly by the mass media, which have hitherto allowed only one-way communication, and by occasional holiday trips to exotic countries.

Examining the possibility of up-to-date cosmopolitan communication at the civic level appeared another big challenge to the project. A group was formed to study the topic and after its meetings delivered the report *Megamedia Centre Bothnia. An IT Vision for Västernorrland.* The group's built-in radar led immediately to the information technology sector. It was found, to start with, that IT in social development was by no means a virgin area. A number of experiments, particularly in the late 80s, have dealt with the subject comprehensively. Nordic researchers have played an important part in the discussions. But interest in these early attempts was primarily focused on IT as a medium in local development and a means of strengthening primary networks.

Developments in the 90s, however, have dramatically enlarged the potential of IT as a communicative instrument. All over the world broadband networks are being built for the multilateral exchange of large quantities of information matter. At the same time numerous multimedia applications are creating ever more lively interfaces worldwide. The main participants in this network are the public and commercial sectors and various private interests. But also the outline of a civic communication structure in a very deep democratic sense is becoming discernable. The new multimedia facilities are shaping the preconditions for a truly global revolution in social communication. The historical organization of the cosmopolitan society in world centres and peripheries may soon be replaced by regional networks, where global simultaneity and compatibility between regions prevail as a result of the convenient access to information, skills and transmission functions. This removes the motives for organizing political territory in a system of metropolises and provinces. Inter-regional democracy comes into being.

The project group's contribution to the discussion on global social communication is the idea of a "Megamedia Centre", strategically sited at the centre of the road and rail network of the Västernorrland region. The Megamedia Centre contains a large number of applications for global multidirectional contact and has been designed for wide ranges of popular use. At the same time the model offers opportunities to regional authorities, business and universities in instituting communication across regional boundaries. The structure has been dimensioned to form a conspicuous regional focal point and to bestow a global aura on the region. By forming the multimedia environment as a collective gathering point it also constitutes an important relay between the global and local spheres of citizens' communication.

The last of the series of social studies touches on the political aspects of a social rearmament of the regions. It is apparent that a programme with broad social ambitions can hardly be implemented without support from the political level. How political interest is to be stimulated is, therefore, a crucial question. But it is also important to discuss the social policy strategies which are required in order to realize the new intentions. One first observation is that the sectoral boundaries of traditional social policy need to be crossed. It is essential to draw up intersectoral alliances and to devise a radical intersocial policy, founded on a scientific knowledge base. But to be in tune with the times, the radical social policy must also step out onto the world stage. It must make world opinion aware of the importance of the social factor in global development issues. The ideological point of attachment is the environmental movement. A synthesis of environmental and socio-ecological commitment is very much asked for now in order to lead society out of the crisis.

It is difficult to find projects which give concrete expression to this vision. But there is at least one: the World Health Organization's programme for an ecological public health. This programme has taken shape at a number of international conferences over the last decade, and was presented at the world conference "Supportive Environments for Health" at Sundsvall in Västernorrland in 1991. The impulse behind the radical health vision is the discovery of a new type of psychosocial sickness, with roots in socio-cultural inadequacies at the very foundations of our society. The conclusion of the WHO experts is that an up-to-date public health policy has to be given a holistic and environmental profile: the creation of good living environments conducive to people's well-being must be the essence of the new agenda. This radical health vision has been launched with great success in the WHO's

"Health for All" programme, in its "Healthy Cities" project and in a series of other activities.

But the new agenda of the WHO has also a broader range. It is a call to the political powers to formulate a strengthened social policy, in which the well-being of the citizens is put at the center of general politics. In this expanded version the WHO programme is a source of inspiration to the regional policy vision which was discussed at the University Group's social conference. A study group was formed with a specific brief to evaluate the effects of the Sundsvall Conference in the region. The results of this study are presented in the essay *Public Health is Social Welfare.* A number of regional initiatives have been taken in the spirit of the conference, but few of them have reached the new level of consciousness set up by the WHO programme. In the regional public health unit there is full awareness of these limitations, which are due to factors over and above its influence. An interest in continuing the work is clearly indicated.

EPILOGUE

The final section of the book is entitled an *Epilogue.* In this the project director has taken the liberty of independently treating two additional themes which, in their respective ways, lead the project to a logical conclusion. One is the role of the economic dimension in an endogenous regional vision and the other is the scientific and academic prospects of the socio-ecological approach.

The economic excursus *Economy in the Viable Region* is undertaken deliberately in general terms to underline that the aspirations of the socio-ecological project extend further than the construction of regional policy safety networks. The crisis of the economy and the inability of economists to deal with it is the theme of an introductory discussion. The destructive implications of the economic crisis for social welfare politics are seen as an especially alarming symptom. Many provincial regions are threatened by the extreme vulnerability of the modern economy, because of their strong dependency on the international economic power-nets.

In an attempt to explain the economic crisis the hypothesis is put forward that one basic cause is the lack of political interest in socioeconomic realities. Economic policy in our time has uncritically elevated the market economy to socioeconomic dogma. In the long run this has aroused major problems, because the market economy is a production economy system. The essential cause of the crisis is not the failure of the market economy but the absence of a consistent "social economy" which assumes responsibility for the historical situation, the permanent infrastructure and the long-term development goals of society.

We now find ourselves in a situation in which the established powers are getting ready to take the ultimate steps away from the essential foundations of the social economy; with this the remains of the civic tradition and the whole historical community are at the mercy of the anarchic forces of the day. In these circumstances a reconstruction of the institutional basis of the social economy in the local and regional community becomes a matter of urgency. The aim of such an effort is to defend the regional community's cultural heritage, its commerce and industry, and its social institutions against imminent economic bankruptcy.

Drawing inspiration from alternative economic tendencies such as the institutional economy, the communitarian tradition and most recently the ecological trend, a socioeconomic model - which we describe as socio-ecological - is outlined. This model presents the social economy as a coherent regional system, possessing three levels: 1. *the natural-resource economy,* which is responsible for the balance in the basic ecological cycle of natural-resource processing. 2. *the institutional economy,* which is responsible for the maintenance of the community's organizational and historical infrastructure; 3. *the production economy,* which deals with daily subsistence and short-term material needs. This final level is organized on well-known market-economy principles and is divided into two sectors: an intraregional for the local market and an interregional for the world market.

Regions which apply a socioeconomic model may gain a better view of the domestic economic situation and of their productive potential. Therefore initiatives can be taken in a more calculated manner. There is reason to believe that even crisis-threatened regions can in this way retain the public welfare that has been gained in the preceding years and entrench it. But the first step towards such a change must presumably be taken by the regions themselves.

In the final pages of the book attention is turned to the scientific environment in which the socio-ecological approach is working (*The Science of Social Ecology*). It is asked to what extent social ecology represents a progressive contribution to the social sciences. In this connection it has to be underlined that the new approach is in several ways in conflict with the established academic tradition at the university. For both historical and ideological reasons the socio-cultural sector - and therefore also the socio-ecological perspective - has become a blind spot in the universe of contemporary scientific observation. Noone who has experienced the harsh conditions of innovative social and cultural research at the university will be surprised at the relatively low level of socio-cultural competence and knowledge in society at large, or at the fact that its societal application is so far behind that of natural science and technology.

Two arguments are put forward which may possibly help to give a socio-ecological perspective its rightful place in the scientific community. The first is the general need of the humanities and social sciences for better theoretical founding, in order to raise the sector to a level equivalent to the natural and technical sciences. The second argument is the world community's acute need for substantial human and social knowledge, to be able to deal seriously with the socio-cultural roots of the human and environmental crisis. The physical location of a science of socio-ecology is also discussed. As an alternative to the national elite university, the regional universities are suggested. By introducing a socio-ecological profile the latter may become new centres of research into socially and ecologically sustainable development (perhaps under the name of *Regional Ecology*).

As a final statement the Umeå project declares its sympathy for the idea of a shift of paradigm which is being launched today by a number of social critics. In the new visions a deep ecological attitude and a cosmology anchored in the life sciences is succeeding the mechanistic rationalism of the Cartesian era. One interesting question is whether this encouraging transformation will also lead to a change in society's perception of the sciences and affect the organization of scientific knowledge.

The view of the socio-ecological project is that the new paradigm opens up doors to radically new thinking, two aspects of which are especially important. One is the establishment of an epistemology of holistic rationality, in which the world is regarded as a systemic whole of interacting processes. The second is the accentuation of the subjectivity perspective, which leads to a methodology based on the observers inescapable involvement in the observed processes. The subjective (or anthropic) and holistic principles together give rise to an endogenous epistemology which goes far beyond exogenous Cartesian rationalism. This is a view of science which particularly invites humanists and social scientists to engage in a search for universal knowledge that has hitherto been strictly taboo.

A new theory of knowledge is on the horizon, opening the way into the soft data of aesthetics, ethics and social empathy - without renouncing the criterion of rationality. And without threatening the spontaneity of the aesthetic experience.

REFERENCES:

Abrahamsson, K.V. et al., 1992: Humanekologi (Human Ecology). Carlssons, Stockholm.

Bassand, M., Brugger, E.A., Bryden, J.M., Friedmann, J., Stuckey, B. (eds.), 1986: Self-Reliant Development in Europe. Theory, Problems, Actions. Gower, Aldershot.

Boutros-Ghali, Boutros, 1994: Building Peace and Development. United Nations, New York.

Bramwell, A. 1989: Ecology in the 20th Century. A History. Yale University Press, London.

Brown, L. et al., 1992: State of the World Report 1992. W. W. Norton, New York.

Capra, F., 1982: The Turning Point: Science, Society and the Rising Culture. Simon and Schuster, New York.

Chase-Dunn, C., 1989: Global Formation. Structures of the World Economy Basil Blackwell,Cambridge, Mass.

Clout, H.D. (ed.), 1987: Regional Development in Western Europe (3. ed). David Fulton, London.

Costanza, R. (ed.), 1991: Ecological Economics. The Science and Management of Sustainability. Columbia University Press, New York.
Devall, B., Sessions, G., 1985: Deep Ecology. Gibbs Smith Publ., Salt Lake City.
Dicken, P., 1992: Global Shift (2. ed.). London.
Dobson, A. and Lucardie, P., 1993: The Politics of Nature. Routledge, London.
Dogan, M., Kasarda, J.D. (eds.), 1988: The Metropolis Era. Vol, 2: Mega-Cities. Sage, London.
Eckersley, R., 1992: Environmentalism and Political Theory. Toward an Ecocentric Approach. UCL Press, London.
Ekins, P., 1992: A New World Order. Grassroots Movements for Global Change. Routledge, London.
Galtung, J., Brien, P. and Preiswerk, R. (eds.), 1980: Self-Reliance, A Strategy for Development. Bogle-l'Overture, London.
Giddens, A. 1990: The Consequences of Modernity. Stanford University Press, Stanford CA.
Goldsmith, E., 1992: The Way. An Ecological World View. Rider, London.
Griffin, R., Rahman-Khan, A., 1992: Globalization and the Developing World. UNRISD, Genève.
Hansson, L.O., Jungen, B. (eds.), 1992: Human Responsibility and Global Change. Proceedings from the International Conference in Göteborg 9-14 June 1991. Göteborg.
Hettne, B., 1990: Development Theory and the Three Worlds. Longman, Harlow.
Holloway, M., 1951: Heavens on Earth. Utopian Communities in America 1680-1880. London.
Hubendick, B., 1985: Människoekologi (Ecology of Man). Gidlund, Stockholm.
Johnston, B.J. and Taylor, B. (eds.), 1986: A World in Crisis? Geographical Perspectives. Basil Blackwell, Oxford.
King, A., Schneider, B., 1992: The First Global Revolution. A Report by the Council of the Club of Rome. Simon & Schuster, London.
Lovelock, J. 1995 (1988): The Ages of Gaia. W.W. Norton & Co., New York.
Mc Intosh, R., 1985: The Background of Ecology. Concept and Theory. Cambridge.
Mc Kinlay, R.D. and Little, R., 1986: Global Problems and World Order. Frances Pinter, London.
Merchant, C., 1992: Radical Ecology. The Search for a Livable World. Routledge, New York.
Merchant, C. (ed.), 1994: Ecology. Key Concepts in Critical Theory. Humanities Press, Atlantic Highlands, N.J.
Naess, A., 1974: Ökologi, samfunn og livsstil. Utkast till en ökosofi (Ecology, Society and Lifestyle). Oslo.
Norgaard, R.B., 1994: Development Betrayed. Routledge, London.
Our Global Neighbourhood. The Report of the Commission on Global Governance.1995, Oxford University Press, Oxford.
Pepper, D., 1989: The Roots of Modern Environmentalism. Routledge, London.
Postman, N. 1993: Technopoly. The Surrender of Culture to Technology. Vintage Books, New York.
Redclift, M., 1992: Sustainable Development. Exploring the Contradictions. Routledge, London.
Robertson, R., 1992: Globalization. Social Theory and Global Culture. SAGE, London.
Råberg P., 1984: Den ekologiska människan (Ecological Man). The Aesthetic Society, Report 3. Univ. of Umeå.

Råberg, P., 1984: Den ekologiska människans ursprung (The Origins of Ecological Man). The Aesthetic Society, Report 4. Univ. of Umeå.
Råberg, P., 1985: Det ekologiska rummet (The Ecological Space). The Aesthetic Society, Report 5. Univ. of Umeå.
Råberg, P., 1987: The Space of Man. New Concepts for Social and Humanist Planning. Almqvist & Wiksell Int., Stockholm.
Råberg, P., 1990: En annan framtid. Ekoregion Nordvästmanland (Another Future. The Ecoregion of Nordvästmanland). Kollegiet för humanistiska framtidsstudier, Umeå.
Seers, D. and Öström, K., 1983: The Crises of the European Regions. Macmillan, London.
Serres, M., 1990: Le Contrat Naturel. Minuit, Paris.
Simon, D. (ed.), 1990: Third World Regional Development. A Reappraisal. Paul Chapman, London.
Skolimowski, H., 1992: Living Philosophy. Eco-Philosophy as a Tree of Life. Penguin Group, London.
Stavrianos, L.S., 1966: The World since 1500. A Global History. Prentice-Hall, Englewood Cliffs, N. J.
Stöhr, W.B. (ed.), 1990: Global Challenge and Local Response. Mansell Publishing, London.
Sörlin, S., 1991: Naturkontraktet. Om naturumgängets idéhistoria (The Contract with Nature. The Intellectual History of Man's Intercourse with Nature). Carlssons Bokförlag, Stockholm.
Taylor, D.R.F. and Mackenzie, F., 1992: Development from within. Survival in Rural Africa. Routledge, London.
The Challenge to the South. The Report of the South Commission, 1990. Oxford University Press, Oxford.
UNDP, 1991: Human Development Report 1991. Oxford university Press, Oxford.
UNDP, 1992: Human Development Report 1992. Oxford University Press, Oxford.
UNEP, 1991: Caring for the Earth: A Strategy for Sustainable Living. London.
UNFPA, 1991: Population and the Environment. The Challenges ahead. London.
UNICEF, 1990: The State of the Worlds Children 1990. Oxford University Press, New York.
Wallerstein, I., 1979: The Capitalist World-Economy. Cambridge University Press.
Waters, M. 1995: Globalization. Routledge, London.
What now - Another Development. The 1975 Dag Hammarskjöld Report on Human Development and International Cooperation. Development Dialogue 1975:1-2. Uppsala.
WCED, 1987: Our Common Future. Oxford University Press, Oxford.
Worster, D. 1994 (1977): Natures Economy. A History of Ecological Ideas. Cambridge University Press, Cambridge.

II REGION AND TERRITORY

"Thoughts of the Mountain". Pencil drawing by Kirstie Ekelund

Regional Development in the New Europe

Walter B. Stöhr

AIM OF THE PAPER

The aim of this paper *) is to put the focus on socio-territorial dimensions of planning, and to stress the importance that European planning agencies should pay more attention to the socio-political perspective in designing strategies for the forthcoming Paneuropean community. To underline the case I will show how the recent phase of European development has brought with it increasing polarization, on several levels, between different areas.

I will also give a summary presentation of main issues connected with modern endogeneous (intraregional) development strategies. These strategies are designed for levelling socio-territorial imbalances which emerge in the wake of economic development, particularly in peripheral territories.

EUROPEAN INTEGRATION

Economic vs. socio-political issues

European integration after World War II initially had been triggered by political motives: a key objective of the Montan Union (European Coal and Steel Community) established in 1951 had been to put mining and heavy industry - as the major bases for waging war - under international control. Soon however with the formation of the European Economic Community (1957) and the European Customs Union (1968), economic motives in the sense of maximizing overall growth and efficiency took the lead and have been driving European integration until, a few years ago, political issues again came to the foreground when the EC moved towards becoming also a political union.

*) Presented in this form at the R&D conference in Härnösand 24 April 1992 under the title *Regional Development in the Pan-European Community*. The geopolitical changes which have occured since then have not altered the principal message of the paper.

Economic aspects

During the period dominated mainly by economic motives the EC had been fairly successful in reducing international and interregional quantitative disparities within its sphere, although its South extension from 9 to 12 countries (1981 and 1986) signified a step back in this respect and required substantial resources to make up for increased disparities (Molle and Cappelin, 1988). Subsequently then about two thirds of the EC Regional Development Fund had to be dedicated to the four Mediterranean member countries (Greece, Italy, Portugal, Spain).

Under the cover of what seemed a rather successful attempt to reduce quantative economic disparities within the EC, however, there emerged a substantial qualitative differentiation between the more innovative areas of the EC and the rest which increasingly became the reservoir for cheap labour, raw materials and branch plant location of the more innovative areas of the EC. Since about 1990 this reservoir has been substantially increased by the haphazard effort of most of the former COMECON countries to transform into market economies and open up to the world economy. Particularly the former COMECON countries are in danger of becoming the low-wage unskilled production reservoirs and the garbage dumps of the more developed EC countries. The integrated areas will increase in future as, in the first instance the EFTA countries, but in a second phase also these Central and East European countries are or will be postulating entry, at first into the European Economic Area, but then also into the European Single Market.

Socio-political aspects

With this rapid cumulation of expansion demand upon the EC, all of a sudden a formerly grossly neglected socio-political aspect broke open to the surface, and significantly enough showed most drastically in both the most highly developed areas of the EC as well as in the less developed Central and East European areas trying to transform into market economies: the demand for ethnic identity and socio-political self-determination at the local and regional scale. A dramatic new restructuring challenge of this increasing and intensifying market economy area emerged, by which both its advocats in the West and its sudden adherents in the former "East" seem to have been taken by complete surprise. Instead of the smooth and frictionless transition to overall wealth and efficiency postulated by "naive" neo-classical economics increasingly en vogue since the breakdown of Communism, it suddenly showed that societies not only needed rationally acting individuals and the State and nothing between and around it, as Margaret Thatcher once put it (Dahrendorf, 1989), but demonstrated the importance of social group identity and Civil Society.

In many of the (economically) most "highly developed" areas of the EC the influx of labour from poorer areas within and outside the EC - attracted and often officially invited during the past decades of economic expansion -

in the subsequent restructuring period became segregated and the objects of hatred and aggression. The local "indigenous" population (i.e. the dominant national culture) saw its own jobs and social achievements imperiled and was not willing any further to share them, given shrinking economic opportunities in the restructuring process. The result was the emergence of right extremist and sometimes even fascist movements in most Western countries and of increasing ethnic regionalism or nationalism.

In most East and Central European countries, on the other hand, the transformation towards a market economy and a democratic political system at the same time also triggered a wave of ethnic regionalism and strife which led to the breakdown of many of the former national units. The background to this in part is also the unwillingness of some of the more developed regions (e.g. Croatia, Slovenia in Yugoslavia) to continue "subsidizing" the poorer parts of their former national units: in part it is sheer ethnic "nationalism" of establishing a homogenous and self-determining national unit (e.g. Slovakia in the present CSFR), a desire which during the past decades had been suppressed by central rule.

On the whole this showed that with unilateral emphasis given to economic objectives of integration, the so far neglected cultural, ethnic and socio-political issues suddenly broke open and today represent a major determinant of the further European integration and development path. The cultural, ethnic and social issues manifest themselves particularly at the regional scale, a level which has thereby received new emphasis and importance. This is expressed not only by the different movements for a future "Europe of Regions" but also by the role local and regional communities have assumed in triggering indigenous development and in meeting global challenges (Stöhr, 1990).

ENDOGENOUS STRATEGIES AS A DEVELOPMENT ALTERNATIVE

Recent European development, as described above, clearly demonstrates two facts: First, that the integrational process has very different effects on the geopolitical terrain in terms of economic sustainability and general standards of living. Second, that the process in its economic onesidedness evokes reactions on the socio-cultural level and provokes manifestations of socio-territorial identity. Both observations lead to the conclusion that socio-cultural and identity-forming processes should be considered more seriously in future phases of the European transformation process. In doing this, endogenous strategies of development should be considered as a supplement, or in several cases alternative, to main-stream macroeconomic policies.

Strategies of the kind already exist, as a result of theoretical discussions over more than one decade. The effort is known as "endogenous development", "bottom-up planning" and under several other names. The concept evolved through an intense dialogue amongst development scholars from several countries and continents. Amongst the pioneers have to be mentioned Dudley Seers, John Friedmann, Stefano Musto, Ernst A. Brugger and, in the opinion of the organizers of this conference, Walter B. Stōhr.

Of special interest to endogenous development strategies are regions, which for different structural reasons are obsolescent or bypassed by mainstream development. In the emerging economic power networks of Europe and the World these obsolescent regions may appear almost anywhere in the geopolitical landscape. However, endogenous strategies seem to remain of paramount interest especially for peripheral regions of the European countries, and more so on a continental scale. I shall here make a summary recapitulation of main points of the endogenous development strategy, as presumably relevant in the forthcoming European development and planning debate in the 90s.

Characteristics of endogenous regional development

Endogenous development is primarily sustained within one specific region, rather than being mainly externally supported. These are some of its key characteristics (see also Stöhr, 1984):

1. Relatively wide differentiation of development strategies applied: As distinct from more centrally steered regional-development policies, which were essentially based on the relatively uniform success model of early industrialization and urbanization, these endogenous programs appear to be much more differentiated. To a high degree they are based on the specific historical, cultural, institutional, and natural conditions of the respective areas and are aimed at the broadest possible mobilization of local and regional resources (natural, human, capital, and so forth) for the satisfaction of basic needs of the regional population.

2. Societal complement to market mechanism: Many of these initiatives give priority to the production of goods and services considered to be socially valuable in the region as well as to the satisfaction of basic needs of specific target groups, both criteria it was felt were not sufficiently taken care of by the market mechanism. The Mondragon Cooperative network in the Basque Country, for example, guarantees its members employment within its regionwide federation and within a radius of 50 kilometers from the member's location of residence. It has also in principle excluded arms and nuclear components from its production program (although the latter principle is said to have recently been undermined).

3. Participation as a necessary but not sufficient precondition: Most programs provide for as broad as possible a participation of its members, both in

decision making and forthcoming benefits. In some cases this participation is primarily firm or sector related and only in the second instance regionally organized (for example, Mondragon). In other cases their organization is primarily territorially based (for example, the community economic-development cooperatives on the Atlantic coast of Canada or of western Scotland). The Mondragon cooperatives, which specialize along product or sectoral lines, have recently, however, found it useful to also form territorially organized subgroups (Grupos Sociales) in order to increase their collaboration. In many of the cases mentioned, this participation and collaboration takes place parallel or outside (sometimes even as a counterweight to) the constitutionally provided local and regional representative bodies (which, in many cases, have traditionally been dominated by small elites or central government representatives).

4. Trans-sectoral orientation: In contrast to the economic monostructures that have emerged in peripheral areas during the last decades (frequently specializing in the exploitation of natural resources or of cheap labor) - and in part as a reaction to this fact - most of these initiatives aim at a more diversified, multisectoral development ("standing on more than one leg") and also at an increased interregional mutual interaction between sectors and economic functions.

5. Promotion of regional economic and financial circuits: Counterbalancing the increasing internationalization of economic and financial circuits, many of these initiatives aim at the strengthening of interregional economic and financial circuits. This is to facilitate the retention of a higher share of value added within individual regions, to safeguard regional investment requirements, to increase the innovative capacity within the respective regions, and to make them more resilient against the direct impact of worldwide economic shocks. The Mondragon Cooperative Federation, with its more than 160 enterprises, has even during the recent crisis years been able to increase employment and generate new enterprises, while the greater remaining part of the Basque economy was in serious crisis and burdened with an unemployment rate of more than 20 percent. Depending on the type of region, a higher degree of processing of regional resources, of regional research and training and more intensive regional interaction among economic sectors and/or between producers and consumers (for example, producer-consumer cooperatives) are aimed at.

6. Innovation orientation, multilevel and not restricted to technological innovation: While the concept of innovation-oriented regional policy, which developed in recent years, was mainly related to technological innovation (either by process or product innovation), many of these programs also include innovation in the organizational and institutional spheres, such as in the forms of decision making and cooperation, in organization of work, and so forth. Most of them contain new (or revitalized old) forms of entrepreneurial regional cooperation and broad democratic decision making. In the sphere of technological innovation, they include the promotion of parallel technologies, for example, of potentially decentralized (human-capital intensive) technologies applicable to small and medium-sized plants along with the traditional promotion of (finance-capital intensive) large-scale technologies. This may also require changes in the organization of work and in the institutional sphere. Of particular interest in this context is the Mondragon cooperative federation in the Basque Country, which as mentioned, provides for an endogenous research-training-production-innovation-financing complex with direct feedback loops. This has in various cases permitted its enterprises a much higher innovation rate compared with corresponding enterprises outside its network.

7. Promotion of territorial identity: Most of these programs are either based on or aim at a high degree of regional identity, be it retrospectively in the sense of ethnic or historical communality or prospectively in the consciousness of a common future fate. The presence of regional identity appears as an important prerequisite both for the cooperation among diverse (often for economic and political reasons divergent) interest groups within the region as well as for the retention or recuperation of initiative and creative personalities in the region.

8. Synergy and integration among regional economic functions, regional identity, and decision-making structures: In most of these programs, a direct linkage between economic (production, service) functions and regional decision-making processes (of workers and/or consumers) in various forms of entrepreneurial or territorial self-determination are provided for. This usually provides also for a high degree of identification of the local/regional population with these programs. Many of them take the form of cooperatives. The eastern Canadian community-economic-development programs furthermore are open for participation to all members of the respective municipalities and, at the same time, subjected to various forms of democratic control by them through contact and coordinating committees. In many cases, the economic crisis situation together with a high degree of regional identity appear to have had a strong mobilizing effect upon the local population, particularly if a high share of younger age groups was still present and no rigid social stratification existed within regions.

9. Promotion (or at least permissiveness) on the part of central authorities of linkage to transregional cooperative networks: In almost all cases, such endogenous development was only possible if the respective central government agencies were willing to either tolerate them or promote them in a way guided not primarily by central-agency interests. Furthermore, it proved essential that such development was supported by (frequently informal, but in any case, not politically dominated by established institutions) cooperative networks or "committed link cadres". These link cadres frequently fulfilled important functions in strengthening the bargaining position of regional groups vis-à-vis central authorities or external (multiregional) economic enterprise, in training and consulting, as well as in increasing the local consciousness of the reasons underlying existing problems and of the required self-organization for overcoming them.

Central government therefore does have a role also in the promotion of endogenous development.

REGIONAL SYNERGY AND FEEDBACK MECHANISMS

I shall use the opportunity to extend the list presented above by mentioning a hitherto not extensively discussed factor in endogenous strategy: the strategy of regional synergy. I will also add a few comments to this theme, which in my belief help strengthen the endogenous planning strategy.

Synergy has been defined as the "effect obtained from the combined action of two distinct ... substances (being) ... greater than that obtained from their independent action added together" (Encyclopaedia Britannica, quoted in Stöhr, 1986). In connection with regional development it can be interpreted to mean that the outcome of networks and cooperation (interaction of regional actors) will normally be greater than the mere (arithmetic) sum of the respective individual regional inputs. In its economic and technological dimension I have called this a "regional innovation complex" (Stöhr, 1986).

But regional synergy has importance also beyond the economic and technological dimension. It includes the interaction between regional culture, the natural and built environment, regional social activities and movements, particularly civil society and NGO's, the regional political structure etc. Experience shows that regional feedback loops need to exist/be established between these different components of regional life in order to facilitate sustainable and endogenously supported regional development. These loops need to operate as negative feedbacks or checks and balances in order to avoid that single (often externally determined) developmental factors (e.g. an externally based mining/industrial company) destroy other regional conditions (e.g. the environment, labour relations, local production and service facilities etc), considered important for the regional life sphere or development by the regional community.

Some of the above mentioned regional autonomy demands of European Regions have been caused by the lack (or frustration) of such regional feedback mechanisms. Cases in point are the present autonomy demands of

Slovakia in the CSFR, who blame their detrimental industrial and energy structure (armaments industry and dangerous atomic energy plants) upon former central government decisions in Prag, or the Polish and Hungarian case studies reported on in the UN-University project "Global Challenge and Local Response" (Stöhr, 1990). - But the lack of such regional feedback mechanisms also causes serious deficiencies in many externally dependent regions in the West, such as in mono-industrial areas or in company towns. In these cases often a dominant industrial sector tends to monopolize local financial resources and political structures to the extent that it prevents more diversified or smaller industries to establish or survive, and is often also a handicap for economic, technological and political innovation. Many present "old industrial areas" suffer from this phenomenon which turned out to be an often serious handicap to their restructuring needs.

In ideal terms, the hypothesis would be that regional feedback mechanisms could prevent that one single developmental factor (or as in the above cases: actor) becomes so dominant that it seriously impinges upon other regional conditions considered important for the life and developmental quality of the regional community. In concrete terms this would mean that local political feedback loops (e.g. in the local municipal council) would act as a brake once e.g. a local dominating sector imperils the local environment as an important basis for local living conditions (e.g. water supply) or for development (e.g. tourist development, attractivity for high-skill sectors etc).

For such regional feedback mechanisms to actually become effective in many cases, however, either the delegation of decision-making power to the regional level or a process of regional self-empowerment (Friedmann, 1992) may be necessary which give the regional community sufficient power to intervene. This will in most cases have to be accompanied by a regional learning process, however, which gives the local population at large and its representatives sufficient information about different developmental facets and particularly about their mutual interrelations and the consequences of different developmental options over time.

The functioning of feedback mechanisms can be conceived of in a systemic perspective in which different sub-systems interact within a regional system and towards the outside. This involves systemic (integrated) rather than the usual partial (fragmented) thinking. Important sub-systems are the natural and built environment, the cultural, economic, social and political sub-systems (Fig. 1).

If the feedback mechanisms between these sub-systems function at the local/regional scale, they can prevent that "over-development" in one sub-system causes excessive damage in other sub-systems to the detriment of the integrated development potential of the local/regional community. It can also prevent that similar negative effects at the local/regional scale are caused by action at higher levels. While all sub-systems interact, the key role no doubt lies in the political sub-system where a democratatic decisions-making process on the development path of the region should occur.

REFERENCES:

Dahrendorf, R., 1989: "Zehn Jahre und kein bisschen leiser. Margaret Thatchers konservative Revolution - ein Frontalangriff auf die britische Gesellschaft". Die Zeit. Hamburg. Nr 19 - 5.5.1989.

Friedmann, J., 1992: Empowerment. The Politics of Alternative Development. Blackwell, Cambridge MA and Oxford UK.

Molle, W. and R. Cappellin, 1988: Regional Impact of Community Policies in Europe. Aldershot and Brookfield, Avebury.

Stöhr, W., 1984: "Changing external conditions and a shift in regional development strategies", in Musto, S.A. and Pinkele, C.F. (eds.): Europe at the Crossroads, pp. 283-308. Praeger, New York.

Stöhr, W.B., 1986: "Regional innovation complexes". Papers of the Regional Science Association. Vol. 59. pp. 29-44.

Stöhr, W.B., 1990: Global Challenge and Local Response. Initiatives for Economic Regeneration in Contemporary Europe. United Nations University. Mansell, London and New York.

Stöhr, W. and Taylor, D.R.F. (eds.), 1981: Development from above or below? The dialectics of regional planning in developing countries. John Wiley and Sons, Chichester.

Fig1. **Regional feedback mechanisms. Systemic sublevels**

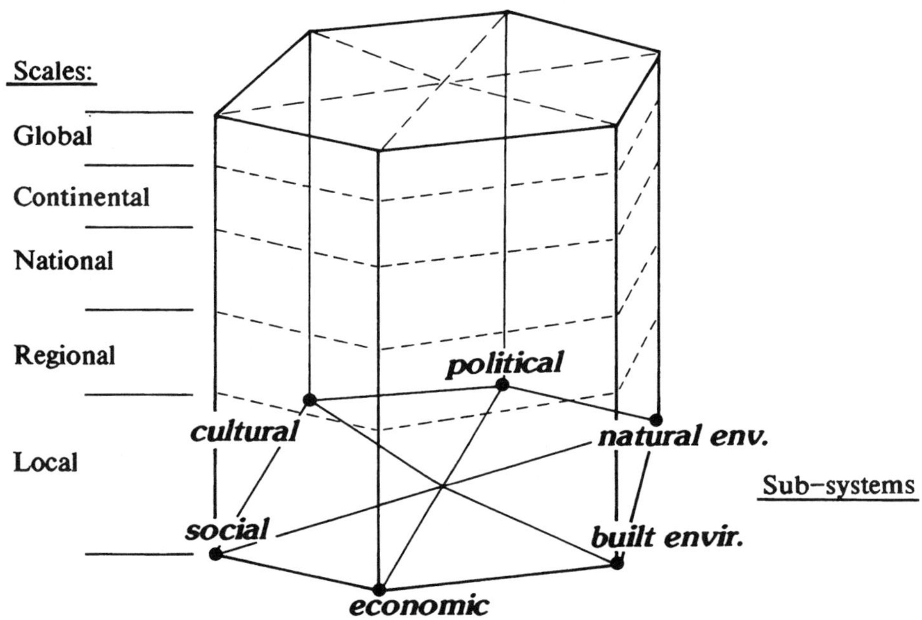

Region, Territory and Identity

Torsten Malmberg

SPACE FOR MAN

Man and environment are the two basic components of human ecology, the long neglected but now fast growing science which in the actual crisis seems to be quite indispensable (Hubendick 1985). It is not seldom forgotten that the two are related to each other in a natural way by behaviour. And this comportment is mainly realized as chains of elements which do not function meaningfully but as inborn, programmed serial systems, interpreted within the frame of ethology, i.e. the modern Lorenzian psychology. Behaviour is therefore very sensitive to disturbance, and so the relation mentioned is easily put out of work. We have here most probably one of the foremost factors behind environmental and other troubles.

The primary ecological interest of man is, of course, centred on vital resources. Among them the most fundamental ones, time and space, are remarkably often left aside or step-motherly treated. They are strongly connected to each other, forming a unit in the geographical school of Hägerstrand at the Lund University. The temporal resource can just be met with passively. But space is strived for actively, which is manifested e.g. in the term "habitat selection", that is the choice of a place to live in, where the expansion of the human settlement and a healthy environment unfortunately often stand against each other.

The representation of "The Space of Man" - which is the title of a valuable book by the director of this conference (Råberg 1987) - in our central nervous system arises from manipulation of objects and from movement in the physical environment rather than from immediate perceptual copying of the environment.

In his foreword to a book by a physicist, Albert Einstein remarked that the concept of space was probably preceded by that psychologically simpler one of place, defined like this: "Place is first of all a (small) portion of the earth's surface identified by a name". He contrasted the historical concept of space as positional quality of the world of objects against that of space as container of material things (Jammer 1954). This last meaning is applicable for instance to the region.

TERRITORIAL BEHAVIOUR

To have roots in a place is to possess a secure base, from where to look at the world, a firm grasp of one's own position. Deep relationships with localities are thus as unavoidable as similar ones with people. Without such connections human existence may be possible but is bereft of much significance. So we simply state that if we can find our place in the world, it must be a cosmos. In a chaos there is no ordered place whatsoever. Territories represent cosmos because they are consecrated one way or the other by human occupation, a symbolic transformation which also suggests a repetition of our evolution.

Strictly speaking, behavioural territories are spatial phenomena to which humans are knitted by action and emotion. It is true that there must always be a possibility of defense in the background, but normally territories do not awake but avert fighting, mainly by means of distribution of available ground and resources between individuals and groups. In fact the ecological-ethological concept of territoriality is so broad and so dynamic that it is difficult to define it incontestably. Awaiting consensus about a good formulation, some key words may be used provisionally: a) knowledge, b) disposition, c) delimitation, d) marking, e) reservation, f) competition, g) defense, and as a kind of summary h) identity.

From this it will be clear that territories effect restriction on unselective movement, that they divide between members and non-members and that they assume some form of recognition. Behavioural territoriality involves an ordering principle with numerous variations both of a rhythmic and a non-rhythmic character, often resulting in hierarchies like chinese boxes. Every territory that has been in function for some time leaves a lasting imprint upon the relevant area. Whether predominantly marked by nature or by culture it tends to create, by its very existence, certain stamps that might be ponderable factors in the future. This is in agreement with experiments by Edney (1972) who found that ground ownership was not reliably associated with defense of land. Instead strength of territoriality seems to be clearly related to length of disposition, and thus the time perspective is always relevant to the study of territories.

Physical threat and implicit threat in the form of status are among the most likely releasers of territorial behaviour. Further, territories and their physical manifestations are maintained only when needed and are generally disclosed just by human reactions. It is suggested that behavioural territories beside their basic task to provide ground for the physical necessities of life have other meanings too. They probably represent environmental entities suited for adequate stimulation of the nervous system, at the same time guarding psychological demands like those of privacy and security. A triune

"Thoughts of the Mountain". Pencil drawing by Kirstie Ekelund

inborn ground of territorial behaviour is well demonstrated by MacLean (1973) and his co-workers. The three centres of the nervous system are located in the upper brain stem (establishing and defending of primitive territories), in the limbic system within the brain (emotional reactions) and in the right hemisphere of the brain (concept and integration of conscious territoriality).

There were hardly any serious studies of territoriality before World War II outside anecdotal evidence back to the Sumerians and among explorers. This might depend on the fact that the phenomenon has been trapped in the chasm between "the two cultures", the natural sciences and the humanities. Thus the only complete international monograph on physical human territoriality seems to be that by the present lecturer (Malmberg 1980).

The Swedish term "revir" (territory) was up till about twenty years ago in human contexts just used as a designation of administrative forest districts of which there were six in the county of Västernorrland. Today the situation has totally changed, and the actual word is e.g. found in nearly every number of the dailies, unfortunately mostly in the form of "revirtänkande" (territorial thinking) and with negative connotations. Without doubt every inborn behaviour runs the scale from minimal to maximal expression, where the extremes may be detrimental. But the normal territorial behaviour seems to be necessary for survival and has nothing to do with thinking but with instinctive reactions. Perhaps it should here be mentioned that we wholly leave aside non-physical territories, for instance competence areas, influence fields and organization levels, which doubtless are as frequent and important as the physical ones.

REGION AS TERRITORY

The human society has a marked tendency for geographic grouping according to special criteria of uniformity. This is a consequence of the fact that man to a certain degree is gregarious and space-bound, resulting among other things in regionalism. The term "region" comes from the latin verb "regere" which means to direct and to rule. Most tenacious are doubtless regions which are based on common resources or a uniform administration. Very genuine are also those kinds of regionalism which have appeared in the form of and inside nation states. Well-known are the subdivisions like province and county which have not seldom a distinct autonomy, like the Swiss cantons and the United States of America, the last ones being of special interest because of the straight artificial boundaries.

In a valuable paper called "Regionalism and Irregionalism" Lewis Mumford (1927) presented some fundamental reflections which seem to be as sound today. He stressed the human neglect and systematic misuse during earlier centuries concerning relevant ground areas. Up to a certain point, he said, economic and social life can be conducted without regard to regional actualities. However, the blind heaping up of population in metropolitan districts, and the equally blind impoverishment through bad marketing and an inadequate distribution of the population in rural areas, cannot continue indefinitely. And he summarizes:

"We can no longer ignore the region; for the day of the pioneer who under the colonial regime blindly laid waist a particular area and moved on, is over; and the period of ruthless exploitation of the land and its people for purely monetary ends, as during the industrial era throughout Western civilisation, is drawing to a close. In every social operation the region lays down certain fundamental conditions. These conditions do not always determine the nature of the operation; but no plan has any prospect of permanent success that is framed and executed without regard for the fundamental natural and cultural history of the region. When we acquire the regional outlook, we think of the region as a whole, and we realise that in each geographic area a certain balance of natural resources and human institutions is possible, for the finest development of the land and the people".

Aristotle gave in "The Ideal State" the first acceptable description of behavioural territories and presented at the same time some prerequisites of territorial conflict. As to the national state he asserted that it should cover an area that is self-supporting, surveyable and easily available for the inhabitants but confusing, difficult to attain and impregnable for its enemies. This is certainly crucial for the relations between behavioural and political territorology, the last term sometimes used as a designation of the subscience of territory. The national state as well as the administrative region and other such entities can be sound just if they display natural collective behavioural territories, and this is, unfortunately, not at all always the case.

"If the law of nations is not to deny itself, it must defend not the actual, more or less accidental territorial status of a certain historical moment but the basic spatial structure as a unit of order and area, that is *nomos*" (Schmitt 1950). This is the Greek word for law, originally suggesting the first ground measurement, the first division and distribution of land, later even identified with the boundary between households and probably covering realities not far from territory.

Friedmann and Weaver (1979) presented a chapter called "Towards a paradigm shift in regional planning" which was explained as "The recovery of territorial life". Here it is asserted that "A new approach to territorial development" is necessary. Basic needs refer to the sum of reciprocal claims in a territorially integrated society, and in such a society everyone is regarded simultaneously as a producer and a consumer. Selective territorial closure refers to a policy of self-reliance at relevant levels of territorial integration: district, region, nation. Such closure is a way of escaping from the fetischism of growth efficiency. It is an expression of faith in the abilities of a people to guide the forces of their own evolution. It means to rely less on outside aid and investment, to involve the masses in development, to initiate a conscious process of social learning, to diversify production and to pool resources. On the whole this is the way we have to go tomorrow.

The actual tendency, not the least in Europe, is a weakening of the nation states and a strengthening of the regions. Compared to other countries Sweden is today a strong central state and has strong municipalities but rather weak regions. This was the general message from KEFU (The Council for Economic Research and Education) at its seminar in Lund, February 5, 1992, where the key words in the context were: identity, distinctive character and active planning. A general ultimatum was formulated, meaning a strengthening of the regional level!

IDENTIFICATION WITH THE ENVIRONMENT

It is remarkable that in territorial connections you can throw out your bait in any water and be sure to get fish, i.e. territories. As soon as a spatial entity is unique enough to attract loyalty feelings from some people and not from others there will arise territorial behaviour. This doubtless means that we have to do with some fundamental reactions which in a simple way could be characterized as the tolerance limit between people. Helmfrid (1974) asserted that the increasing knowledge on continuity and identity of territories mediates a better and more profound perspective concerning the history of cultivated land.

The term "topophilia" is a neologism, particularly useful to include all of the human beings very different affective ties with the material environment, to which the response may be primarily aesthetic. It can vary from the fleeting pleasure one gets from a view to the far deeper sense of beauty that is sometimes revealed. Doubtless, the response can even be tactile, a delight in the presence of air, water, earth. More difficult to express are permanent feelings toward a place, because it is home, the locus of memories and the spot of gaining a livelihood (Tuan 1974).

The notion of identity is a basic one in our everyday life. Thus we recognize the identities of people, place and even nation. Possibly, identity is a phenomenon that evades simple definition, although some of its main characteristics are apparent. In particular, the difference between "identity of" and "identity with" should be noticed. The identity of something refers to a persistent sameness and unity which allows that this something can be differentiated from another thing with which there is no identity.

Influence on place identity resulting from changes in the individual's physical world can be traced to more than simply technological development. Demographic and ecological alterations in a community, themselves the result of economic, political and social impact, may have important consequences. The intrusion of unwanted groups, the evidence of crime in the area, beginning signs of physical decay may all precipitate stronger emotional attachment to the home and the surrounding region.

Similarly, groups who are dislocated from their residential "turf" often feel lost in their new setting and long for their old home. Depending on the stage of the life cycle and the discrepancies between the old and the new location, a sence of belonging to the new neighbourhood may or may never appear (Fried 1963).

The primary mental troubles connected with absence from the well-known, original, more or less imprinted territory are, of course, those summarized as homesickness. Several authors have stressed that typical homesickness is particularly frequent among inhabitants of mountains, which explains an old German name of the disease: "Schweizerkrankheit". Probably it depends on the conspicuous differentiation of the landscape, above all of that spot on earth where one has lived. Broadly speaking, there is, of course, no reason why mountain tribes should love their home district more than peoples of the plains. But evidently humans tend to be more impressed by the strongly structured mountains where every point is unique than by the plains where all the points are rather similar (Simmel 1908).

Clearly, individuals do know who and what they are in terms of such affective ties to house and home, neighbourhood and community. It is, however, important to add that definitions of the physical world of a person's self-identity extend far beyond this. What emerges as place-identity is a complex cognitive structure that is characterized by a host of attitudes, values, thoughts, beliefs, meanings and behaviour tendencies which are far more than just emotional attachment and feelings of belonging to particular places.

It is hardly possible to find a more persuasive sign of the strength of a person's love of his home than that manifested by the Icelander Gunnar of Lidarände in the saga of Nial, who about a thousand years ago was killed by the enemies, having refused to go into exile. Indeed, he was on his way to leave for Norway but turned around to have a last glance of his place. He caught the whole scene from sea to fjeld suddenly lit up by the sun. He then uttered the famous words. "How fair is the valley. It never was so beautiful before. I will go home and never leave again"!

THE CASE OF VÄSTERNORRLAND

In his fascinating biography of Patrick Geddes, the Scottish pioneer of human ecology as well as of several other innovative disciplines, Philip Boardman (1978) reports how young Patrick went up on a hill outside his home town Perth and regarded the rivulets running down the slopes. In his imagination he magnified the scale into large river valleys and started thinking of them as natural regions for study and administration. During the rest of his life Geddes built out these theories and distributed them not so much in writing as in lecturing, paying visits to different parts of the world

Even though there is no real proof it seems to be highly probable that the Tennessee Valley Authority (TVA) and other famous projects are offsprings of the brilliant Scot's ideas which made him the worlds first real planner. The Swedish term for territory, "revir", comes from the latin "ripa" (shore), which is also identified in the words "river" and "rival", competitor for example of territories. All this signals the important function of water courses in the characterization, bounding, marking and identifying of behavioural territories.

It is well-known that the northern Swedish river valleys since old times have functioned as central settlement and communication axes in the landscape, leading from the sea in the south-east to the fjelds in the north-west. The recommendable ecosystem approach has already been applied for example to the Lule River Valley in the north by Åkerman and Lundholm (1990) in their monograph about humans and resources in the said valley from 1300-1600. Ångermanälven is the main river within the Västernorrland County and has the largest basin as well as the largest water quantity among running courses in the northernmost parts of Sweden. Doubtless this gives a strong territorial mark to the region.

The Västernorrland County originally covered an administrative district comprising the provinces of Hälsingland, Gästrikland and Härjedalen and got the Härnösand County as an addition in 1654. Later on for long periods even the provinces of Jämtland and Lappland belonged to this district. However, in 1762 the very large area was divided into two counties, Västernorrland with the provinces of Medelpad, Ångermanland and Jämtland and Gävleborg County with the rest. Later Jämtland was constituted as a separate county together with Härjedalen. The actual county of Västernorrland has a land area of 21.786 square kilometers and is sparsely populated by just 267.935 inhabitants.

Unfortunately it has not been possible for the present lecturer to make an extensive investigation of the Västernorrland County as to territorial characters. However, the mountains which protrude between the river valleys toward the sea should have the suggested effect of making a landscape imprint on the population. Of special interest in this connection is "Höga kusten" (The High Coast), the only example of its kind in Sweden. It is true that there are two very large rivers, Indalsälven in the southern and Ångermanälven in the central parts of the region. Doubtless, however, the latter and its valley, Ådalen, is as we have found the dominating one. As to towns, Sundsvall in the south has by far the largest population and industry, but Härnösand is anyway the primary centre of the region for administrative, cultural and social reasons, being the residence town of the county governor and having an episcopal chair as well as a university college.

These suggestions may represent important factors for the identity and territoriality of the Västernorrland County. But it is hardly decisive if we can classify this region as a natural collective territory or not. Experiences not least from the United States, where boundaries drawn by ruler in offices have functioned well, seem to point in this direction, at least in connection with judicial and similar matters. The important thing is, of course, if the people of the Västernorrland County, regardless of reasons, show some unitary feelings for their home territory. However, the longer the district and its boundaries have been intact, the more we can calculate with territorial reactions. Common characteristics to language, literature, customs, livelihood and common history and culture in general may be strongly influential.

Much is in favour of the fact that a large northern region like the Västernorrland County with its economy predominantly built on natural products and a population very interested in hunting and fishing should be a good behavioural territory. We also have proof of a strong attachment to home and unwillingness to move, when lack of employment and support speak for emigration.

As already stated, to feel at home is to take part in a milieu that has first of all a special physical character. At the same time nobody can be conscious about his home-feelings without having been away from home, intellectually, emotionally and physically. Therefore, it is not surprising that the most penetrating accounts of regional identity have been written by authors in exile: Henry James on America and James Joyce on Ireland are two of the best examples.

In contradistinction to the stream of people from the northern to the southern parts of Sweden, particularly to the urban regions, we have now use for a stream in the opposite direction. In fact we have a beginning in the form of retired people longing for home. To give an interesting international example, most Norwegian sailors who during this century lived for decades on the Galapagos Islands returned to their birth places in Norway at high age (stated by Gillsäter, 1986). In a lecture at the annual conference of "Sveriges Hembygdsföreningars Riksförbund" (The National Federation of Swedish Home District Associations) in Östersund 1986 I recommended sparsely populated communities to ask old people born in the area if they would like to go back home and stay there. This proposal has also been made by others and it has started to yield positive results.

On the European continent the exploitation of the landscape and its flora and fauna has in recent times reached devastating proportions. Under such circumstances a still rather natural region like the Västernorrland County should be looked at with great expectations from the urban and industrial people of the south, seeking adventure, wilderness and recreation. It is true that also regions like Västernorrland are badly hurt by unwise forest clearing, nuclear contamination from the east and so on. But a change of the economic direction, which for different reasons cannot wait much longer, could put us in another and better situation. Thus a consistent and radical approach in the name of ecology, management and environment protection, could doubtless make the Västernorrland County a model territorial region mediating to its people a new self-respect and identity!

REFERENCES:

Boardman, P. 1978: The Worlds of Patrick Geddes. London.
Carlstein, T. 1982: Time Resources, Society and Ecology. Lund Studies in Geography B 49. Edney, J.J. 1972: Property, Possession and Permanence. Journal of Applied Social Psychology 2: 275-282.
Fried, M. 1963: Grieving for a Lost Home. In: The Urban Condition. Ed. L.J. Duhl. Newport-London. P. 151-171.
Friedmann, J. and Weaver, C. 1979: Territory and function. London.
Gillsäter, S. 1986: Urtid i nutid - Galapagos. Höganäs.
Helmfrid, S. 1969: Bruk och missbruk av regionbegreppet. Ymer 89: 17-30.
Hubendick, B. 1985: Människoekologi. Stockholm.
Jammer, M. 1954: Concepts of Space. Cambridge, Mass.
MacLean, P.D. 1973: A Triune Concept of the Brain and Behaviour. Toronto and Buffalo.
Malmberg, T. 1980: Human Territoriality. The Hague-Paris-New York.
Mumford, L. 1927: Regionalism and Irregionalism. Journal of Sociology.
Råberg, P. 1987: The Space of Man. Almqvist & Wiksell International, Stockholm.
Schmitt, C. 1950: Der Nomos der Erde im Völkerrecht des Jus Publicum Europaeum. Köln.
Simmel, G.1908: Soziologische Untersuchungen über die Formen der Vergesellschaftung. Leipzig.
Tuan, Y.-F. 1974: Topophilia. Englewood Cliffs. N.J.
Åkerman, S. and Lundholm, K. 1990: Älvdal i norr (River Valley in the North). Almqvist Wiksell International, Stockholm.

The Forces of the Social Life-Space

Gunilla Lundahl

We are living in a time of cataclysmic changes. We can see them in progress round about us at an ever-increasing rate and with staggering results. The time has come for us to realize that we are compelled to set off on the voyage towards a new world picture and new forms of living. The World Watch Institute has given us just over a decade to change our ways, if the earth as an ecosystem is not to collapse. It is not a new technology that is called for. It is a new style of living.

Terror, suspicion and uneasiness follow in the trail of changes in the familiar pattern of things. We have to be prepared for this. But the changes, the decampments, also lead to a breakthrough of wonderment, curiosity and creative force. This we can hope for. Build on.

A shift of figures of thought is now taking shape more and more clearly. At the centre of it all is our view of nature. The 17th century saw a contest between two scientific approaches. That contest was waged in philosophy and the academies. But also in a war against craftsmen and peasants, alchemists and witches. It was a contest between a meristic/holistic approach and a mechanistic, dualist one. The mechanistic approach was victorious and has left its mark on science, technology and planning in the west down to the present day.

The meristic/holistic approach tells us that the parts can only be understood in terms of the whole. Understanding is founded on a dynamic relation between subject and object. Knowledge develops dialectically. Opposites only become intelligible as different sides of a whole. Empathy and intuition are implements in the search for knowledge.

According to the mechanistic world picture, the whole is understood in terms of its parts. The world is an atomistic meccano which can be disassembled and built up again. It is governed by laws which a rational mind can analyse. The world is explained in terms of hierarchic pairs of opposites - spirit-matter, male-female, body-soul, nature-culture, subject-object. Reason can explain everything.

The paramount exponent of this view of science, and its founder, was René Descartes, the French philosopher who froze to death in Stockholm in

1650, having been summoned there by Queen Christina. He employs the image of clockwork to describe the world and reason as its ordering principle. Francis Bacon was his rival in the construction of intellectual models, significant partly through his New Atlantis, in which he formulated the theses on which the Royal Society in England came to be founded. The contemporary Finnish philosopher Georg Henrik von Wright gives an exciting and perceptive description of these events in his book "Science and Reason". Concerning, among other things, the background to the Royal Society, he says: "One cannot help noticing that the language in which both Bacon himself and the founders of the Royal Society describe their programme is shot through with sexual symbolism. Nature is the woman whom the scientist must conquer. Methodically and systematically he is to unveil Mother Nature, exposing her secrets. Penetrate her and in this way force her into complete submission. Thus, right from the beginning, the new scientific attitude was given a powerfully masculine image - something which the feminist movement in our own time has had cause to notice."

Ruling and dominating nature, then, was the foundation of this new view of science, which accordingly took upon itself the right to exploit, and so it has continued.

The vanquished intellectual image had been fashioned by alchemists and hermetics, who were now ridiculed, expelled and forbidden. It had been formed by the witches, with their knowledge of life. They were burned. 1.000 in a week at Torsåker, not far from here, during the 17th century. It had been formed by the craftsmen and peasants, with their experience and their hands as a knowledge base. They were crushed in the rebellions of artisans and peasants which acquired a foothold in the towns. Opposing them was the rational scientist, the one and only possessor of true knowledge.

During the 20th century, we find science making discoveries which foul up the objective, rationalist, mechanistic conceptual apparatus. The theory of relativity. Heisenberg's uncertainty principle. Quantum mechanics pointed to incompatibilities. The electron was both wave and particle, both movement and matter. The dualism between the two was dissolved. The atomic event could not be captured in both time and room simultaneously. The beholder came in between. Chaos research unrolled the world fabric in all its immensity. The new physics is looking for parallels in oriental Taoism.

In the present-day philosophical debate, the great projects of modernism, derived from the Enlightenment, are being called into question. The ego is deconstructed by Julia Kristeva and François Lacan. In Pierre Bourdieu's descriptions, man becomes an actor, a chance projection of history and society.

The rational scientist is threatened. Mechanical nature coughs and rattles. The fog is spreading.

The overtly sexist view of truth and rationality has shaped a culture in which feminine thinking and female work have been subordinated and obscured. It is above all Evelyn Fox Keller who, in "On Gender and Science", has analysed the traditional science from which women have been discarded because they are subjective, emotional, spontaneous, preoccupied with relations and processes, whereas men stand for objectivity, distance and sense of purpose and are therefore given a high ranking position. Men's caring ethic focusses on justice, while women's focusses on needs. Women's rationality comprises the capacity for holistic integration, understanding, listening to the material. That rationality, consequently, is perpetually in conflict with the male version.

Concepts like nature, work and the division of labour have been defined with a male horizon in view. Women's research calls for a revaluation. Nature is not an object. We ourselves are part of it. Nature can be recreated and cherished, that is a profound feminine experience, routed in responsibility for home management and survival in virtually every culture, stored in Praxis as defined by Bourdieu. Work has been defined as a disciplined, paid activity in the production of goods or services. For women, work also means physically and spiritually renewing humanity, sustaining life, what we usually term reproduction. The guarantee of that life and culture will continue. The division of labour between the sexes is described as a division whereby each has taken his/her half of the burden. But this division of labour is based on exploitation, on women's unpaid work, with child birth, caring activities and the running of the household looked on as something belonging to nature and therefore discounted, made invisible, defined away out of economics and political decision-making. Half of all material production takes place within the subsistence economy but is not counted, is by definition excluded from the system of National Account. Nature, similarly, is not seen as a resource or economic asset. Which makes all economics an absurdity in a survival perspective.

In one dimension there is a theoretical discussion of the nature of the world. Parallel to this there are the practical experiments, the applications of new insights into the order of things. The two fertilize each other. Today the environmental movement and the women's movement stand for social utopias translated into everyday living and values. The eco-city, the eco-village are new social experiments which are looking for a firmer foundation and to which many of the environmentally conscious are committed. Women's researchers see in collective housing and co-operative organization solutions to problems which women with caring responsibilities are confronted by in their everyday lives. The idea of democracy and of influence over one's living situation is always getting knocked about. Once again we have to show and test its meaning in practice. This is done at local level, where self-determination, co-determination and co-operatives can flourish. For experts, and for architects and planners not least, it implies a new professional role, in which capacity for empathy, listening and pedagogical action become important qualities. The regional movements are an expression of this.

I would like to quote a few examples of hope-inspiring phenomena, possible positive specimens which can point the way towards the renewal which we have to inaugurate.

I will start with experiments with an ecological focus. Because they are experiments, self-management, co-operation, autonomy and social involvement, the attitude to nature and life, play an important part. Several examples come from Denmark, where there is social curiosity and trust in the informal.

My first example concerns a *multi-family dwelling in Baggesens gade, Copenhagen,* and about 10 households in the building, which dates from the end of the 19th century. It was dilapidated and needed repairing. The owner was more than willing to sell it to the tenants, and the people living there very much wanted to limit their expenditure by using their own labour. The residents also included an architect, Floyd Stein, who was strongly committed to ecological building. The courtyard façade needed repairing. That became the nucleus of their project. First, though, they got started on the courtyard. Using compost, they developed a flourishing flower and vegetable garden facing south. They decided to put a glass screen on the outside of the façade. They found a builder who was ready to work with them and on their conditions. A framework of old ships' masts was erected. Windows were collected from demolition sites. The floors were insulated with crumbled frigolite from the radio dealer. Simple ventilation was installed in the floors. Further minor repairs and decorations were carried out by the residents together. A greenhouse with a flow form water basin was built. This gave the residents not only cheap repairs, 30 per cent less energy consumption and a number of aesthetic and architectural benefits, but also greater self-reliance, a perception of their collective competence. They became closely acquainted with each other and also with their surroundings. They acquired a new sense of the interrelationships of use, consumption and ecology. They also acquired a new, semi-private arena - the narrow rooms inside the glass wall. There an inner life is reflected, vegetation is cared for, washing is hung up to dry and both play and rest occur. From the inside they can experience the life of the courtyard at close quarters. All this adds up to an exciting spatial increment which transforms social patterns. Otherwise the semi-private zones are almost obliterated in the modern city.

Another example is the renewal of *Rundradiogatan in Göteborg.* A 1950s district was to be given additional insulation. Architect Christer Nordström asked to be put in charge of one of these buildings. The outer wall was retrofitted to this building as to the others, except that an air gap was left between the façade strata. Air solar collectors were put on the roof. A very simple fan system was installed in the attic, and this circulates the warm air in the space between the two walls. The result is a 40 per cent reduction of energy consumption. But this was not the main thing for Christer Nordström.

To manifest the sun as an energy source and productive force, he had a greenhouse built the full length of the south façade and divided it into sections for the 20 or so households. He then handed it over to them, with the proviso that nobody was to give away their holding during the first year. To start the growing, the residents formed a purchasing association for implements and materials, and so the first community came into being. Those who were used to gardening inspired the others. Before the year was out, everybody had put their holdings under cultivation and the tomato plants were prospering. Some of the greenhouse near the entry was turned into sitting rooms with grapevines in the ceiling. Rag rugs were eventually put down on the basement corridors. Additional social rooms were fitted out. The social life of the building flourished. The normal entrances to the building were abandoned in favour of that passing through the greenhouse. During the second spring, a number of enthusiasts began digging up the lawn in front of the greenhouse. New carrot patches were laid out. Fruit trees were planted, for pleasure and shade. "So nice for the old boys to have something to do instead of hanging about at home and grumbling," said the elderly women pensioners. And the men, foot on spade, made extensive social contacts. The word "social control" took on a new meaning. And the residents improved their eating and living habits. But social inertia is formidable, and so far no housing manager has followed the example thus set.

Many people are afraid that ecological living will cause inconvenience and add to the workload about the home. On closer inspection one finds that time is both lost and gained, but above all, what happens is that a great deal of necessary work is brought out into the open and people are made aware of it. Practical work often becomes one side of social emancipation. Elderly people have noticed this, at least in Denmark. *Mariendalsvej 14 in Copenhagen* is one example. A pension fund organized a questionnaire and study project for its members on attitudes to and the content of ecological living. This met with a great deal of interest, and a first building has been completed - the one in Mariendalsvej. Solar heat reaches the communal swimming pool, creating a good climate in corridors and social areas. Composting and care of the garden are the vitalizing responsibility of the residents. And here the residents become a major support for each other.

For children, an ecological day nursery can mean that awareness of the world about them takes on a palpable dimension. Even so they can all look very different. As part of the *urban renewal of Kreutzberg in Berlin,* an abandoned factory has turned into a day nursery with a great deal of growing and solar heat - an environment with generous space, many happenings and sensory impressions to assimilate, and a lot to explore. Another *day nursery* is *in Wales.* Christopher Day has designed rooms which very palpably illustrate natural form, circadian rhythm, light as both a mystical, mythical and explanatory force. A building for fantasy and intimacy.

Ecology demands pedagogics. Curiosity and wonder are aroused and

Senior collective. Mariendalsvej, Copenhagen.

Ecological day nursery in Wales. Architect Christoper Day.

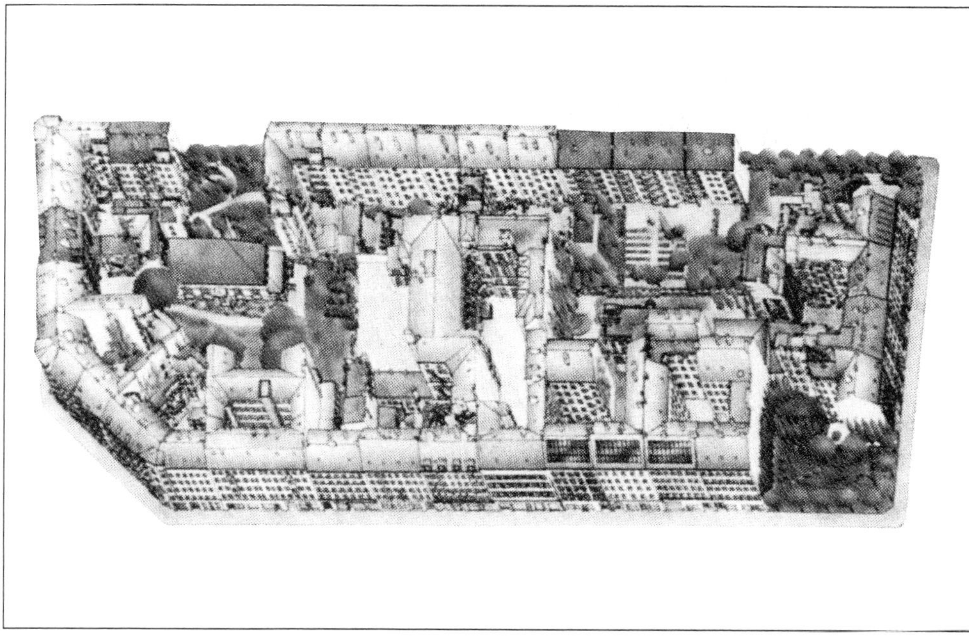

Ecological city renewal. Block 103 in Kreuzberg, Berlin

Detail from ecological city renewal. Slagelse Green City project, Denmark.

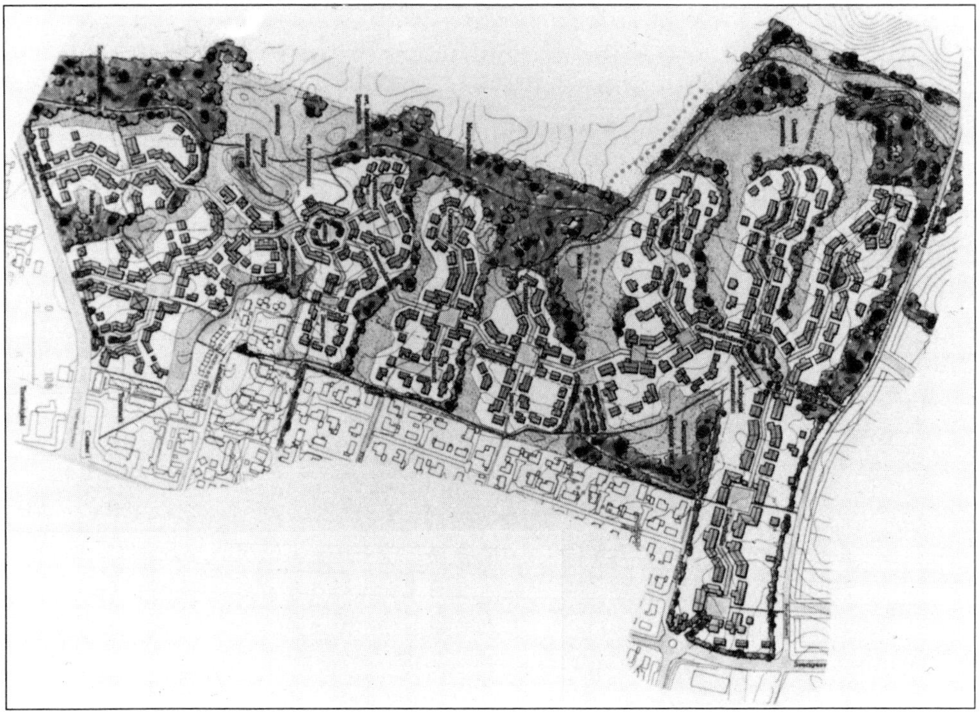

**Torsted Vest district, Horsens in Denmark.
Associated with the WHO Healthy Cities project.**

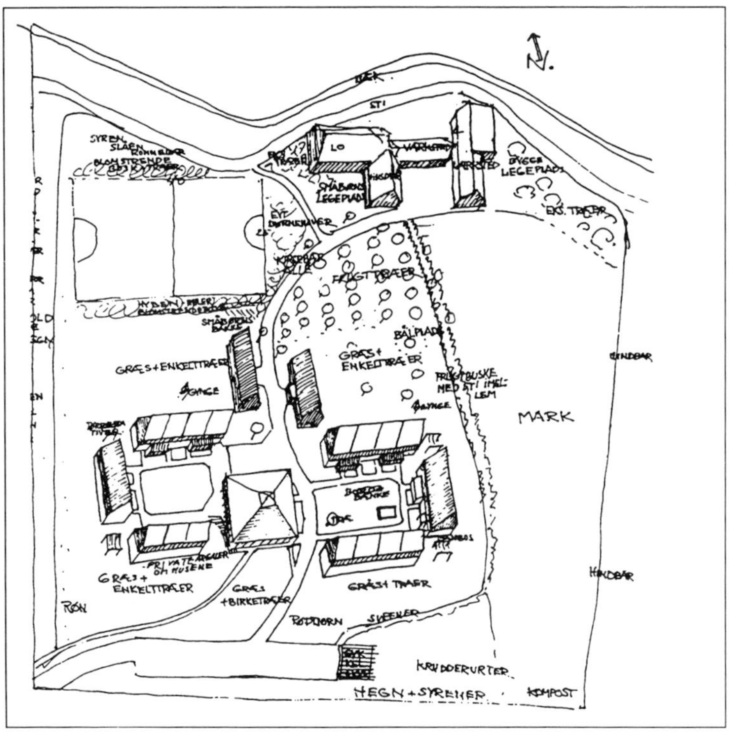

Eco-cooperative Overdrevet. Hinnerup, Denmark.

followed by insight, knowledge and capacity for action. Who takes responsibility for this? Not the traditional school. In Copenhagen, at all events, the drawings and a firm scheme for an ecological building exists - a kind of alternative to all the Science Centres which have been created. An effort was made in Karlstad to build one up on a minor scale, but it foundered on the rock of local government finance. Perhaps the first *town farm in Skövde* is a beginning. People must be enabled to meet in a place where questions can be asked and knowledge exchanged.

Workplaces are also good breeding grounds of ecological insight. Here we can take as our example Ton Albert's *NMB bank in Amsterdam.* Part of the philosophy - as few lifts as possible, stairwells instead, open, well-lit rooms for meetings - important for the dynamics and pulse of the workplace.

Ecological buildings have often been said to be ugly or unnatural, but ecology has a great potential for endowing architecture with intelligibility, sensuality and an engaging content. Look at *Frei Otto's building in Berlin.*

Does ecology exist at the level of urban development? Levels vary from the block to the entire town or city or urban region. One can see, for example, how the city of Berlin has formulated far-reaching demands concerning ecological infrastructure and ecological urban renewal. In *the project Healthy Cities,* the WHO is trying to elevate the ecological ambitions of planners. In Denmark, people are taking things into their own hands and planning ecological rural communities for extensive self-sufficiency. In the Netherlands, Lucien Kroll has launched *the Ecolonia project near Utrecht.* A total of 300 buildings are to be erected over a protracted period. A new development is to acquire an identity of its own through an analysis of the site, in-depth participation by the residents and the efforts of about ten interacting architects. For Lucien Kroll this means laying the foundations of creative processes. Local culture, local non-toxic, energy-efficient materials, local resources and planning on a dialogue basis - that, to Kroll, is ecology, but not just words. Practice which develops. The architect has to make himself invisible. That is success.

In Sweden we have all the fifty or so eco-villages which have been dreamed of or made to come true. Living projects and experimental workshops from which a lot can be learned. There are many ways to choose, many mistakes we will not now have to make.

Are there ways into an ecological pattern of action with more organic growth and closer links with existing communities? I would like to point to *Tävelsås in Småland,* near Växjö. That is a village which refuses to be obliterated simply because it has merged with the Municipality of Växjö. People have joined together to preserve important community functions and a social identity of their own. They also had the good fortune to find a building contractor, Hans Andrén, with a commitment to architecture, ecology and local culture. He retained Markens architect office for the projects which have so far been realised in the village - homes for 18 households, a school, a leisure centre, caring amenities for the elderly and a meeting hall. Lars Danielsson at Marken follows a tradition of organic architecture from the turn of the century, from Van de Velde and Frank Lloyd Wright by way of Aalto to Erik Asmussen in Järna. Artists and gardeners from Järna have also been involved in the project. The dwelling houses occupy a lake-side villa plot between two older houses. In an interesting way, they reiterate themes of the buildings round about them. A complementary blue strikes up a conversation with the surrounding yellow houses. The garden provides space for individual plots, an artificial stream carries rainwater from the roofs to the lake. Solar collectors and an accumulator tank save energy. Concerned with detailing, planning forms, scale and meetings with light and human beings give this development its ecological identity. The same goes for the other new development near the church, which ties in with the natural scenery, the lake shore and its bathing, history (that of the school and church), public meeting traditions and the pre-existing architecture. The new development is a manifestation of the village and its determination to survive.

The Anthroposophical community in Järna is also worthy of reflection. One readily admires its architecture, the beautiful positioning of its buildings in the landscape, the gardens and the tranquil, sometimes festive mood. An ivory tower utopia, one might think. But it is not only the philosophy which holds this community together. It is also the manifold practice of that philosophy. There is a diversity which is so vitally necessary to a community - with pedagogics, art, healing, growing, social and living philosophy, research, water purification, food production, food manufacturing, flowers, medicinal plants, production of dyes, furniture, a bank, conferences, periodicals, books, a banqueting hall and housing scattered throughout the community. A nucleus of this kind can mean any amount for the spiritual growth of a region. And by spiritual I mean, not religious, but pertaining to the sphere of the spirit.

The household is the level at which ecology becomes intelligible to everybody. And ecology, to coin a phrase, is a matter of "householding", of "oikonomia". But the household is concealed in the shadow land of women where there is also a great feeling of powerlessness. Women are fighting their campaign to bring their life-space, their rationality, their diurnality, into the light. Not only for egocentric reasons but also in the conviction that a society built on equality between the sexes will be a better society, that children as they grow up need male as well as female role models in their surroundings. Besides, the way domestic work is most organized today, it is being irrationally conducted in progressively smaller nuclear families, where a great deal of important knowledge and opportunities of influence are lost. A working community, by contrast, means greater happiness and more social experience.

In the Nordic region there is a research group based on 12 years' annual seminars on the subject of "Building and Living on Women's Terms". This group has developed a concept of *the intermediate level* - a level in between the individual household and the planning society, a women's public domain. A level at which a great deal of reproductive work can be accommodated, a level which can provide a completely new point of departure for planning.

The cohousing blocks have long existed at this intermediate level. Today there are something like 50 of these communities in Sweden, most of them based on labour-sharing. That is, the residents do their cooking together, manage their building and develop a communal culture. The rather exciting thing about these buildings is that they are so different. The collectives form individuals with great differences of character.

Most of these buildings have been erected by public housing utilities, some of them by housing co-operatives, and a few are privately owned tenant-owner flats. Some buildings have between 12 and 15 apartments, other something like 50 - except for the collective villas which accommodate only a few households each. Some have very generous spaces for communal activities, while others are less well off in this respect. Some are accommodated in converted buildings, some in new ones. Some share a building with pensioners' flats, others do not. In nearly all these buildings, there is a baby boom soon after people move into them. In some buildings there is good home cooking every weekday, while in others it comes less frequent. In all these buildings there is a great deal of openness towards the world at large. Many people from the outside are drawn into them.

In these buildings a large fund of social competence develops among children and adults. In the case of *the Hässelby collective housing unit,* things went so far that when children from the building started school they were put into different classes because they were too strong together. Their competence is translated into a variety of activities which create community sense in wider circles. Networks and knowledge are built up. The informal sector, as it is commonly called, is strengthened. Zoning and specialization in housing and employment are broken up. Here we have the embryo of the housing transformation prompted by the changing family structure and the gainful employment of women.

"The Grand Domestic Revolution" is the title of a book written a few years ago by the American architect Dolores Hayden. It deals with the widespread women's movement in the USA which, during the 19th century, campaigned for women's emancipation from domestic slavery in the family by means of co-operative enterprise and facilities. A long line of utopian communities were founded in which household work became a shared concern. The extent and vigour of it all can be surprising. What is above all surprising is that it left so few traces. These women wanted to make domestic work visible, bring together the private and the public and make democracy begin in the nursery and the kitchen.

Today we are still waiting for the Grand Domestic Revolution, which could become a powerful re-creative force and an important basis of an ecological society. Housing collectives are at all events a beginning.

One consistent feature of the ecological and collective projects is the important part played by self-management. This is perhaps the most important reform in planning for a sustainable region. Planners and architects must modify their role. The genius loci, the spirit of the place, is not only to be found in nature and history. It lives in people and the community between them. By way of conclusion I would like to describe a German project - a *youth centre in Stammheim.* With very slender economic resources, the youngsters have acquired a place which is in every sense their own - dreamed, built and brought to life by them personally. The architect Peter Hübner gathered youngsters, residents and others concerned to a meeting to outline ideas. He took upon himself the role of listener, translator, educator. In this way he became the choreographer, conducting and fashioning the performance. The actual process was kept going by the cast. He saw it as his task to make the performance an enjoyable experience - it is when people think something is fun that they want to join in it. The youth centre, then, was constructed with a good portion of voluntary work input. Using many people's enthusiasm as a motive force was no problem. Utilizing dreams was self-evident. Newspapers and local inhabitants were pulled into the process, and the project became theirs too. The use of the building was the important starting point. This has generated an ongoing process. Building neither begins nor ends at a predetermined point in time. The design expresses ideas and wishes. The utilization is part of the design and the design part of the utilization. It is only when opportunities of a choice begin to come into view that the notion of what quality is can evolve.

The Ecological Life-Region

Per Råberg

PLANNING WITHOUT THEORY

Territorial planning is one of the weightiest sectors in the political and social transformation which our society is undergoing today. Still, regarded as a theoretical discipline, it is a highly overlooked field of debate. Such an inconsistent state of things is nothing which surprises the experienced critic of modern society very much. As a matter of fact, it is often a good starting point for a closer analysis of the inner contradictions which permeate the contemporary phase of modernity. That is also true of the actual paper, in which we are taking a closer look at the social implications of regional policy and planning.

Let me start by listing some facts. Territorial planning is weighty as a productive enterprise, in terms of the activities and number of persons employed in building, civil engineering and industrial construction. The sector also commands a large administration consisting of government departments, national and local authorities, research councils and trade associations. The many expert groups in architecture, urban and regional planning and national development make up a stratum in their own right. In addition we have the educational departments at universities and colleges.

The weight of territorial planning also becomes fully clear when we consider the role of infrastructure in municipal politics. Architectural space forms the framework for many of the dynamic processes in the communiy. Maintenance of the infrastructure entails vast expenses over and above the primary construction costs. Unlike the functional dynamics, it represents a slow-moving factor whose impact remains visible in the environment for a long time, often for centuries. Directly and indirectly, the spatial frames affect our view of reality. By its visual vigour the architectural environment is probably the strongest factor of all for giving a sense of reality to our inner conception of society.

Considering the factual importance of territorial planning, the paucity of relevant theory in the field is remarkable. Please note, I am not claiming that territorial planning is devoid of theories. But the theories used are mostly fragments derived from other sciences more firmly rooted in traditions of primary research. Geography, regional economics, political science, geopolitics, urban sociology, environmental psychology and mathematical theory have all supplied the planners with a host of theoretical arguments. These fragments form an eclectic and interchangeable frame of reference for the practice of planning

There also exist a number of modern planning schools, mostly inspired by the social sciences. They can be roughly divided into the social-reformist, the policy-analytical and the social mobilization tradition (John Friedmann's classification in his book "Planning in the Public Domain", 1987). These directions of thought have derived inspiration from the entire spectrum of intellectual currents from the last decades, such as positivism, structuralism, neo-Marxism, neo-liberalism and post-modernism (for a closer description, see R.J. Johnston "Philosophy and Human Geography", 1983). All these are well reputed names and trends of the time.

What I maintain is that there is a conspicuous shortage of theoretical approaches emanating from the expertise of territorial planning itself and dealing with its genuine subject, which is *the spatial dimension.* The frequent apostrophizing of Walter Cristaller's and August Lösch's locality classification patterns (from the 1940s) in recent literature on regional planning, illustrates the moderate level of aspiration which prevails. This is an odd state of affairs, considering that within the spatial field there exist a number of epistemological issues which should be subjected to intellectual and philosophical study in order for a professional level to be properly established in the field.

Let me to quote a few examples. At the profoundest level, we are confronted with the ontology of space: which properties characterize the spatial niche between planetary gravitation and the cosmic time dimension? The insights which we can attain from such analyses are by no means of purely theoretical interest. Then follows the conception of space as an element of bioexistence and ecological circulation. This directs our attention to problems connected with the life-space as an operative field: what can be said as to the vital point of departure of the operative field and the energies needed by organisms to overcome the inertia of distance? As a subdivision of these themes we then discern the human living space, regarded as a medium for biological and cultural activities. Further, the polarity between man's quest for wholeness of his living system and his urge to transcend its boundaries is a decisive subject when we study the evolution of human society. The themes I have mentioned here are just a few of all those which a theoretically comprehensive spatial planning practice should carefully take into account.

When confronted with all these eminent questions, however, we are left

"Thoughts of the Mountain". Pencil drawing by Kirstie Ekelund

almost entirely in the lurch by contemporary planning expertise, and we cannot help asking why. On closer reflection we find that this ignorance is hardly a coincidence. Implicitly or explicitly it expresses a decided standpoint, and the higher the level at which we probe the planning hierarchy, the more distinct that standpoint becomes. It is fully clear that the spatial dimension is regarded by its own expertise as a fairly inessential aspect of the knowledge field of social planning. Among experts, instead, it is the material structures filling the topographical space which are regarded as the determinant factor. The leading organizational principle for the material structure, according to post-spatial planning, is the time dimension, not space. Time/space maps, in which the territory is deformed according to the friction of distance, are considered to convey a more accurate picture of reality than the topographical map. Persistent belief in space as an active planning factor is jeeringly dismissed in some circles as "spatial fetishism".

And yet it is with spatial planning and nothing else which these experts deal in practice. Also, in general community planning routine, the geographical dimension is being ascribed an increasingly important role today, especially at the makro-level.

PLANNING WITHOUT PEOPLE

The theoretical vacuum of territorial planning is especially noticeable in tasks relating to the human sector. The notion of society as an existential territory of man is met with raised eyebrows and quickly ruled out at planning expertise level. Planning for human needs is regarded as one single sector - the socio-cultural - along with other planning sectors, e.g. the production sector, the communication sector, the technology and political sectors. The planner of our times profiles himself as a neutral social engineer. His task, as he sees it, is to solve technical problems of distribution within the material structure in accordance with the demands of various clients. This is a carry-over from early 20th century functionalism which still, despite minor attempts to break loose, represents the main trend.

The result of the defensive expert attitude is a pragmatical planning practice which, for lack of guidelines, makes common cause with the strong forces of development - that is the market, technological progress and political trends. It has obtained a prestigious niche in the widespread technocratic power structure of the modern mass society. Unfortunately, in this way the entire sector of civil and communitarian life is edged without professional protests into the margin of mainstream development. We strongly sense the icy undercurrents of the present rationalist age, at the same time realizing the difficulty of eliminating the threats which this paradigm poses to society.

The consequence of technocratic pragmatism at expert level is that the spatial appearance of the human community today is anybody's guess. There are no principles governing its dimensioning, delimitation or infill, and no organizing rules except the different sectors' demand for allocation. Small or gigantic, either will do, planning for homes or traffic machinery is a matter of taste, and beautiful or ugly is somebody else's business. The currently most powerful interest wins, whatever the long-term interests of the community. The pragmatic planning machine of our time has provided us with a gigantic infrastructure, but the social outcome of the efforts is a community with no heart and no working brain. The result is there for everyone to see in the metropolitan environment so typical of our age: chaotic remnants of inherited social residues within the mass of anonymous or self-asserting new planning. The monstrous mega-cities springing up like mushrooms all over the planet are the logical culminations of this ultra-pragmatic style of planning.

But, somebody may object, we still have the architects. Does not their work bear witness to the social concern and the artistic sensitivity we strive for? Yes indeed, if only we put on blinkers first. For the truth is that the architect's niche in community planning is so narrow that only a tightrope walker can make a go of it. We need only rise to the level of zonal and urban planning to see that architects' arguments have anything but the last word in present-day planning. And yet it is obvious that the decisions made at the higher levels have very palpable effects on the lower, architectural level. The awful thing is that the architectural level also represents people's actual life-space.

One important aspect of physical planning during this century is, in actual fact, the shift of the powers of decision to higher and more abstract levels of organization (political decentralization has not been able to offset this trend). The decisive long-term plans for the local community are drawn up today by departments of state supported by regional researchers with qualifications in political science, economics and even mathematics. In the abstract rhetoric of regional research, certain localities are nominated creative centres while others are sentenced to be slowly garotted. Productive corridors are drawn across the continents according to the forecasts of logistics, and wide support zones are hatched on the peripheries. In the expert debate, entire countries are closed down with the exception of a few creative regions which, through a rallying of national forces, are to be made internationally competitive.

Mathematical logic remorselessly explodes the historical man-made landscape; in the society of the footless economy the socio-cultural space seems more and more to be an anachronistic leftover.

THE USERS PERSPECTIVE

But let us take another approach. How does this powerful but at the same time elusive physical planning appear from the user's point of view? In the role of citizens and members of the civic community many of us have difficulty in accepting the humble part ascribed to us by the planning elite. The planners' division of society into a number of spheres of interest, with the citizen's interest representing only one segment of the whole, frustrates our innermost wish to actively take part in and to survey that space in society which we consider our own. We feel rather that a systematic de-humanization of the primary living environment is taking place. We face a deconstruction of the central values in life, which seems difficult to influence, since it is part of a more general de-humanization of contemporary society, linked to the dominating paradigm.

It is true that for a long time we have been carried away by the politicians' propaganda in favour of a society of flexibility, international mobility and extensive specialization. Decades of mass-media indoctrination have taught us to hold in contempt those values embedded in the holistically oriented local community. We have come to regard its moral concepts, its norms and aesthetic values as chains and restraints, the decayed remains of a past epoch. As members of a cosmopolitan culture, we have made ourselves believe that the electronic network, reaching around the globe and using a time-controlled system, is also the ideal organization model for our society.

However, our positive expectations of the social constructions of the technical society turned to disenchantment once it became apparent how illusory were the gains in actual living quality, and what fundamental human values had been lost. A humanist revaluation is now taking place provoked, amongst other things, by the reduction in the existential quality of the environment experienced in the physical social space around us. Beholding the fragments of a humane environment which territorial planning leaves behind it in our metropolitan cities, we eventually acknowledge that the socio-territorial frame represents a value, indeed an indispensable one, in our existence. We begin to realize that cosmopolitan living out of suitcases, at air terminals and in hotel bars does not entirely satisfy our need for a dignified frame of living. We realize that we need a place to return to, a spatial centre for our life. This place does not have to be a private oasis or a tourist paradise, but it should possess a soil in which we can put down roots. We also need sufficient time here to restore the links between the family generations and with our friends. Yet what is at stake is perhaps not so much our personal security as the possibility of experiencing a fuller life than modern nomad living has to offer.

Contrary to all dogmas declaring the death of the local social space, we are suddenly realizing the indispensable importance of spatial frameworks in our vision of the good society. This importance does not stem primarily from social space regarded as the sum of a number of specific functions, but from its role as the arena where our most profound human needs are acted out. Only through a new respect for the spatial framework of society, can guarantees be established that will allow such values as human closeness, social warmth, true solidarity and humanitarian compassion - so profoundly missing today - to flourish and guide us. All those fundamental concepts for a civilized society which are encountered today only as empty phrases in political rhetoric.

PLANNING THE HUMAN SPACE

How can we as critical citizens defend convictions like these against today's planning élite? What strategies will we have to devise in order to assert a humanist vision against the market rationalism which has grabbed hold of the living organism of our society?

In the following pages I will try to formulate a possible answer to these questions. I shall present a new theoretical approach which may contribute to transferring the debate about spatial planning to the overriding level of discussion which its human and social content deserves. It originates in socio-ecological ideas expressed by the Study Group for Humanist Futurology in Umeå. The intention has been to try and fill the theoretical vacuum which characterizes social and political planning at the heart of its competence and to help put a humanist theory of planning on its feet. But I shall start this discussion by first pointing to a few important and inspiring contributions written within different disciplines and fields of expertise beside main stream physical planning. They are all contributions to the elementary building up of knowledge and competence which spatial planning will have to submit itself to in order to deserve in an alternative future the name of expertise.

THE REGIONAL TERRITORY

Outside established planning theory there exists an unused fund of experimental knowledge. Some small portion of this knowledge is to be found in the very field of expert planning. What I have in mind are visions such as Ebenezer Howard's Garden City project from the turn of the century, Community Centre ideas from mid-century and Britain's New Towns movement. Here we have a whole bundle of *communitarian* planning ideas which have long been highly respected in architectural and planning history. But they have never played any decisive role in geopolitical reality, because they have invariably been outflanked by more powerful trends.

A more important approach, with similar ideological profile, is presented by the *regionalist school of planning*. This school experienced its heyday during the seventies, due to its involvement with the developing countries, which attracted great interest on the part of western researchers at that time. John Friedmann, Dudley Seers and Walter B. Stöhr are three wellknown exponents of this approach, which also has a well documented previous history. One needs only mention the names of Paul Vidal de la

Blache in France, Patrick Geddes in Britain and the Americans Lewis Mumford and Howard Odum. All four of these pioneers are rooted in a geographical scientific tradition.

The ideas of the regionalist school are summed up in a book by John Friedmann and Clyde Weaver, characteristically entitled "Territory and Function" (1979). In the book the authors describe the regional planning crisis at that time. They note that regional planning in modern times has pursued two disparate lines of development, namely the functional and the territorial, and can be viewed as an ongoing dialectic between these two. The functional doctrine represents an economic network planning in which society is looked upon as a resource space for rational exploitation by the market forces. The geographical territory is interpreted as an abstract communicative space of unlimited extent, subject only to the laws of localization policy and communications technology. Local development in the community is governed by exogenous forces, whether in the form of the state or different enterprises.

The *territorial* doctrine of planning emanates instead from the needs of the rooted population in a local area, seen against the background of that area's historical conditions. Territorial planning is regarded as an indigenous activity and is based on a certain local autonomy. The primary aim of planning is to improve the living conditions of the population within the autonomous area. Activities are conducted with simultaneous consideration for historical continuity and the productive qualifications of the place.

Friedmann and Weaver show how functional planning theory, most recently in the guise of the growth-pole doctrine, has resulted in the transfer of territorial power from the local population to the market forces. Initially this happened under the aegis of the nation state and with governmental approval. The state, not the local community, was made the socioeconomic production unit, and the schemes of economic production and welfare distribution were placed into separate policy fields. In this way the states acquired profound in-built geopolitical imbalances. At a second stage the market forces leave the state level and establish a transnational corporate structure, for which the planet itself is the limit. What the market is now asking is for the central power also to break down the national barriers and in this way surrender national territory to the economic forces. The polarization which has already begun into national centres and peripheries is being accentuated and is assuming international dimensions, even in the privileged world.

But the undermining of territorially based popular power at local and national levels and the surrender of social power to the transnational corporations are, Friedmann-Weaver maintain, a threat to democratic values, and will lead to the disintegration of society. As they see it, one-sided development along these lines is a cultural regression which will have to be modified. As a healthy alternative intended for both local and national development, the authors present a regionalist planning model which they call *the agropolitan approach.* They base their theory of regional policy on a hypothesis of elementary human needs. Their hypothesis, admittedly, is roughly sketched, but it has a great value as a declaration of intent.

It is interesting to see how the authors then proceed to transform the ideal requirements of their scheme into a territorial model. They describe the social territory as a combination of three spatial spheres superimposed on each other: the common cultural, the common political and the common economic sphere. The three spheres do not exactly coincide spatially, and each of them has its unique frames of reference. The social territory, however, constitutes the central zone in which the three spheres meet in an interactive schema. *The agropolitical district* is referred to as the smallest territorial unit in which these spheres can be made to interact. The area of the district should be capable of accommodating a working population of between 20 and 100.000. Friedmann and Weaver assume the district to include one or more urban communities with a population of between 5 and 20.000 inhabitants.

The authors state that the agropolitan model has mainly been devised with a view to tasks of development planning in Third World crisis regions. At the same time they argue that, suitably adapted, the model is also applicable to the industrial countries of the North and can thus acquire general validity. The agropolitan model expands into a comprehensive geopolitical vision, the purpose of which is to direct world development towards humanly and socially more sustainable lines.

THE EXISTENTIAL SPACE

The regionalist school of planning touches down plumb in the middle of the territorial discussion. An equally central contribution in terms of ideas can be sifted out of a different tradition, namely 20th century radical philosophy. I am thinking of two kindred lines of thought, the phenomenological with Edmund Husserl as its dominant figure, and the existential represented by Martin Heidegger. Both directions are far removed from the reductionist trend of modern philosophy, with its demonstrative turning away from existential questions to problems such as logical analysis and scientific methodology.

Both approaches begin with pointing at the dual foundation of social reality on both a subjective (anthropocentric) and an objective (territorial) perspective. Edmund Husserl formulates the conditions for an anthropocentric pursuit of knowledge by launching the idea of transcendental subjectivity. That concept implies that every subjective act of consciousness contains an essence which is quite independent of the personal colouring, and also that subjective registration is not a private (solipsistic) process but is constitutive and universally valid. Herein lies the objectivity and usefulness of the act when building up a fund of knowledge concerning human behaviour. In Husserl's postulate we have not only the point of departure for a new theory of knowledge but the seed of an anthropic theory of society and planning.

Heidegger's philosophical analysis focuses on the outward sphere of the world of phenomena. In his essay "Bauen, Wohnen und Denken" (1951) he clarifies the connection of human existence with the spatial dimension. Out of the dynamics of existence the life-space is created and expands, and it is during confrontation with the spheres of the life-space that existence acquires its emotive content. Similarly, space forms a basis of human existence and is a precondition of its spiritual stature, Heidegger tells us. Man's life is a lifelong journey through the terrestrial space under the influence of human and transcendental forces. In this immense ambulatory space, however, we distinguish points of crystallization, places where the human, terrestrial and metaphysical forces intersect. These are the places of man's settlement in space, of his society. The place-bound community represents man's home in this earthly life.

The most exhaustive contribution to the discussion of dwelling as an existential category has been made by a Scandinavian researcher, the Norwegian Christian Norberg-Schulz. Significantly, Norberg-Schulz is an architect by profession; in him, theoretical and psychological interests merge with clear visual observance of the designed space. In a sequence of books beginning with his thesis "Intentions in Architecture" (1963), he charts the conditions of human dwelling and its forms of expression. Dwelling, to Norberg-Schulz, is not only one of many life functions in society but represents a fundamental existential category.

Norberg-Schulz begins by defining the existential meaning of dwelling. Dwelling means, most profoundly, having a personal identity of place and emotively identifying with a particular territory in geographical space. Dwelling corresponds to a profound pursuit within the individual of an existential fixed point in the unbounded surrounding world. Only when the individual has accomplished his quest for a fixed anchorage does he achieve a state of existential balance and acquire his full identity as a human being. It is only through identification with existential space that his life acquires its true meaning.

But in order for the individual to experience full identification, the outward organization of the place has to meet a number of elementary requirements. The primary character of the existential space is that of a natural space whose topography and vegetation are framed by the sky and horizon. The form and content of the naturally given space vary in their details from one region to another, but their basic features make up an archetypal element in the life-space of all individuals. The individual's spatial identification means, first of all, that he experiences mental participation in the life-sphere of the natural space.

A second fundamental motif of existential space according to Norberg-Schulz is the man-built house, meaning the home or, in a collective sense, the dwelling structure. Place-bound dwelling is the frame of the socio-cultural sphere of existence. The home is the scene of the creative inventions of human feeling and thought, which it also symbolizes. The home, like the natural space, is an indispensable element of place-bound life and of the structure of the place. The two aspects of the human home are not isolated from each other; they belong together and are expressively intertwined. Man inhabits both spheres and effortlessly moves between them. Together they give the place its specific emotive content, its *genius loci*.

Norberg-Schulz' essays contain a number of sketchy, but acutely perceptive, attempts to define the spatial and architectural components and their interaction. His observations lead to what may be called the beginnings of a theory of architecture for the existential home. In view of their summary character, they are best described as a contribution towards the phenomenological analysis of space.

THE NEW LIFE SCIENCES

The speculations of existential philosophy, by virtue of their intellectual substance, are a major contribution towards an anthropic discussion of space and planning. The models of social practice derive by this welcome support from the sphere of intellectual theory. In the circle of sciences, however, this contribution belongs to the soft human sciences, whose pronouncements on crucial matters of planning do not carry anything like the same weight as the facts of natural and technical science. For this very reason, it is interesting that, even in a number of scientific disciplines, we can point to discoveries and established theories which support the ideas of the life-space as a decisive existential factor. It is, above all, in the biosciences that approaches of this kind are to be found.

The biosciences have undergone a magnificent development in recent decades. The volume of research has exploded in a number of specialized fields, ranging from biophysiology to behavioural ecology. We can point to epoch-making advances in both primary research and various applications. Many observers believe that today it is the biological sciences which occupy the central, catalytical position in science which during the 18th and 19th centuries was held by physics and chemistry.

Apart from the advances of biology itself, developments in modern physics and chemistry have supported a shift of emphasis towards the organic life sciences. The discoveries made by Niels Bohr, Werner Heisenberg and Ilya Prigogine have shaken the foundations of classical science and its mechanistic ideals. The dynamic, apparently organic character of the fundamental physical processes, but also their tendency to self-organization into large-scale systems, have changed our view of micro- and macro-cosmic reality. This change can be described as a growing biologization of the world picture, and it seems to be opening up new and dramatic perspective both in natural science, the theory of science and our entire view of the world.

Another aspect of the expansion of biology is the entry of the biosciences into the field of cultural and human science. The traces of this influence are discernible well inside the human sciences, and, indeed, all the way into the qualitative sphere of the humanities. Modern social anthropology, as represented by Edgar Morin in France, is one example of the way in which biological approaches are beginning to influence the attitudes and methodology of human and social research. In systems theory, the biological view of nature and science has acquired a theory of science - or the embryo of such a theory - which may prove more capable than reductionism of describing the new reality which is taking shape on our retina.

The implications of this prospective synthesis, however, are not yet clear. Many in the sphere of human disciplines fear that the synthesis will mean mechanical reductionism's conclusive assumption of power on the holiest ground of human science, and thus the abandonment of the last residue of humanism in civilization. There is some justification for these fears, at least in the short term. The development of socio-biology presents a number of eloquent examples.

In the long run, however, hopes can be pinned on the possibilities opened up by the biological approach. These possibilities, as was emphasised by Konrad Lorenz in his late days, can only be realized if the onslaught of biology is met with a creative bid from the humanist and cultural sphere. The fruitful situation might arise when an evolutionary biological and a cultural approach start to cross-illuminate each other. A situation of this kind calls for an offensive human science relieved of anachronistic dogmas and actively involving itself in development issues of the future.

My own consultation of the biosciences for support of a socio- territorial theory starts from the humanist vantage point. Admittedly, I have not found, nor can I refer to, any complete, holistic solutions. On the other hand, interesting systematic observations can be derived from a number of disparate subjects, primarily ecology, ethology and evolution theory. It should be noted that the libertarian methodological approaches in the new biosciences have been just as great an inspiration as the empirical results of the specialized disciplines.

THE HABITAT OF ECOLOGY

Let me start with the ecological contribution. Ecology usually defines itself as the science of interaction between living organisms and their habitat; the scheme includes both the biotic and abiotic elements of the environment. By definition, ecology is a science of space, more exactly of life-space. Life-space is not defined as any occupied territory whatsoever, but as a frame for the flows of energies and nutrients through which the biological life of the organisms in the ecological space is sustained.

In the framework of ecological theory, the individual organism is the starting point of the life processes. The individual, however, is viewed, not in isolation from his surroundings, but as an open system involved in a bilateral exchange of energies and information with the environment. It is the individual who has the initiative here, but this does not necessarily give him dominion. For his survival he is entirely dependent on his habitat and on the external resources at his disposal. This makes it easy to shift perspectives and to regard the ecological environment as the ontological point of departure. From this alternative point of vantage, the individual organism is seen as a part of a population which in turn forms a segment in the ecological space as a whole. The ecological researcher studying dynamic interactions in nature oscillates between these two approaches.

In many of their definitions, the ecological theorists state problems in terms which are relevant to processes in the proximal life-zone of the organisms. The important question of the organism's adaptation to its surroundings is perhaps the foremost example of this, and a central field of study. By studying adaptation it has been shown how symbiotically interactions develop in the life-space of the organisms. In this interaction, it is essential for the individual to be adapted behaviourally as well as functionally to the ecological niche in which his life unfolds.

On the other hand, the constancy and predictability of the local environment are an important prerequisite of smooth coexistence. If there is a change in the outward conditions, this impedes the instinctive activation of the organisms until reprogramming has taken place, which itself is a time-consuming process. If the outward conditions are more fundamentally transformed, the organisms cannot achieve the adjustment and the population succumbs.

Empirical ecology often employs an ecospherical macro perspective. The study of the cyclical processes in the biosphere then is the natural point of departure. But even when ecology devotes itself to large-scale system thinking, the questions of spatial definition soon return and so too, do questions concerning the local life-frames of organisms. The overriding spatial entity described on our planet is the ecosphere, which incorporates all elements and conditions for life, but which also forms a gigantic system of its own. Within the ecosphere we have the ecological macro regions and their functional units; these are large-scale systems forming the planetary elements, and their terrestrial and maritime subdivisions.

Within the macro regions we distinguish the micro regions and their ecological niches. The micro regions are of many different kinds and contain the prerequisites of life for the different species. In an instructive model, the ecologist J.W. Valentine refers to this type of space as bio-space. The smallest unit in the spatial hierarchy, finally, he calls eco-space. This designation refers to the small-scale space constituting the life-territory of a local population and, ultimately, of an individual member of that population.

THE TERRITORY OF ETHOLOGY

We now turn to consider ethology. Ethology works within the same polarized scheme as ecology, but from the opposite, subjective vantage point. Ethology, it will be recalled, is the study of animal behaviour - in its natural surroundings, be it noted. Although refinement of situations and laboratory experiments do occur, it is this rider which above all characterizes the discipline of ethology and sets it apart from the study of human behaviour. Unlike some parts of psychology, ethology also aspires to be a natural science.

Ethology's connection with the discipline of biology is also manifested by an evolutionary approach being allowed to impregnate its studies and discussion of methods, above all on a theoretical plane. Furthermore, an ecological approach comes naturally to ethology, with its interest in the role play of animal species in the social and topographical environment. Another sector of ethology looks more to biological physiology and pursues an ingenious study of the perceptive apparatus and the neural processes. Thanks to these interfaces, ethology is amply equipped for a committed study of its main research territory. However, as in several comparable research fields, practice as yet presents a modified picture. The study of behavioural processes clearly occupies the centre of attention, and a descriptive, behaviouristic approach frequently predominates.

The study of the spatial behaviour of animals has long been an established field of ethological research. Fundamental to the spatial attitudes of animals are the schemata of mobile behaviour. Through their mobility, animals are capable of mastering topographical space, they can move from place to place and in this way derive benefit from the assets of the space. With the aid of outward signs or of signals which are species-related, the animal orients itself in space and can move in specified directions. Repeated use of the signalling system lays trails which wind invisibly ahead but which facilitate animals' journeys to particular destinations. In many cases regular communication systems develop which are readily observable even by a human being. These at the same time bear witness to the mobility of the species and to its fixed habits and stationary location.

In the great majority of cases, and especially among the more highly developed species, the communicative and migrative patterns are supplemented by stationary schemata of movement. The stationary schemata are attached to the use of the space as a permanent dwelling place for longer or shorter periods. Unlike the extensive mobile schema, the stationary schema represents a concept for intensive space utilization. In its most humble form, the stationary space is only one of several resting points in a far-flung network. At the other extreme, the place takes the form of a clearly defined territory for life-long habitation. Here the communicative network has been integrated and serves as an implement for mastering the stationary structure.

A THEORY OF ECOLOGICAL PLANNING

I have presented three scientific approaches, all of which underline the importance of the territorial dimension as a dynamic factor in the life of the individual and the species collective. The surprising thing is the unanimity of accents and conclusions, in spite of the wide variety of approaches to the subject. In all contributions, the spatial dimension is highlighted as a prerequisite of the functional life processes. All of them underline the intertwined character of the interaction between the living subject and the life-space; in this spatial interaction, the dual conditioning of human existence acquires manifest expression.

All approaches also underline the topical basis of the life dynamics; place-boundness is a characteristic of organic life on our planet (and, accordingly, of man). Topical attachment by no means precludes mobility and migration, but it is in the nature of things that range is limited. Despite the openness of geographical space and the individual's freedom of action, the existential territory tends to take the form of a limited domain, a habitat. The specific properties of the habitat, however, vary from one species to another, as does the extent of its space.

Implicitly or explicitly the three contributions can also be said to represent a systematic ecological approach to vital human processes. All three appeal to interactive biodynamics as an important basis for social organization, and they all emphasize the importance of the subjective, anthropic approach as a reference point for the organization, inner complexity and scale of the functional space.

None of the three approaches depicted here is of such a kind that it can serve as a basis for articulated planning; nor has any of them shown such ambitions. They do, however, represent a rich and fruitful source of knowledge and arguments which can be drawn on in the new efforts to introduce humanist methodology into spatial planning. Above all their import lies in that they address the ecological approach. We now realize that the basic expressions of human dynamics are subordinate to the same laws of ecological systematics which have organized the lives of all other species into a balanced interplay within their own system and with the surrounding biotic environment. When we describe the de-humanization of society we are often implicitly referring to various crimes against these fundamental ecological principles.

Thus we may conclude - given today's problems - that a solid humanist planning theory should be firmly rooted in a systemic ecological view of man and society. Society is a human living space, which is created in the dynamic interaction between two poles: the human subject and the local biotic living space. The idea that such a society must necessarily assume the form of a confined rural community is overplayed. Despite the need for territorial limitations, a community can take on cosmopolitan proportions and its natural ecological aptitude includes the capacity to interact with many other habitats in the global surroundings.

But what, then, is still lacking? Which problems will have to be solved and what strategic measures will have to be taken in order for the social and humanistic targets to be achieved?

Allow me to attempt a reply. We need an immediate *increase in competence within the human sciences,* in all sectors and at all levels of spatial social planning. Hopefully, together with this increased competence will come the ability to take responsibility for directing social planning towards more distinctly qualitative human goals than is the case today. But to achieve this new competence and to enable it to occupy a proper position, new foundations of knowledge must be laid. We are in fact in dire need of research in the human sciences which will map out the fields of interaction between the human subject and the existential space, with the aim of formulating an ecological humanist theory. To a large extent, this field is as yet unexplored by the social and cultural sciences.

In effectuating this proposition, two specific tasks emerge. First a scientific exploration is needed of man as a spatial being, which means defining his ecological identity. An important part of this consists of the sensory and qualitative needs, which traditionally we define as humanistic. Second, we have to determine the life-space which forms the habitat of ecological man. That task, however, is not at all as simple as my wording may suggest, for the interaction between individual and life-space involves the entire social and cultural, productive and technological content of life at the basic existential level. Only with a creative and holistic approach can a realistic and, accordingly, competitive methodology of humanistic and ecological planning be attained.

A network of researchers affiliated to the University Group for Humanist Futurology at Umeå has ventured to take up the challenge I am outlining. An analytical and experimental study has been in progress along both lines for several years now. It derives inspiration from the regionalist and philosophical directions mentioned above and has been moving towards the ecological and evolutionary fields of theoretical biology. On the other hand there has been absolutely no compromise on qualitative requirements as regards the humanistic and cultural intention. This approach is the opposite of reductionist (cf. E.O. Wilson), it remains anthropological and cultural-scientific (cf. E. Morin).

We have dubbed the Umeå contribution a socio-ecological approach in order to mark its broad span between humanist and scientific interests. Its theoretical outcome, accordingly, should be termed a socio-ecological theory of planning. This is, of course, a tentative approach on unknown and uncharted ground, which partly accounts for embryonic and imperfect features. I will round off this essay with a brief presentation of the socio-ecological theory draft as it appears at the time of writing in the publications of the Umeå Group. I have divided my presentation into two parts: first an account of the individual's ecological scheme of interaction, and then a description of the corresponding life-space model.

ECOLOGICAL MAN

How do we describe ecological man in a scientific outline? There are important differences between our behavioural model and the views held by a majority of the schools of modern psychology. Social-ecological theory e. g. regards man holistically, as an existential being, not as an isolated reactive system. Furthermore, this being is constantly involved in a meaningful interaction with the world around him. He is shaped by the constant interplay with the environment, but at the same time he himself remodels his life-space. Thirdly, ecological man's identity is composed of several systemic levels. This division into levels is interpreted as the result of a behavioural evolution in which, by stages, the species has expanded and streamlined its life-space, partly in order to derive greater benefit from the assets of the resource environment.

The most vital aspects of the ecological personality theory are captured by a series of simple pictorial diagrams showing the system levels of ecological identity (Fig. 1). Each of these levels reflects a paradigm shift in the organism's interaction with the living environment and in its understanding of reality. They represent, actually, increasingly advanced stages in the organism's functional interaction with the surrounding world. And yet, throughout the course of evolution, the more original levels remain the foundation of the newer, more sophisticated stages. In this respect, the behavioural development of man conforms to a familiar evolutionary principle in living systems. At the human stage, the original schemes - in transformed qualitative shape - still form the constitutive frame of operative behaviour.

In man's ecological identity, according to the theory, we distinguish four genuine levels of existence. We have dubbed them the levels of *natural identity, species identity, ego-identity* and *logo-identity*. I will try to characterize each of them in a few words.

Natural identity

The deepest level of eco-identity, in the theory, is made up by the natural identity. The significance of the natural eco-identity is the individual's awareness of being a creation of the biospace and the latter containing the basic preconditions for life. This is an awareness which we describe as organic rather than reflective, and which is probably deeply rooted in evolutionary antecedents.

At the level of natural identity, the individual interprets himself as *a crystallization point of dynamic life* in the biotic space. But he does not apprehend the space of nature as an antithesis; if anything, he regards himself as a part of natural processes, a piece of nature brought to life. From this attitude we deduce the individual's profound affection for the biotic environment, as well as his recurrent bid for empathetic identification with nature.

The task of natural identity in man's life-system is to regulate, at various functional levels, the elementary interaction (the metabolic processes) between the individual and the world around and to sustain the equilibrium (homeostasis) between the subjective and objective poles of eco-existence. This is a vital task which it is particularly essential to safeguard at the complex level of modern man.

Natural eco-identity also comprises the basic ecological reason, which still remains an indispensable foundation for social and political reason at the advanced level of human existence.

Species identity

The second level in the model is occupied by the species identity. The systemic importance of this identity rests upon the fundamental role which the species population plays in bioevolution, the human stage included.

The safeguarding of the genetic pool and the maintenance of optimum living conditions within the unique bio-community are an absolute prerequisite for the survival of the individual, the species and, ultimately, the whole of the biosphere. The processes of species identity also range over the entire scale of eco-awareness, from the basic sensory to the highest reflective level. Through species identification, a host of vital processes (genetic, communicative, pedagogical etc.) are consolidated which promote the long-term survival of mankind.

The carrier of species identity is the individual consciousness, not some kind of collective awareness. In this role the individual regards himself as a typical specimen of a local population. As a result of daily association with neighbours, specific behavioural patterns arise and a sympathetic attitude to one's own species is consolidated. It is an instinctive drive in the individual to try and achieve empathetic identification with the population of the local habitat. In this way he fulfils the implicit goals of species identity: *to promote the well-being of the living collective* and the prosperity of the whole species.

Fig.1. **MAN'S ECOLOGICAL IDENTITY**

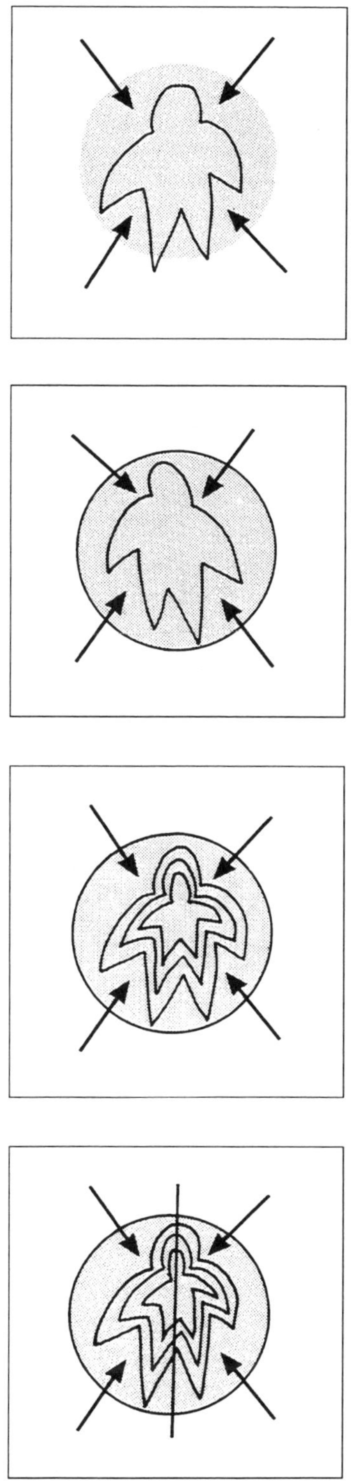

Natural Identity

Species Identity

Ego Identity

Logo Identity

Deficient species identification is experienced in a strongly negative sense and as a lowering of the quality of life. This triggers compensatory actions, and hereby the balance of the individual's ecosystem is restored.

Ego-identity

As the third systemic level we distinguish the ego-identity. We think of ego-identity as representing a later stage of evolution than the two levels already described.

Whereas the latter represent constitutive levels of living, ego-identity is oriented towards operative functions in daily life. The behavioural schemes are characterized by flexibility and distinct activity orientation. But daily activity on an advanced level also demands powerful motivation on the part of the individual. *Egoistic needs and egoistic satisfaction now enter as the motivating force* by which the interactive dynamics is maintained.

We should notice that ego-identity involves several types of interaction. Its behavioural schemes are also more readily accessible to contemplation and manipulation by the individual. Three of the operative schemes are of a more fundamental kind, namely the cultural, the social and the productive. These archetypes interact in many ways, but probably originate from different stages of ego evolution. In daily community life they tend to concentrate within separate action-spheres.

The eco-cultural sphere of action

forms the nucleus of the triad. It includes projects attached to the everyday sensory experience of the world around. Its field of interest reaches from the organic experiences, via social empathy, to the domain of environmental aesthetics.

Cultural ego-identity achieves its objects through impulsive and creative action with a great sensory impact. But there is also a contemplative side. The goal of eco-cultural contemplation is to obtain aesthetic satisfaction on the profoundest levels of aesthetic awareness. The core of the aesthetic dynamics is the comprehensive experience of situations, environments and finally the whole ecological habitat.

The eco-social sphere of action

takes the place of an intermediate sphere in ego-awareness. Its objective is social communication with a view to acquiring information for participation in local community life.

This is an activity which is egoistic in the sense of strengthening the individual's capacity for competent action, but it unfolds within the social community sphere. The family may be the nucleus of the ecological commune, but it is supplemented by more large-scale social networks, such as the neighbourhood and the urban community.

The eco-productive sphere
stands out as an outermost deposit among the action spheres of ego-identity. As the term suggests, it is oriented towards resource utilization and local basic provision.

The instrumental skills required for effective resource management are the core of behaviour. Continuous training of these skills, including use of the technical complements, is an important task together with productive work. Self-interest, as in the earlier sub-schemes, is a crucial motivating force for maintaining a high level of efficiency.

Logo-identity
As the last, supreme level of the identity system we distinguish the logo-identity. In an evolutionary perspective, it is the latest and most advanced of the systemic levels. Among the earthly species, we find it only in man.

Whereas the other sub-identities (in their original state) are realized in a sensory life-space with a local extent, logo-identity is based on a temporal spatial awareness with a far-flung field of application; in the ultimate stage its range is global. *Rational reflection is the appropriate tool of logo-identity.* With the aid of systematic thinking, the immensity of temporal reality can be analyzed and organized to serve human purposes.

Logo-identity thus represents a genuinely human level in our ecological awareness and expresses itself through a host of advanced functions and techniques. Human language, technology, industrial production and science are all aspects of the action-field of logo-identity. The breakthrough of logo-identity in the evolution of the species is a breath-taking leap in the expansion of organic awareness. Ideally, existential consciousness now extends not only to the global limit but far out into a cosmic domain.

THE ECOLOGICAL COMMUNITY

If the scheme which has been sketched here of man's ecological identity is plausible, it provides us with useful tools for understanding the current crisis of civilization and its socio-cultural roots. But at the same time, it indicates crossroads and new possibilities which scientific thinking can try when reflecting on strategies for overcoming the crisis and for shaping humanist and ecological alternatives. I will turn now to describing the model of an ecological habitat with humanist implications, which has been outlined by the Umeå group in the logical extension of the theory of ecological interaction. This is the second basic step towards constructing a theory of ecological and humanist planning.

In designing this model we have been guided by the notion that the ecological habitat should function as a hall of mirrors for the ecological personality; it should form a harmonious element for the behavioural patterns of interacting man. The creative task therefore is to transpose the action schemes from different identity levels to existential environments and finally into a holistic spatial form. The result of our effort to define, along these lines, appropriate environments for the existential levels of eco-identity is presented below (se also Fig. 2).

Natural (green) community plan

Creating a harmonious element for nature-identity is the first task when we articulate man's life-space according to ecological principles. Our method is to reserve, on the site of the settlement, a natural territory with a distinctive landscape character. It comprises a mix of topographic and organic elements and takes the form of a slightly cultivated landscape park.

This green zone can be described, in terms inspired by bioscience, as the eco-organism's natural habitat. Typical of the natural habitat plan is its openness towards the geographical space, but also its penetration of the centre of the habitat. Hereby the undisputed role of the natural plan as a basic platform for ecological life, even at the level of human existence, is forcefully demonstrated.

In the natural habitat we mark the presence of man by indicating a vital crystallization point in the centre; this is done by sculptural or architectural means. This crystallization point symbolizes the presence of nature-identity, and serves as a catalyst for man's instinctive pursuit of empathetic identification with ecospherical space. By indicating a central point indifferent geographical space is transformed into a topical space, a singular living place for the basically singular life-identity.

For pedagogical reasons, we have named the basic plan *the green community plan.* Its existential importance in human society is very much on a symbolic and inspirational plane.

Species (blue) community plan

The second task of ecological planning is to articulate the living space as a territory for the species identity. The space of species identity serves primarily as a unifying frame for the local population and the long-term survival processes of the species.

The primary task of the plan is to reinforce the feeling of communal unity and closeness which is a characteristic of elementary species life. At the same time the plan should possess the spaciousness which is required for various communicative objectives to be generated.

We tackle this task by designing a communicative community plan which is fairly broad, but delimited from its topographical surroundings. Its basic shape is the local territory, which forms an architectural surface covering the settlement. On this spatial surface lines of communication (streets) and meeting places (squares) are marked. A further reinforcement are public buildings for the overarching community functions (schools, caring

Fig.2. **MAN'S ECOLOGICAL LIFE SPACE**

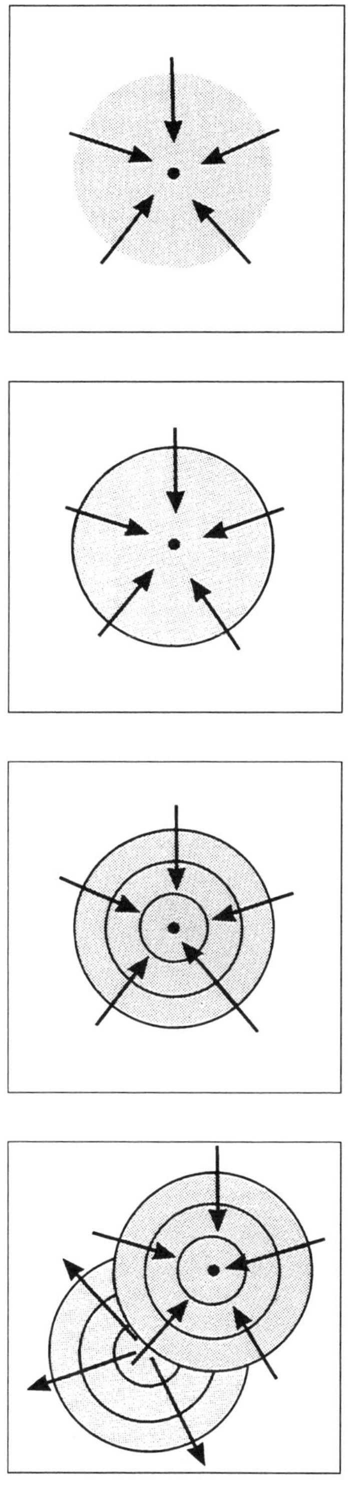

Natural (Green) Community Plan

Species (Blue) Community Plan

Ego (Red) Community Plan

Logo Community Plan

institutions, public administration etc.).

This we call *the blue community plan.* We place the blue plan as an overlay on the green one, but take care not to destroy the clarity and aesthetic expression of the former. By indicating a local habitat the natural community plan is transformed to a secure breeding place for the basic species processes.

Ego (red) community plan

The third task is to design a community plan for ego-identity and its behavioural archetypes. Here we are primarily concerned with imparting form to operative functions which unfold in a three-dimensional action space and creating characteristic ensembles for the cultural, social and productive subidentities involved.

The plan should be designed so as to overlay the blue and green plans described already. For the sake of simplicity we have chosen the designation *the red community plan.* The planning task here is of a decisive architectural character. It is concerned with designing buildings and blocks which will accommodate the operative activities of the local community.

In a consistently designed ecological community plan, however, one also has to define the functional spaces of the sublevels of ego-identity and, if possible, to localize the activities in specified areas. Desires of this kind lead us to divide the red plan into a number of concentric zones. In the innermost of these we place *the cultural functions,* in an intermediate zone *the social functions* and in an outer zone *the productive activities.* The purpose of such a spatial division is to try and reflect the evolutionary stages which the behavioural archetypes mirror.

POLARIZED DEVELOPMENT

One characteristic, which the spatial modules of eco-society sketched so far have in common, is their accommodation within a local habitat scheme. This means that they are primarily designed for access by the instruments of sensory and physical communication which are man's equipment for controlling the proximal surroundings. This is a scheme which is radically broken when we come up against the final task, that of drawing a *community plan for the system level of logo-identity.* With logo-identity man, as we have shown, takes the radical step from a sensory to a reflective awareness of existence and at the same time from a local to a global action space. Designing a community plan which meets the advanced requirements of logo-identity is the second principal task of ecological planning.

This is a theme on which we can speculate widely, but at the same time we recall the demand of evolutionary theory, namely that the advanced systemic levels ought to be subordinated to the more fundamental ones, for the preservation of the organic unity of the system. But how can the global action perspective of logo-identity possibly be incorporated in the local sensory space of nature-, species- and ego-identity and be integrated with its basic existential values? Is it at all possible, in practice, to link the disparate system levels together into a single unitary space, into an integrated life frame which corresponds with the ultimately indivisible ecological individual?

We are confronted here by a massive organizational dilemma of evolutionary scope. At first sight, the incompatibility of the demands made respectively by local sensory identity and global logo-identity upon social organization appears to be a monumental incontrovertible fact. Are we not forced at this point to relinquish the ecological aim of trying to construct a social space uniting both the life spheres into a harmonious element for man? Is this the end of the road for human-ecological calculation?

Before answering this important question, let us indulge in a brief diversion. To put the problem into perspective, we shall take a look at the archetypal history of human society. An examination of the socioterritorial development of man's civilization may provide interesting confirmation that our way of formulating the problem is plausible, while at the same time revealing how the actual evolutionary dilemma has been handled in previous epochs.

The early history of mankind, like that of many other species, took place in a local habitat (the camping place of the hunting community, and, later on, the village of the agrarian culture). Before long, however, the development of a reflective, temporal awareness made possible a drastic expansion of the effective life-space. The great cultural advances along the new lines of logo-identity take place, presumably, against tough resistance from the local habitat level. In this way a situation of conflict is created between the culture of the local identity and that of the advanced logo-identity.

A virtually Gordian solution is found to this conceptual conflict by the culture of the logo-identity breaking free of the inherited patterns and creating its own habitats. The emergence from the rural environment of the early urban system, with its much wider horizons, can be interpreted as a result of this deliberating process. In this way a polarized society comes into existence in which certain groups are still referred to local, spatial life, while others boldly enter the global, temporal arena. Hereby the preconditions are established for a fatal schizophrenia in the mental development of the human species.

This is also the embryo of several thousand years' development leading to the consolidation of the urban system in progressively steeper hierarchies and, at the same time, to a growing polarization of the provincial and urban regions. And to this development we can trace back the accelerating urbanization in progress all over the world in our own time and the deepening regional crisis which is its shadow.

Now, in our time, the polarization between the local and global cultures is as strong as ever, but the pendulum of general opinion has swung radically from the reluctant and backward attitudes in man's earliest history. What we are experiencing today is a complete slanting of positive expectations towards the global value sphere and a downgrading, indeed virtually a denial of the local life sphere and its fundamental forms of awareness.

Oddly enough, when studying the consequences of polarized development today, attention is caught by the many signs of socio-cultural insufficiency which characterize both the metropolitan cities and the provincial living space. Both types of environment in our time have become highly specialized, though in separate directions, with the result that they are both ecologically frustrating for their inhabitants. There is good ground for arguing that both forms of community, because of their limitations, are profoundly alienating and consolidate feelings of powerlessness in the individual. Alienation and powerlessness, as is well known, can easily trigger two defensive strategies, namely resignation and compensation. Codified in routine behaviour, they constitute a factor hostile to society. Through their activation on a large scale, the whole of society and its course of development can easily be put off balance. A perverse development may begin which, soon enough, will result in a crisis of civilization and perhaps, ultimately, in the collapse of the system.

I am unable here to give a fair description of even the main outlines of the problem tentatively sketched. I must make do with hinting at plausible conditions. There is good reason for supposing that a structural connection exists between polarized development and the crisis of civilization (including the environment crisis) of the present age. As a result of the unnatural cleavage entailed by the civilizing process, man is now developing a form of culture which shows clearly pathological traits. That culture today is a growing threat to the planet and to man's own existence.

The polarized society, in consequence, is more and more appearing as an evolutionary dead end, or rather, as an evolutionary makeshift of limited validity. It is high time now that the formula of the polarized society was replaced, or at all events supplemented, with a model for *concentric development* - a model making integral provision for man's craving for both local and global self-realization. There is much to suggest that the inward socio-cultural balance thus recreated might also result in the restoration of balance to our outward eco-systemic relations. In this way, maybe the crisis of our civilization could be overcome.

LOGO-COMMUNITY PLAN

After this excursion into civilization analysis, it is time to answer the question I asked earlier about the odds for achieving a fusion of the two polary schemes in man's ecosystem. The answer is that, in the long run, there may be no other alternative. If the crisis of civilization is to be overcome at all, the polarized development engendering the crisis will have to be modified. This can only be done in the long term by *concentric development alternatives* being fashioned and introduced into societal practice.

Of course, this is not an easy task, but on the other hand the difficulties should not be exaggerated. The problems connected with the introduction of concentric schemes are in fact more theoretical than practical. The international community of our times presents enormous areas which have everything to gain by introducing alternative schemes of social organization. I refer to the growing regional problem areas in nearly all countries and continents of the planet, but also to the dilemma of unequality in development standards between the continents of the globe. Considerations in this and still other directions have encouraged the Umeå University Group to undertake, in defiance of dominant trends, the task of trying to develop a concentric human society model.

From the scholarly exertions in the field during several years one main alternative has eventually crystallized which we would like to present for scrutiny by a larger public. We have named it *the ecological life-region*. The life-region model displays a territorial unit roughly the size of a Nordic county. Its ethno-geographical base consists of a number of local urban communities scattered over the cultivated and natural territory. The ecological alternative community plan which we then project on this regional territory is taking shape as a dual projection. It consists firstly of *a regional plan for the local living level*. In this the concern is the articulation of the region's urban environments and inter-urban contacts for basic sensory life functions; the local community is at the center of interest here.

Secondly, as an overlay to the local regional plan, there is *a regional plan for the global action level;* the latter contains economic, social and cultural functions of an inter-regional impact. The activities at the global level are of a more abstract-cognitive kind and have a clear cosmopolitan touch. They are complemented by a cosmopolitan meeting point, which is preferably located at the geographical centre of the region.

Once the principal model of the ecological region has been sketched, the great implementory task will be to profile activities in various fields distinctly in both directions, towards either local or global needs fulfilment. In addition, forms and expressions must be found in architecture, culture, social life, technology etc. which make visible the separation of functions within the - still remaining - overall regional unity. All these cares should be seen as parts of a pedagogy for dual ecological existence.

In this way a kind of community is created in which the individual can oscillate between the genuine local dimension of living and the equally genuine global dimension without losing unity and control of the panorama of ecological life processes. And this, in actual fact, is the very core of ecological theory. *Only a society which in both its dimensional aspects remains fully intelligible to the individual is under control of an ecological judgement and capable of developing along socially well balanced lines.*

To keep things straight in our own minds, however, it is important to realize that the global action level of the region is not to be seen just as a symbol or substitute for the true global space. In regional practice it is indeed a dynamic world arena, and includes a centre which radiates cosmopolitan dignity. Still, the vision of the eco-society does not stop short at the territory of the home region; it includes interacting ecological regions both near and far. It is - in principle - only the sum total of the interacting regions which makes up the global action space of ecological man. In order to confirm this, the life-regions are linked with each other through federative networks and ultimately perhaps constitute a worldwide mosaic.

However, speculations concerning the global organization does not entice the eco-regions to enter a new phase of human empire building. In the global network of the regions, the single life-region (according to bio-ecological principles) remains the organic, operative module. This insight lies at the heart of the eco-political consciousness, which now enters the scene. In the global house, man remains his own master.

I will have to make do with this rhapsodic presentation of the socio-ecological model. Various parts of it are sure to change as time goes on, and before being accepted the model will have to be subjected to close empirical and experimental testing. Allow me just to emphasize, for the last time, the basic intention. The purpose of the Umeå Group's regional studies is not only to uncover a number of plausible facts about man's nature and natural existence. The ultimate aim is to establish a well-founded humanistic planning methodology. An initial indication of the design of such a methodology is provided in the applied exercises presented in the contribution *Life-Region Västernorrland* which will follow next in this publication. It contains a sequence of sketchplans for the regional eco-community, drawn in collaboration with an expert group of architects.

TOWARDS A RADICAL REGIONALISM

Looking at the results of my own efforts and those of the university group's network, I see quite clearly that the Umeå approach comes close to the approaches in the other contributions in this sequence of papers. The human pathos is the same as in the civic initiatives depicted by Gunilla Lundahl. The scientific perspective coincides with Torsten Malmberg's human-ecological view, which does not fight shy of biological and even evolutionary explanations. The concrete model of society to which our studies lead corresponds closely to the model of endogenous regionalism into which Walter B. Stöhr has given us an insight.

If the various radical approaches were to be linked together, then perhaps the leap could be made from a debate on the margin of society and social science to a new paradigm of social planning. The technological utopia of hyper-urbanism would at last come up against competition from humanist and ecological schemes of society, which try to conceptualize man's existential space. Schemes which, unlike the segregating visions of the present, make simultaneous provision for man's craving for participation both in a local and a global culture.

The basis for such an effort would be a new theory of spatial planning, built on solid knowledge of man's ecological needs. And the main motivating force behind would be the new insight of the central part taken by spatial planning in realizing a humanistic vision of society. Both by its human import and organizational role the field stands at the centre of an alternative devolopment vision. In a society where Cartesian atomism has been replaced by humanist holism spatial planning may even come to occupy the souvereign part in political consciousness today upheld by the economy.

If, however, there is to be a breakthrough for a new planning theory, additional groups of agents will have to take part beside those already mentioned. Even if a theory is socially and scientifically well-founded, it will have no impact unless the social situation is ripe for it. Practitioners at both the civil and political level must be ready to accept the ideas and willing to translate them into action-plans. They have to experience the social crisis strongly enough to venture into the unknown. At the same time, the deficient credibility of the classical schemes must stand out with full evidence. Do these conditions exist today?

This question directs our observance further towards possible fields of application for the new theory. In this, our interest is primarily caught by the role played by the worlds provincial regions as organizational entities. The provincial regions, often in the administrative form of a county, contain latently many of the ideal preconditions which a socio-ecological theory wishes to develop further. Its seems logical and reasonable that a humanist theory of planning makes these regions the main object of its practical and ideological efforts. How are the prospects for such a strategy?

I believe that the time is ripening for radically innovative models in regional planning. I can give three reasons for this optimistic forecast. The first is the dramatically expanding scope of social territories beyond the interest of the metropolitan power structures. In a country like Sweden this means 70 per cent or more of the national territory, if we are to go by the hegemonic megacity visions that have been marketed lately. In a continental European perspective we discern a circle of marginalized national regions surrounding the central European metropolitan network. The regularity of polarized development becomes still more conspicuous if we turn our gaze to the countries of the South and the East and consider the effects there of metropolitan magnetism on the provincial regions. *There need be no doubt concerning the necessity at world level of regionally profiled development models, now and in the foreseeble future.*

The second reason for regional optimism is the drastically declining quality of the international metropolitan cities as a living environment. The big cities of the latest turn of the century - Paris, London, Berlin and New York - still possessed an irresistible charm created by the tension between the wind of cosmopolitanism and residual gusts of social intimacy. These are qualities which have been lost almost completely in the communications desert of the contemporary metropolis. Credible alternatives might perhaps be able to reverse the human tide, or at least lead to the creation of a balance, desirable for both parties, between the world's metropolitan and provincial regions.

The third reason is the growth, for a couple of decades now, of a global awareness of the crisis and the alternatives. Slowly, and not without serious setbacks, a new body of theoretical and practical knowledge is emerging, a counterpoise to the official dogmas which hitherto have domesticated the regions and paralysed their will to action. This counterveiling knowledge will soon make it possible for the regional agents to act independently, because it emanates from the very interest of the region and legitimates regional competence. Remote-controlled exercise of authority and economic prophesying can then be replaced by civic assumption of responsibility. Creative life-regions will emerge from the developmental shadows. As a part of this changes, perhaps, also a longed-for shift of paradigm in planning can take shape.

REFERENCES:

Altman, I., 1975: The Environment and Social Behavior. Monterey, Calif.

Berg, P.G. (ed.), 1993: Biologi och bosättning (Biology and Settlement). Natur och kultur, Stockholm.

Bertram, W.J., 1982: Dimensionen des Landschaftserlebens. Bonn.

Björkman, M., 1978: Ekologisk psykologi (Ecological Psychology). Stockholm.

Bollnow, O. von, 1963: Mensch und Raum. Stuttgart.

Boughey, A.S., 1975: Man and the Environment. An Introduction to Human Ecology and Evolution (2. ed.). New York.

Castells, M. and Henderson, J. (eds.), 1987: Global Restructuring and Territorial Development. Sage, London.

Clapham, U.B., 1981: Human Ecosystems. New York.

Egli, E.,1959-67: Geschichte des Städtebaues, 1-3. Zürich.

Friedmann, J. 1987: Planning in the Public Domain: From Knowledge to Action. Princeton University Press, Princeton, N. J.

Friedmann, J. and Weaver, C., 1979: Territory and Function. Edward Arnold, London.

Heidegger, M., 1954: Bauen, Wohnen, Denken. Vorträge und Aufsätze II. Pfullingen.

Ittelsen, W.H. et al., 1974: An Introduction to Environmental Psychology. New York.

Johnston, R.J., 1986: Philosophy and Human Geography (2.ed.). Edward Arnold, London.

Jonsson Britta, 1983: Alternativa livsformer i sjuttiotalets Sverige (Alternative Forms of Life in the Sweden of the 70s). Uppsala.

Lorenz, K., 1973: Die R̥ckseite des Spiegels. München.
Lorenz, K., 1978: Vergleichende Verhaltensforschung: Grundlagen der Ethologie. Wien.
Lynch K., 1981: A Theory of good City Form. Cambridge, Mass.
Malmberg, T., 1980: Human Territoriality. Mouton, The Hague.
Mumford, L., 1961: The City in History. London.
Mumford, L., 1975: Architecture as a Home for Man. New York.
Månsson, T. (ed), 1992: Kretslopp. Basen för en hållbar stadsutveckling (Ecological Circulation. The Basis of Sustainable City Development). SOU 1992:42. Allmänna förlaget, Stockholm.
Norberg-Schulz, C., 1963: Intentions in Architecture. Oslo.
Norberg-Schulz, C., 1971: Existence, Space and Architecture. London.
Potter, R.B. and Salau, A.T., 1990: Cities and Development in the Third World. Mansell, London.
Rapoport, A., 1977: Human Aspects of Urban Form. Oxford.
Rokkan, S., Urwin, D.W., 1983: Economy, Territory, Identity: Politics of West European Peripheries. Sage, London.
Råberg, P., 1972: Funktionalistiskt genombrott. Radikal miljö och miljödebatt i Sverige 1925-31 (Break-Through of Functionalism. Radical Architecture and Architectural Debate in Sweden 1925-31). Norstedts, Stockholm.
Råberg, P., 1987: The Space of Man. New Concepts for Social and Humanist Planning. Almqvist & Wiksell Int., Stockholm.
Råberg, P., 1990: En annan framtid. Ekoregion Nordvästmanland (Another Future. The Ecoregion of Nordvästmanland). Kollegiet för humanistiska framtidsstudier, Umeå.
Sach, R.D., 1980: Conceptions of Space in Social Thought: A Geographic Perspective. Minneapolis.
Sale, K., 1980: Human Scale. London.
Sommer, R., 1969: Personal Space. The Behavioral Basis of Design. Englewood Cliffs, N. J.
Stöhr, W.B. and Taylor, D.R.F., 1981: Development from Above or Below? Wiley, New York.
Wallace, R.A., 1973: The Ecology and Evolution of Animal Behavior. Pacific Palisades, Calif.
Weaver, C., 1984: Regional Development and the Local Community. Wiley, Chichester.

Life-Region Västernorrland

Jan Henriksson and Per Råberg

INTRODUCTION

Many of the Nordic regions are wrestling with severe future problems, due to the ongoing restructuring of the economic map of Europe. The leading development models of today clearly favour the big-city regions, while the fate of the provincial regions is written in the sky. In this way, very large areas of the countries' territory risk being left in the shadow of development.

The dominating macro models of regional policy greatly need to be supplemented by models which deal with the future of the threatened regions in a more inspired manner. Above all, models are needed which show a deeper interest in the regions' own social and economic potentials. In the debate on regional policy, this is usually termed strategies for endogenous (*or self-reliant*) development.

This study presents an experiment in endogenous regional planning. The experiment was conducted with the geopolitical map of the County of Västernorrland in Sweden as a factual base and source of inspiration. The focus has been on the territorial community plan. Our aim was to outline a plan whose emphasis would be on an ecological and socio-cultural perspective. We have termed our sketch a *socio-ecological* community plan. The theoretical aspects of the socio-ecological plan are described in more detail in the previous chapter of the book.

The study forms part of the project *The Ecological Life-Region* described in this book. It was undertaken in co-operation with a group of architects headed by Professor Jan Henriksson, Stockholm. Regional consultants taking part were the County Administrative Board Planning Unit in Härnösand (Per Sjöstedt, Head of Unit) and the Town Building Office in Härnösand (Hans Thunell, Town Architect).

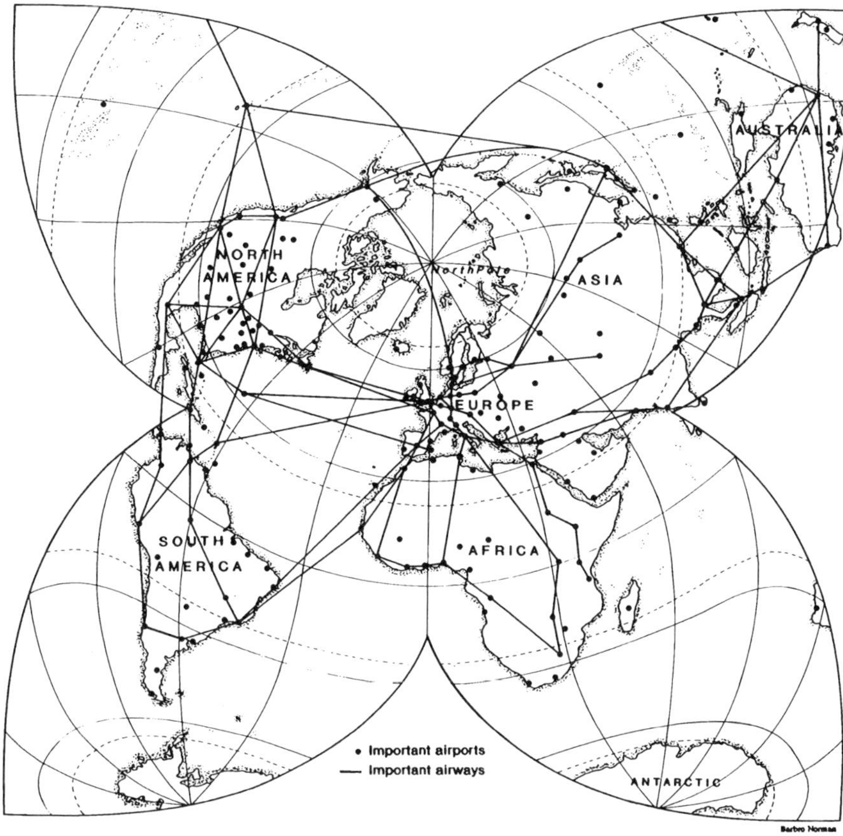

The global network of international airways.

THE GLOBAL PROBLEM PICTURE

The global urbanization process is leading to increasing imbalances between the world's metropolitan regions and the intermediary or peripheral regions. Many historic settlements in the world's industrialized nations even appear to be under sentence of death as a result of the ongoing process of concentration. This is a process whose effects are to be seen on several scales; marginalization has a local, but also a national and a continental face.

At the same time, the survival and prosperity of the metropolitan regions is becoming more and more dependent on efficient hinterlands and peripheries; the provincial regions constitute the ecological and economic precondition for the life of the expanding centres. Another dilemma is that the regression of the peripheries augments the burden on the metropolitan regions, owing to the regional and social supports which then become necessary. And yet, in future the avoidance of social tensions between different countries and parts of countries may become still more difficult than at present.

The creation of better living conditions for the provincial regions is therefore a matter of pressing concern for the international community, whatever one's ideological standpoint in the debate on regional policy. The problem is that we are left in the lurch by the leading theories of development, designed as they are for the victorious metropolitan centres. Models will have to be tailored for the intermediary regions, giving them a chance of developing on independent terms alongside of the metropolises. Perhaps these models should also be trained on other objectives, more closely bound up with the new visions of ecological and social sustainability.

THE ECOLOGICAL LIFE-REGION

Here we present, in the form of preliminary sketches, a model plan for the sustainable social and ecological region. We have dubbed it "The Ecological Life-Region". This sketch is primarily a contribution to the discussion of the problem regions of Europe, and has been prepared to fill the planning vacuum by which those regions are surrounded. But the ideological content of the model also makes it applicable to the fateful debate on a balanced global development, and to discussions concerning the need for an alternative, qualitative lifestyle in world society. The socio-cultural ideas underlying the ecological life-space model can be summed up under the following heads.

Endogenous development

The life-region has control of its human and economic base resources and shapes its strategies primarily on the basis of responsibility for the survival of the home region.

Qualitative goals

A high quality of life at local living level is the direct planning goal of society. Qualitative goals include insistence on dignified work, a stimulating living environment, control of the life situation and good social relations.

The dual life horizon

Public planning is aimed at meeting the demand for local and, simultaneously, global social activity for everybody in regional life.

The ecological imperative

An ecological systemic balance is aimed for, both in the community's relations with the external natural environment and as regards its inward, socio-cultural relations.

The individual as yardstick

In the ecological life-region, the welfare of the individual is the primary goal of social policy. The individual's quality of life also remains the relevant yardstick for assessing the living value of the whole community.

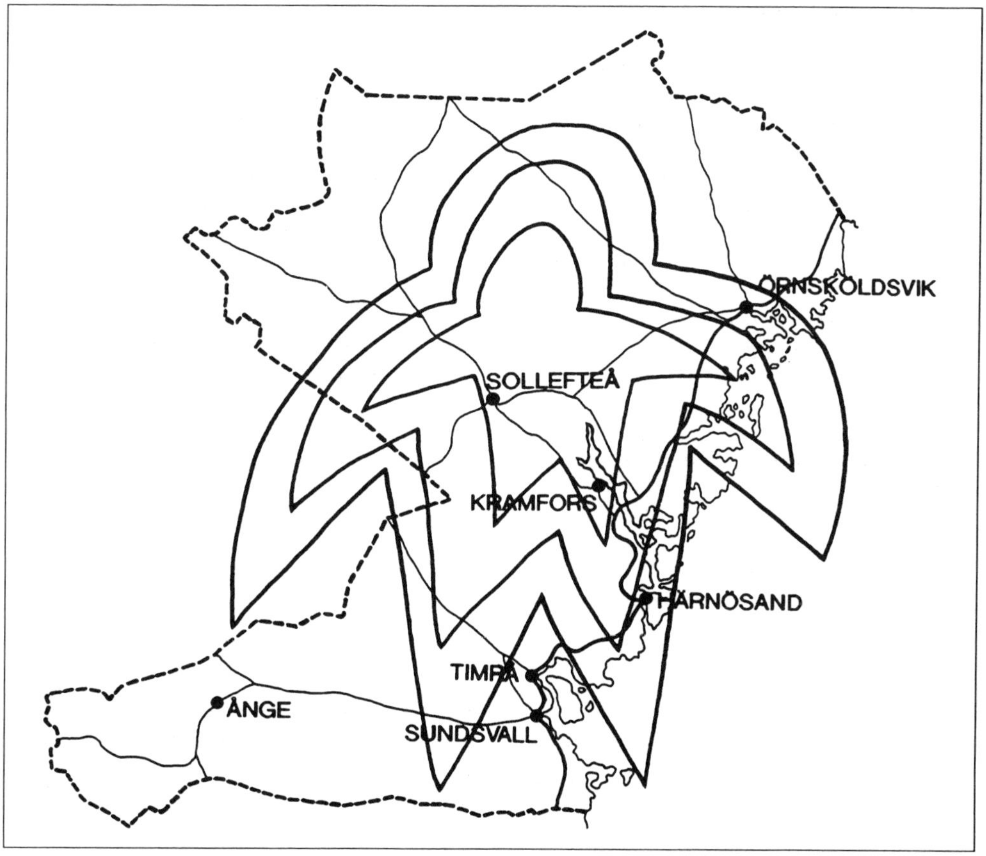

County of Västernorrland

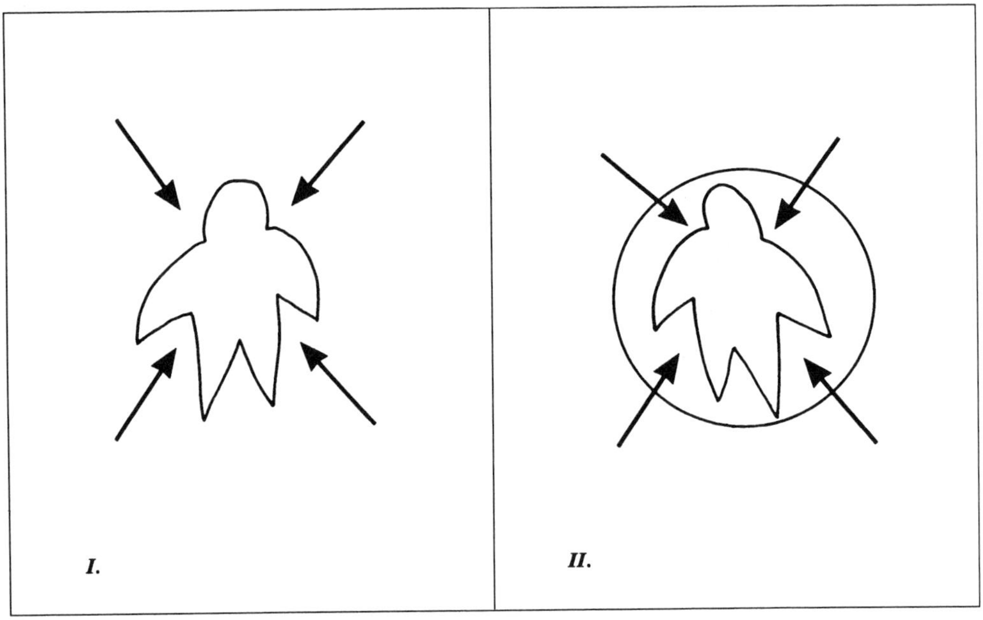

THEORETICAL STARTING POINT 1: ECOLOGICAL MAN

The conceptual basis of the ecological model plan is a theory of ecological man. Man is regarded as an environmentally related being, whose biological and mental constitution has developed through interaction with the surrounding habitat in a long process of evolution. Within man's ecological identity we distinguish four systemic levels, all of them with an evolutionary prehistory. All levels are still making themselves felt with undiminished force in our existential behaviour.

I. Level of natural identity
The individual regards himself as a vital crystallization point in ecospherical space, as a piece of nature brought to life. His impulsive wish is to achieve emotive identification with the ecospherical life source, using that term in its widest spatial sense.

II. Level of species identity
The individual regards himself as a member of a population of the same species. This population is framed by a habitat with defined boundaries. The species instinct impels him to look for identification with the local collective and to safeguard the functional whole which the ecological territory forms roundabout him.

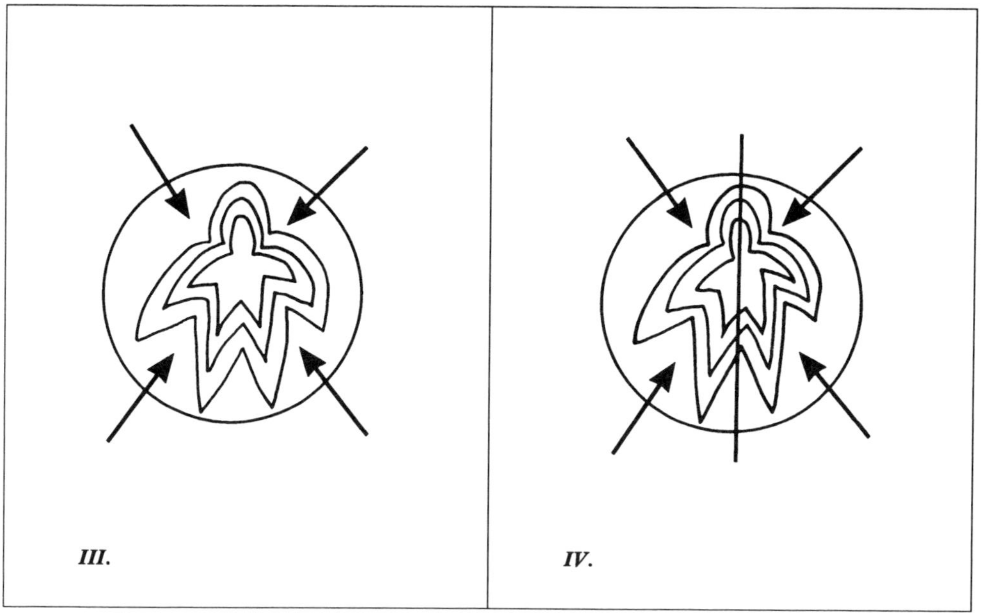

III. Level of ego-identity

The individual regards himself as a solitary being whose aim is to satisfy egoistic needs and interests. This is done primarily through actions in the local life space and within the local collective. The self-realization of ego-identity takes place on three separate operative levels. Deepest down, those levels also hark back to three stages of the evolution of ecological behaviour:

- **a. The sensory (aesthetic) action level.**
- **b. The communicative (sociative) action level.**
- **c. The instrumental (productive) action level.**

IV. Level of logo-identity

The individual regards himself as a logically calculating being. He undertakes rational operations in space and time, guided by his cool intellect. The world with which ecological rationality identifies is no longer the local territory but a sphere of thought with global and cosmic extension. Creating an existential synthesis of the action spheres of logo-identity and the deeper systemic levels described above is necessary as long as man remains an indivisible organic whole.

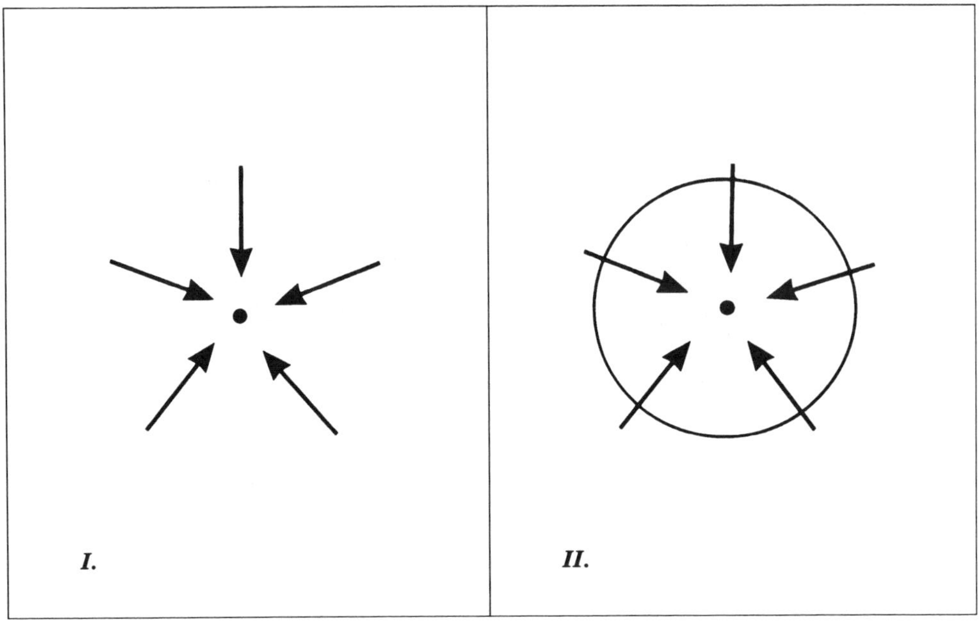

THEORETICAL STARTING POINT 2: ECOLOGICAL LIFE-SPACE

The ecological theory of planning postulates that man instinctively organizes society in order to provide good living conditions for his ecological identity. The first task is to fashion an appropriate spatial frame for the ecological life functions. We distinguish between four elementary levels of spatial organization in the ecological life-territory. Each of them corresponds to a systemic level in man's ecological identity.

I. Level of natural space, (Green Community Plan)
The task is to concentrate the organic life-space into a point of crystallization within the open ecospherical space. The distant vault of the heavens serves as a symbolic form descriptive of the whole, while the topography of the ground level makes up an action-space. A life-space for the individual's most elementary natural identity takes shape.

II. Level of the species space, (Blue Community Plan)
The task is to form the life-space into a territory for the local population of fellow-members. The individual belongs to this population as a typical representative of the species. A social habitat of homogeneous extent and with fixed boundaries against natural space is articulated. This is a life-space for the generations and for the mediation of society's cultural heritage.

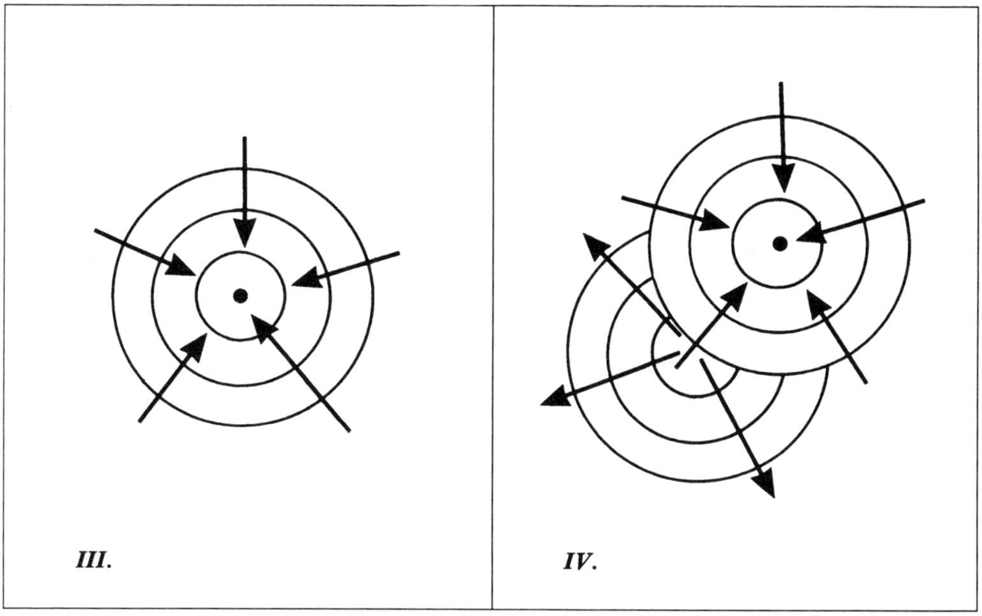

III. Level of the ego-space, (Red Community Plan)

The third task is to organize the life-space into an arena for the self-realization of the ego-identity. A diversified three- dimensional action space now arises, forming a frame for the operative functions of community life. We distinguish between three distinctive operative zones:

a. **The cultural (sensory) central zone**
b. **The social (communicative) medium zone**
c. **The productive (instrumental) peripheral zone**

IV. Level of logo-space, (Global Community Plan).

The final task is to organize the boundless rational space of the logo-identity into an action space. The hard problem to overcome here is its linking to the place-bound primary levels of eco-identity. The cosmopolitan eco-region may stand forth as a natural compromise, realizing in its shape the vision of a bipolar community. In format and content, the eco-region provides an intermediary link between the local organic life and the global cognitive level of existence with its networks.

Centre and periphery in future Europe?

REGIONAL REALITY 1: SWEDEN AND EUROPE

The debate on European integration inspires great hopes but also awakens fears. One thing is clear: this is a process of economic concentration which entails considerable socioterritorial changes. A centrifugal process has already begun which might result in a heavy continental polarization between centre and periphery. One plausible scenario is that the metropolitan regions on the continent form a European power network of hyper-standard, while the periphery is marginalized in terms of competence and creativity.

Although the pattern shows many overlaps in both directions, this forecast should give the peripheral nations cause for reflection in their choice of strategy for the future. Perhaps the centrifugal forces cannot be offset by copying the continental models. Alternative ideas and models, better suited to the periphery of Europe, should be tested. By and large, however, alternative visions are lacking at present.

REGIONAL REALITY 2: CENTRE AND PERIPHERY IN SWEDEN

Faced with the alluring prospect of European integration, regional researchers and national politicians outline new visions of Sweden's economic future. One model which is gaining ground today pleads for a concentration of strength in the country's historic big-city regions (the Mälaren region, the Göteborg region and the Malmö region). This, it is supposed, will give rise to competitive European regions.

It is hard to see how such a concentration can occur without further capital, competence and creativity being drained off from regions of the country which are threatened already, thus diminishing their prospects of survival. The obliteration of the historical agrarian landscape is already going on in large parts of the country and may soon be followed by the closing down of entire communities and regional road systems.

70 % or more of the settled territory of the country is threatened with closure, given the theoretical visions now entertained by many national development experts. Through their traumatizing effect, those visions can very well become self-fulfilling prophecies, whatever their long-term credibility. But are the uncertain gains for a small number of metropolitan regions worth the immense price we will all have to pay for a social transformation of this kind?

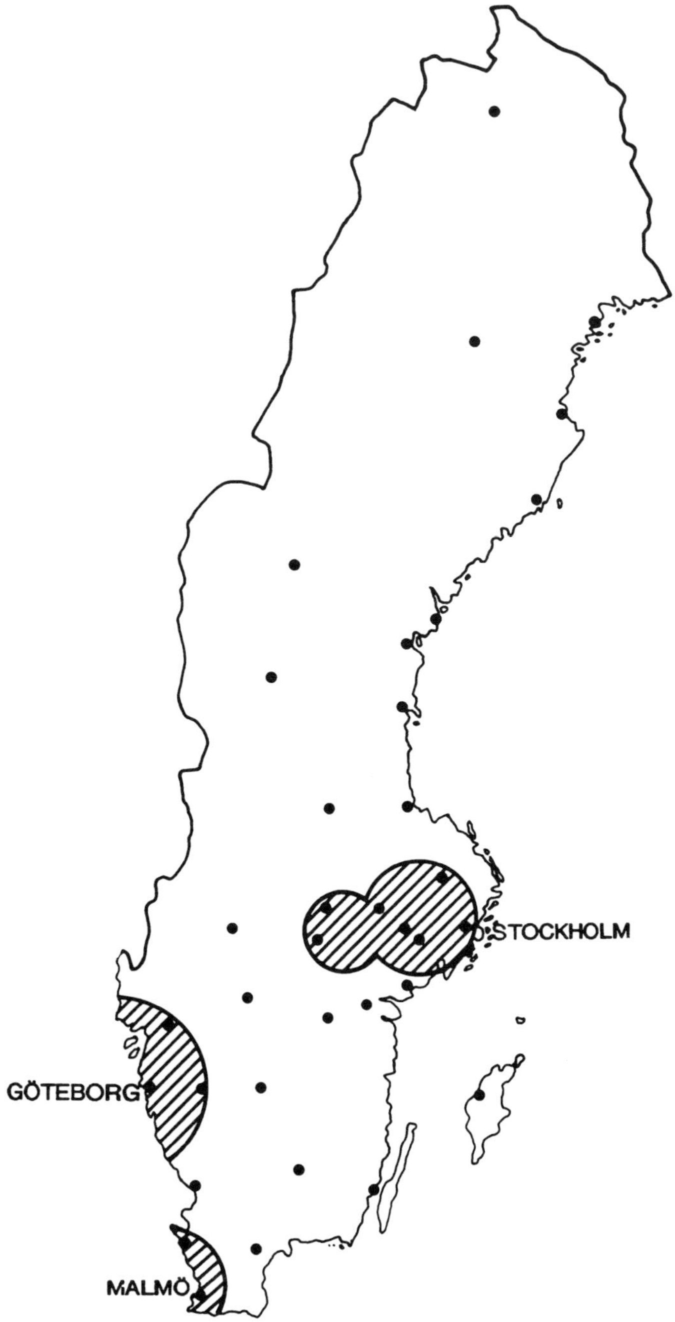

AN ALTERNATIVE: A SWEDEN OF LIFE-REGIONS

An alternative map of Sweden can be drawn, using the life-region as a model of regional development. In the map shown here, the municipalities outside the big-city regions have been organized into regional units which, in scale and structure, agree closely with the ecological life-region. Within the new regional units we imagine a functional reorientation in a human-ecological direction. The aim is to create a dynamic interaction between the socio-cultural and economic forces of the regions. This may also create opportunities for an endogenous development capable of resisting the forces of disintegration.

A less menacing picture of the country's regional future emerges, characterized by a better balance between the highly and less urbanized regions. Considering the many questions and doubts connected with the metropolitan utopia, the alternative model can serve as a much-needed economic and social safety net for the entire country.

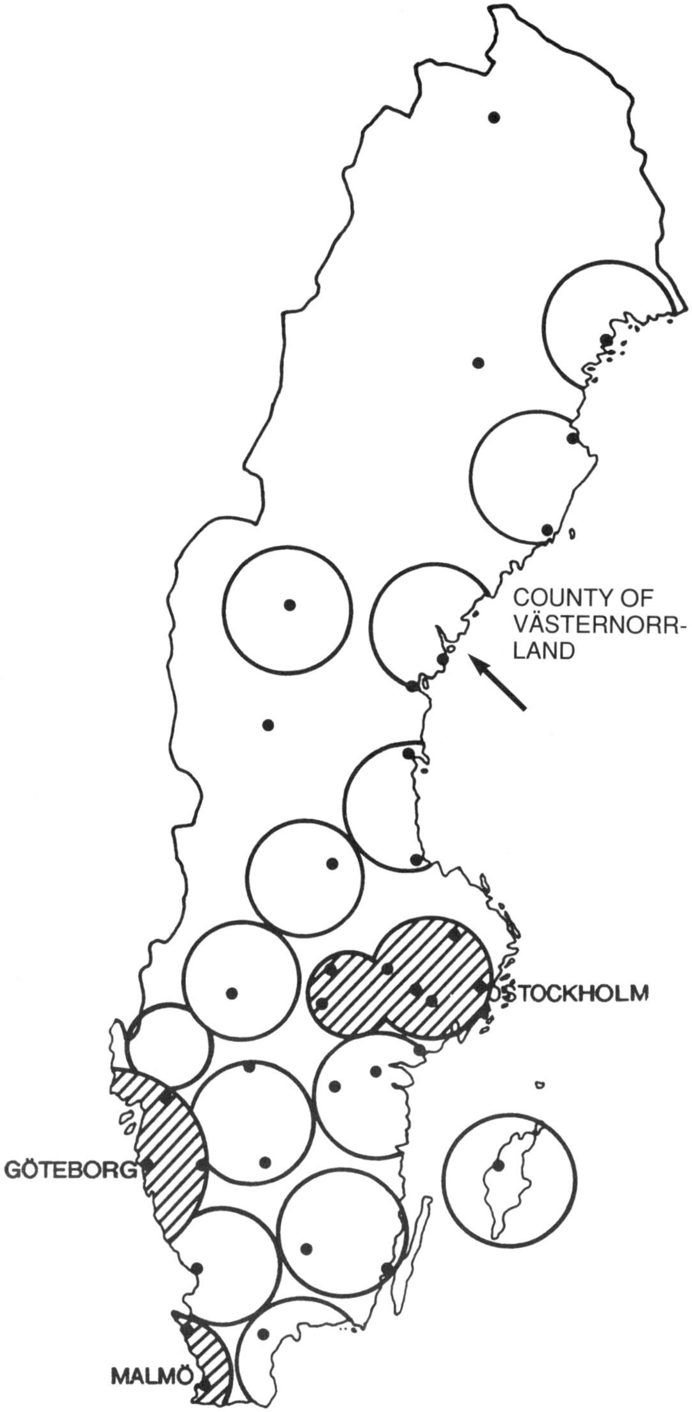

LIFE-REGIONS AS EURO-REGIONS

We imagine the life-region to represent a qualitative life territory, the integrity of which it is important, for socio-cultural reasons, to preserve. In the international economic perspective such regions are relatively small-scale units, but they can strengthen themselves by forming sectorial inter-regional coalitions in economic, cultural and social matters.

By forming inter-regional coalitions, the life-regions can become competitive alternatives to the metropolitan regions on the international market. Their socioterritorial stability helps to attract businesses to the municipalities. In addition to their superior quality of life, they can procure for themselves a technological standard equalling that of the metropolitan regions.

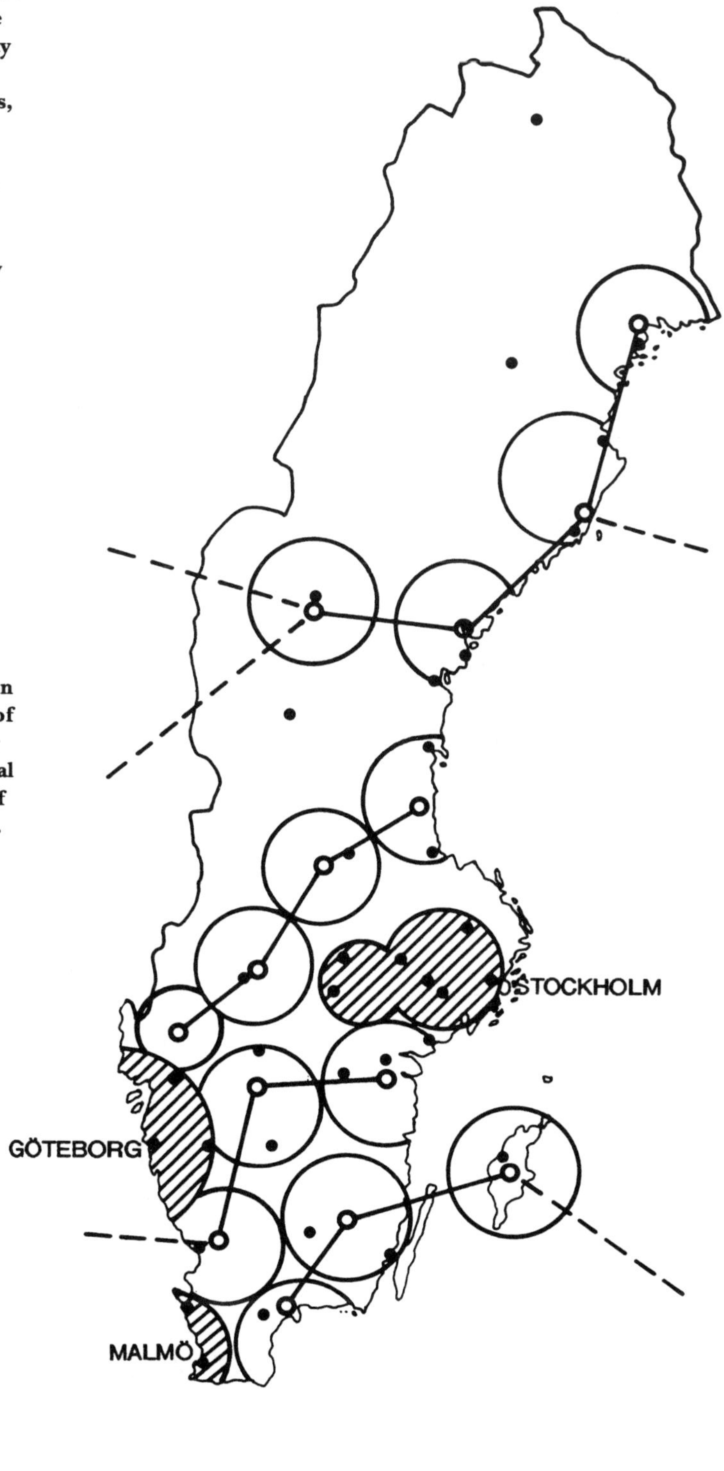

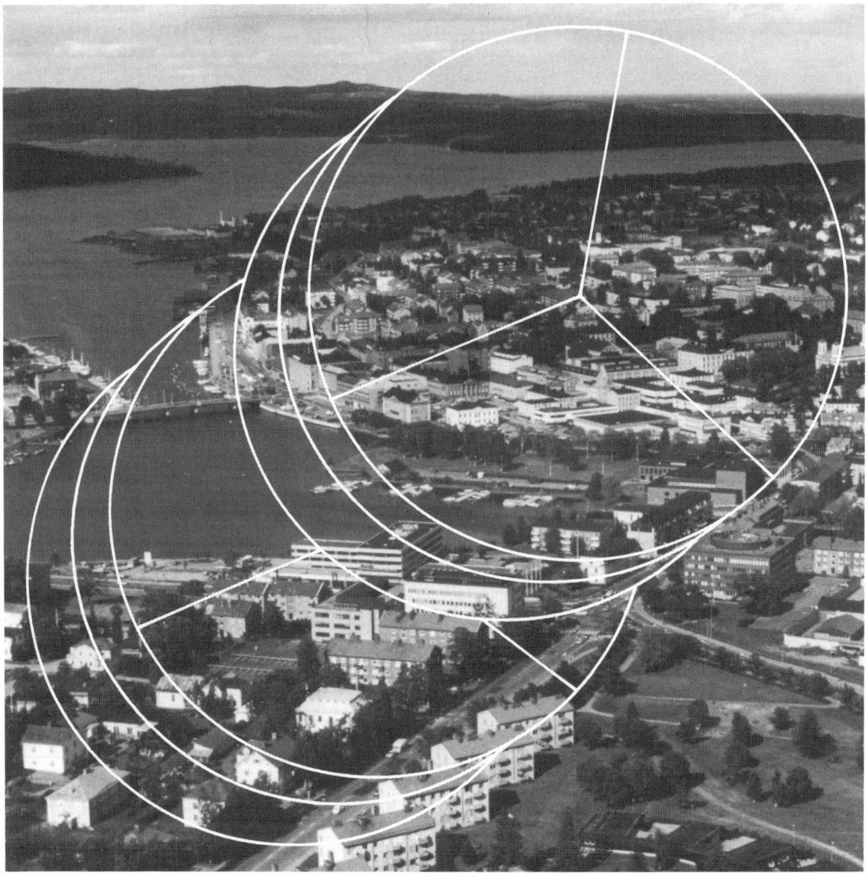

A SWEDISH LIFE-REGION: VÄSTERNORRLAND

The socio-ecological region is not intended as a rhetorical myth for the constellation of regions whose existence is threatened. It is a community model for practical implementation in municipal and regional planning. The design of an ecological planning methodology is an important stage in making the radical model practicable. This matter has been studied by the Umeå University Group for Humanist Futurology, and the work has included applied regional studies. A series of tentative community analyses has been performed in the County of Västernorrland.

One of these studies deals with the territorial planning dimension. The following pages show the result of the Group's experiment in territorial planning. The experiment was conducted in collaboration with Professor Jan Henriksson's architect's practice in Stockholm. Contacts with county planning specialists and several visits to the site have helped to enhance the authenticity of the sketches.

The study is organized in two sections. In the first the region's global action level is analysed. Some of the main aspects of an ecological plan of renewal are presented in a sequence of regional plans. The second section outlines, in somewhat closer detail, a plan of renewal for the local level. It should be noted that the renewal plans are confined to the spatial organization of the community. They represent an adaptive planning methodolgy aiming for adjustment to the existing settlement and infrastructure.

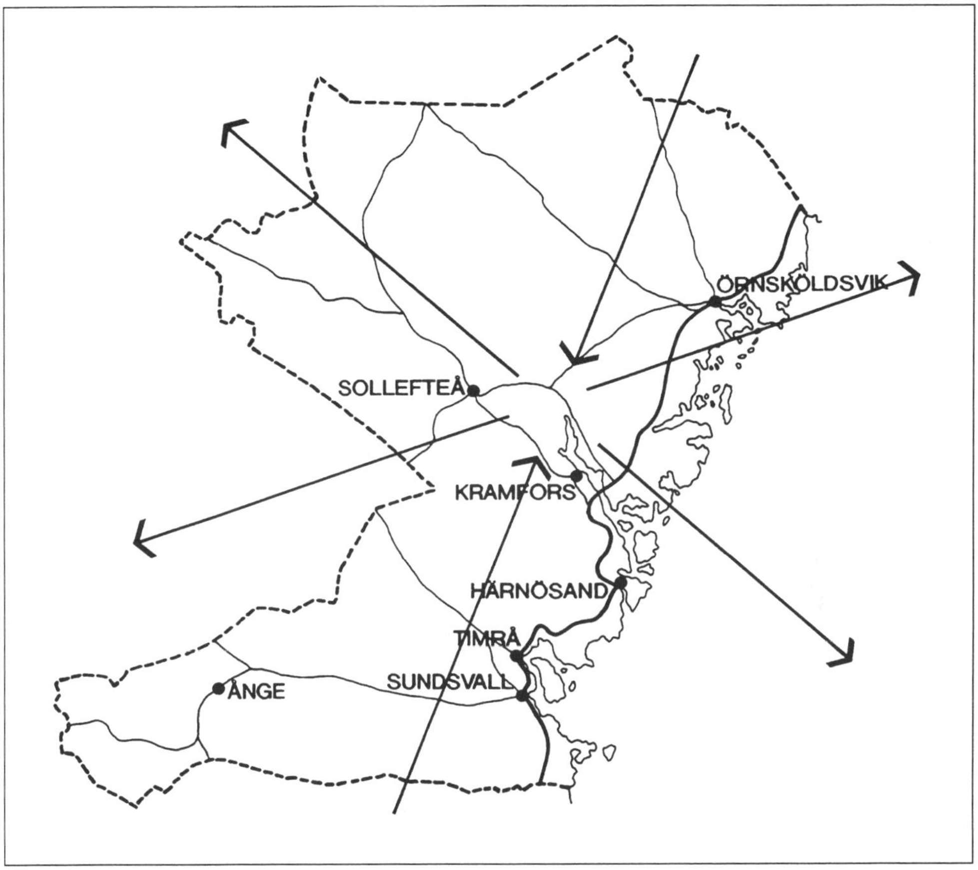

RENEWAL PLAN FOR THE GLOBAL LEVEL OF THE ECO-REGION

In the global plan for Västernorrland, we treat the county as a powerful actor on the international scene. Economic, social and cultural processes are calibrated for interaction with other regions on the front line of a planetary culture. The infrastructure of the entire region has been organized so that the region can effectively discharge its cosmopolitan role.

Where possible, global cultural identity is articulated in the architectural landscape. But the overall unit of the regional territory is marked at the same time. It should be remembered that the global arena retains its moorings in a local and - ultimately - individual life perspective. The organic unity of the bi-polar life-space is preserved by regarding the global regional space as an overlay on the local level of regional space.

Contents:

Regional structure plan: green (natural) level
" " blue (species) level
" " red (ego) level

Inter-regional meeting point: green level
" " blue level
" " red level

County of Västernorrland
Ecological regional renewal
Green regional plan

- **Inter-Centre**
- **Natural landscape zone**
- **Area of outstanding natural beauty**
- **Natural Protection Area**
- **Panorama**

REGIONAL STRUCTURE PLAN I. GREEN LEVEL

The green regional plan forms the basic level of the ecological renewal plan for Västernorrland. The designation shows that it is the natural landscape space of the region that is referred to. Also in the global perspective the geographical landscape remains the foundation of the individual's activity. What is asked for is not the natural space as a field of economic action but its aesthetic qualities and its value as a symbolic life frame.

Safeguarding the natural space as a symbolic community frame is the prime task of territorial regional planning. The task includes the marking of "Areas of outstanding natural beauty" and designating "Viewing points" or "Special protection areas". The political purpose of the green regional plan is to protect the landscape values of the region from destructive interference by various interests.

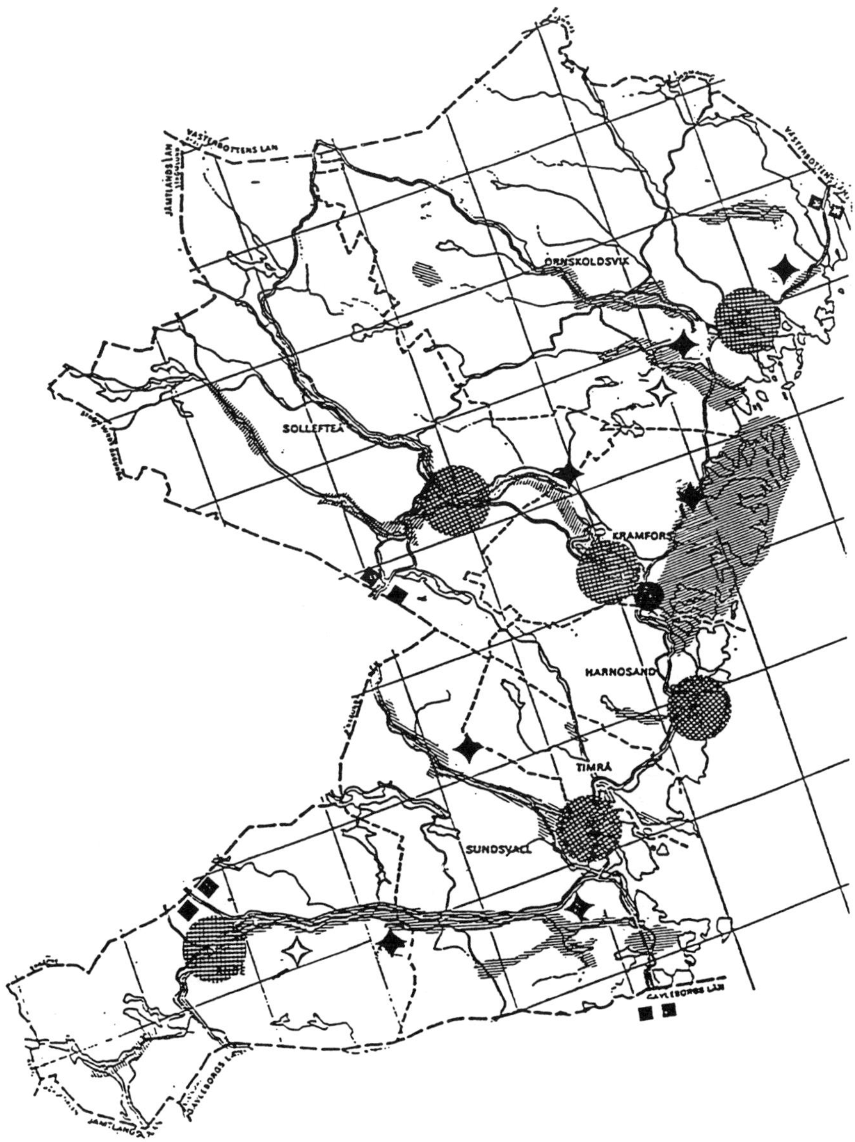

County of Västernorrland.
Ecological regional renewal

Blue regional plan

Inter-Centre

Urban centre

Zone of cultivated landscape

Cultivated landscape of outstanding value

Regional entrance

Regional park

REGIONAL STRUCTURE PLAN II. BLUE LEVEL

The second ecological planning level we term the blue (species) level. The blue regional plan frames the institutions and infrastructure for the common concerns of the regional community (such as schools, the administration of justice, medical care and planning). The public institutions are responsible for the continuity and long-term stability of the community. Their societal frame is a space for the generations and the species rather than for individuals and group interests.

The public space, therefore, can also be termed "the cultivated environment". The cultivated environment is formed by the inherited agrarian landscape with its historic settlement and transport routes, its open vistas and borders with the natural space. The historical infrastructure is partly still in use and has partly been elevated to monuments. But the man-made environment also includes urban centres, and it is foremost here that its task is to frame the community institutions.

The task of the blue regional plan is to articulate the public level as a constituent element of a comprehensive community structure. We should bear in mind that it is on the blue, public level that man's activities are elevated to the level of an organized society, with a history and a permanent population.

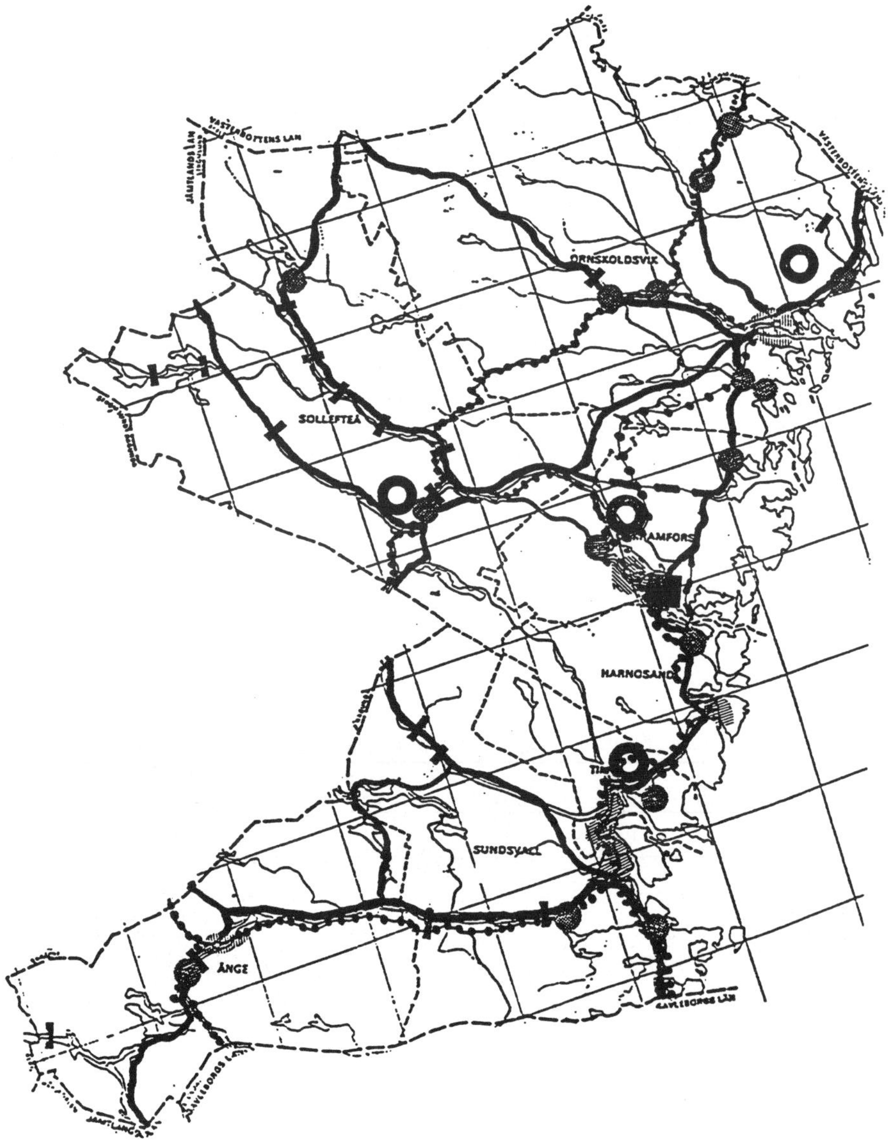

County of Västernorrland.
Ecological regional renewal

Red regional plan

- ■ **Inter-Centre**
- — **Through-road**
- ••••• **Railway**
- ● **Interregional production**
- | **Power station**
- ○ **Airport**

REGIONAL STRUCTURE PLAN III. RED LEVEL

The red level forms an uppermost layer of the ecological planning hierarchy described here. The task of the red regional plan is to co-ordinate the operative structures of the living community, insofar as these are of cosmopolitan dignity. It is the infrastructure for production facilities, for the social networks and the dynamic cultural institutions which is the subject of planning measures.

The operative structure naturally falls into two sectors: the executive functions, located near the local communities, and the administrative functions at a cosmopolitan meeting point in the geographical centre of the region. One important point is that no permanent population is living at the cosmopolitan meeting point; that place serves as a common stage for the inter-regional ambitions of the municipalities. An important role is also played by the region's communications system. The main emphasis should be on the trackbound network, to which future settlement will have to be adapted. An intra-regional circular line links the local communities of the region with the cosmopolitan meeting point. High-speed trains then connects the regional centre with other regional centres at inter-regional level.

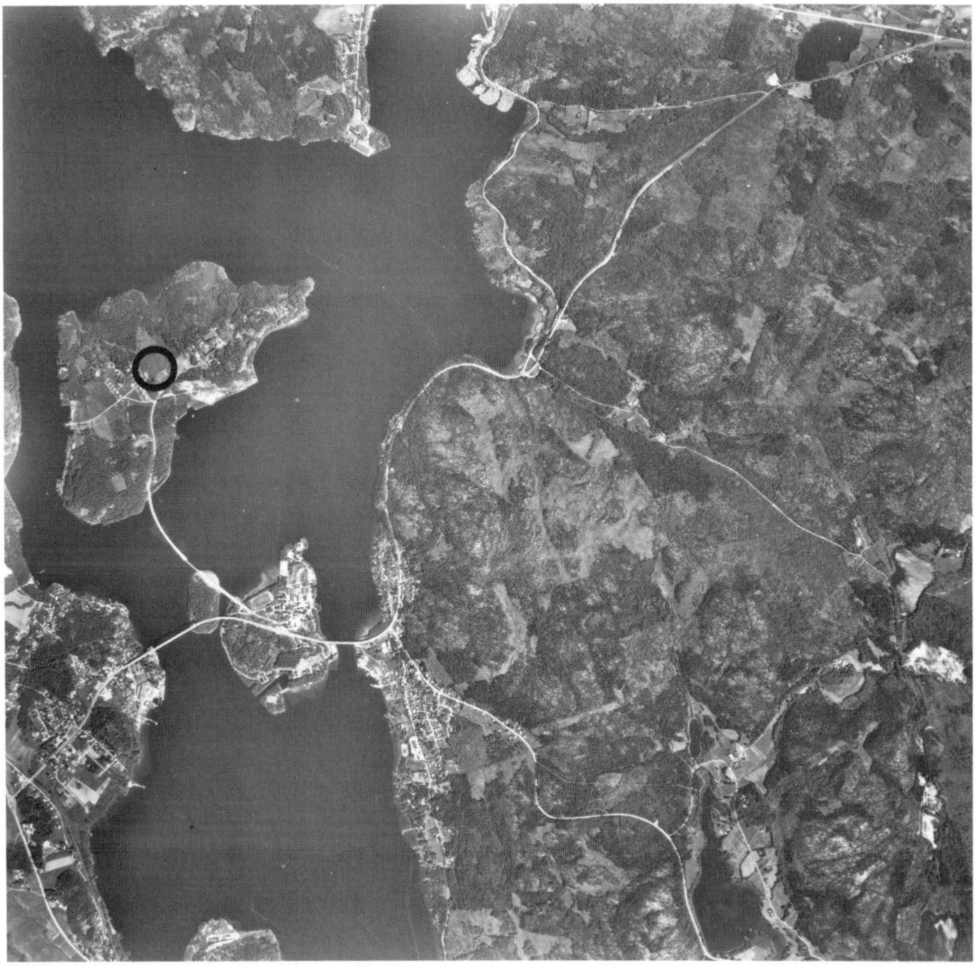

Ångermanälven river
"Inter-Centre", Svanön

INTER-REGIONAL MEETING POINT
A GLOBAL INTERFACE

Planning of the cosmopolitan meeting point is a task demanding careful consideration. The "Inter-Centre", as we have called it, is the region's main interface with its global surroundings. This image has to be emphasized both in the functional plans and in the articulation of the centre's architecture. The Inter-Centre's positioning at the point of intersection between the local communities in the region is one of the tasks. Another is the specification of the scale of the settlement. With its moderate monumentality the centre deviates from the local level. The architectural vocabulary also differs from that of the local environment, by its bold universality.

In the ecological plan for Västernorrland we have located the Inter-Centre to the isle of Svanön, which is placed at the mouth of the Ångermanälven river. Svanön is a dramatically situated spot in the coastal area of Västernorrland. It is situated near the centre of the county's communications system.

INTER-REGIONAL MEETING POINT I: GREEN PLANNING LEVEL

Our first step applying ecological principles to the Inter-Centre is to work out the green - natural - plan for the meeting point. We solve the task by preserving, in the centre of the settlement, a green corridor which forms flanking stretches of parkland on the periphery. To heighten the dignity and urban character of the natural plan, a symbolic monument - which we have chosen to call "Cosmotopos" - is erected in the centre of the plan. The Cosmotopos maintains a dialogue with a group of buildings for intellectual contemplation of the natural dimension; this has been dubbed "Ecosophicum". Profound ecological life values are associated with the green urban plan and govern its central position in the Inter-Centre.

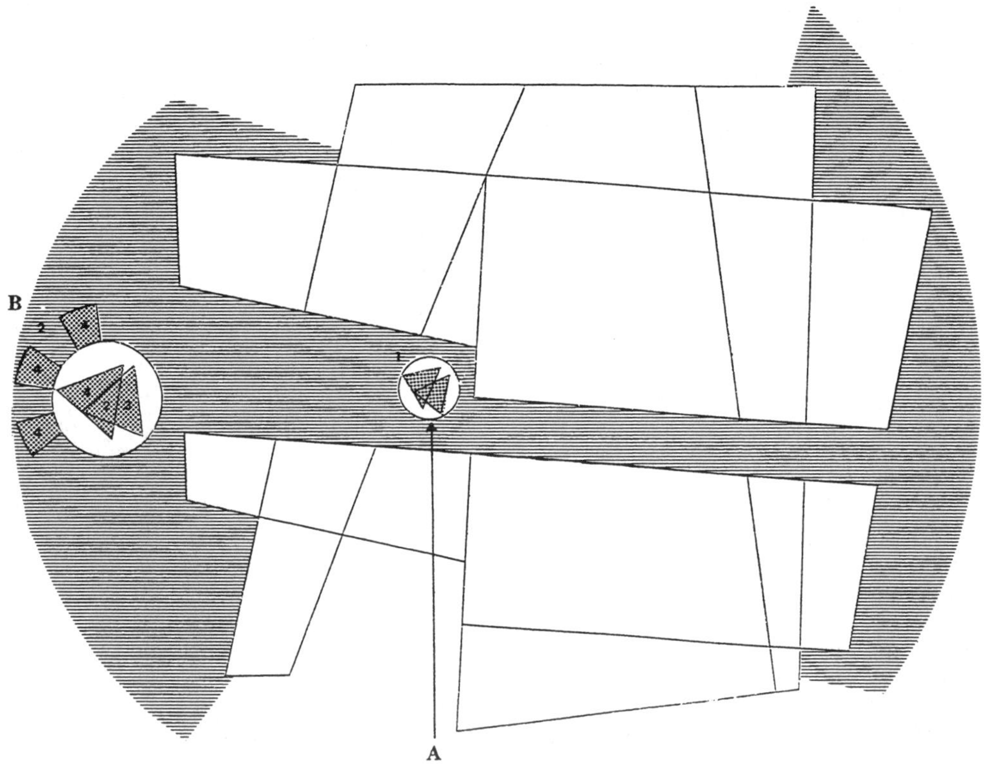

Inter-Centre. Global meeting point of the Eco-region

Green central plan

A. Cosmotopos
Place for ecospherical identification

B. Ecosophicum
Eco-philosophical centre of the region
1. Auditorium
2. Library
3. Seminar
4. World Eco-University, regional division

INTER-REGIONAL MEETING POINT II: BLUE PLANNING LEVEL

The second planning level in the regional centre comprises the plan for the public buildings of the meeting point (here with certain service adjuncts). These are facilities of common interest to the municipalities of the region, and they supplement present-day county planning for the public sector. The hotel and municipal hostels make it possible for the population to stay in the place for longer or shorter periods (though not permanently!).

A regional parliament denotes the summit of regional politics. In the socio-ecological vision, its establishment is associated with hopes for a deepening of civic democracy. But facilities for global tourism have also been included and given an attractive environmental setting.

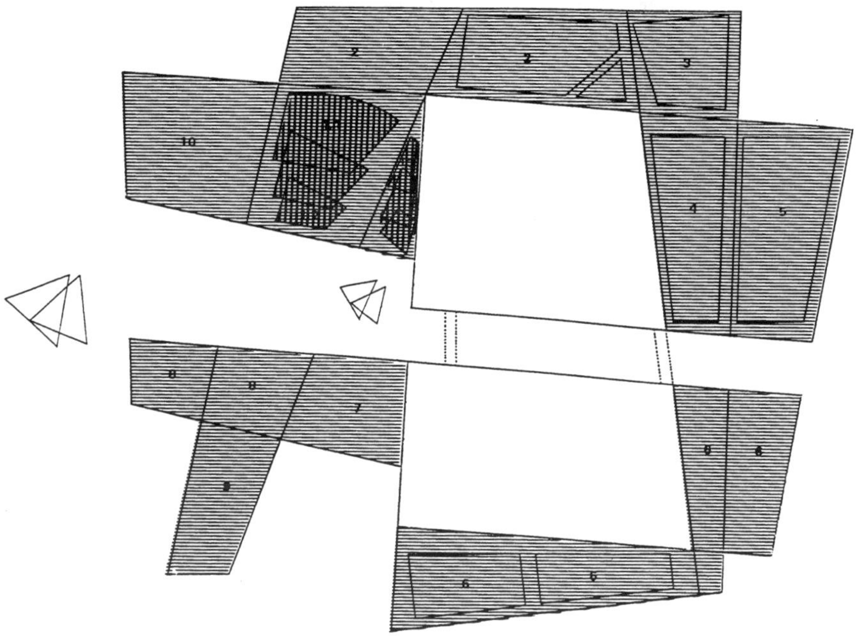

Inter-Centre. Global meeting point of the Eco-region
Blue central plan

1. Regional Parliament
1.1. Civic Forum
1.2. UN Section
1.3. Secretariat

2. Regional/Inter-regional administration
3. Regional R&D institutions
4. Archives and library
5. Regional goods depots
6. Hotel and municipal hostels + service
7. Regional caring services
8. Botanical garden
9. Spectator sports
10. Amusement park

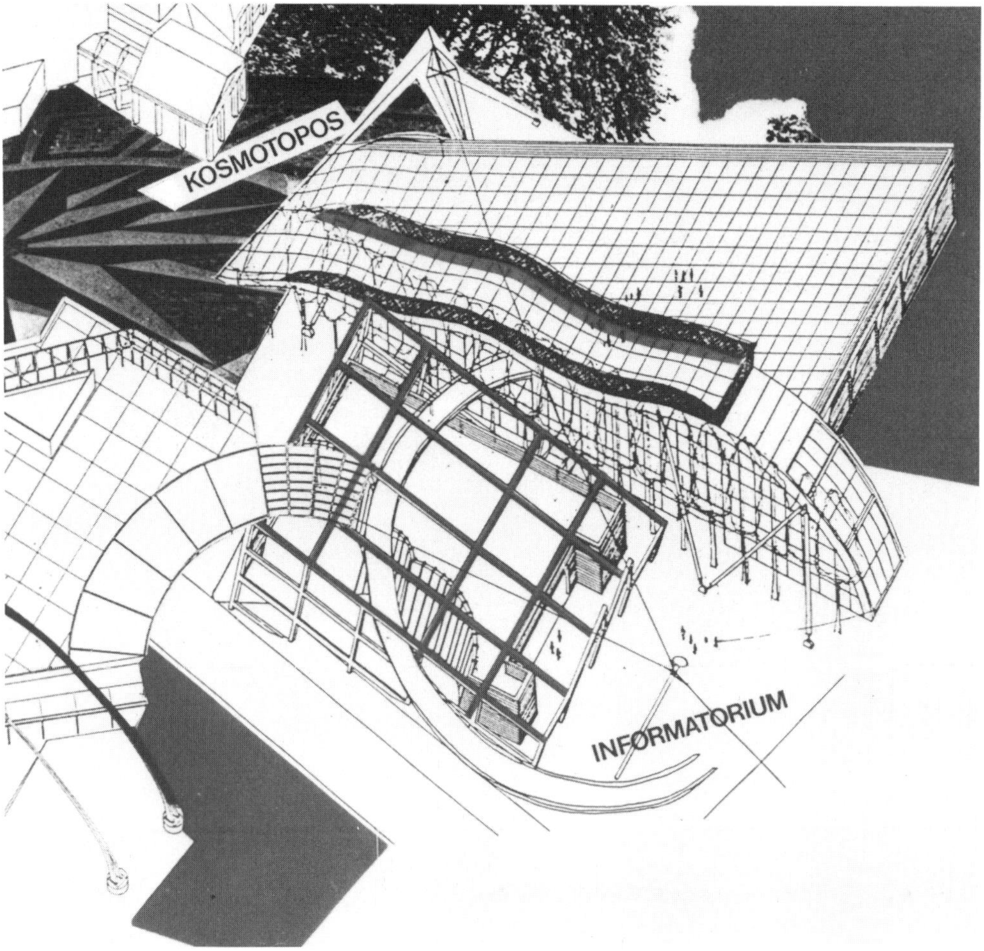

INTER-REGIONAL MEETING POINT III: RED PLANNING LEVEL

The dynamic activities of the regional centre have been brought together in the operative red plan. Its area occupies the central zone of the meeting point and is an overlay to the green and blue plans. The operative functions of the red plan are divided between three sectors: the cultural, social and economic sector.

The operative buildings are on a relatively large scale and are characterized by the oscillation of cosmopolitan aesthetics between cool abstraction and utopian fantasy. The central functions have been designed to make the meeting point an attractive venue for conferences, courses, international sporting and cultural events, school outings and intra-regional entertainments. Inter-Centre imparts a completely new, cosmopolitan dimension to the life of the regional community. The trauma of provincialism gives way to a sense of the region's full global presence!

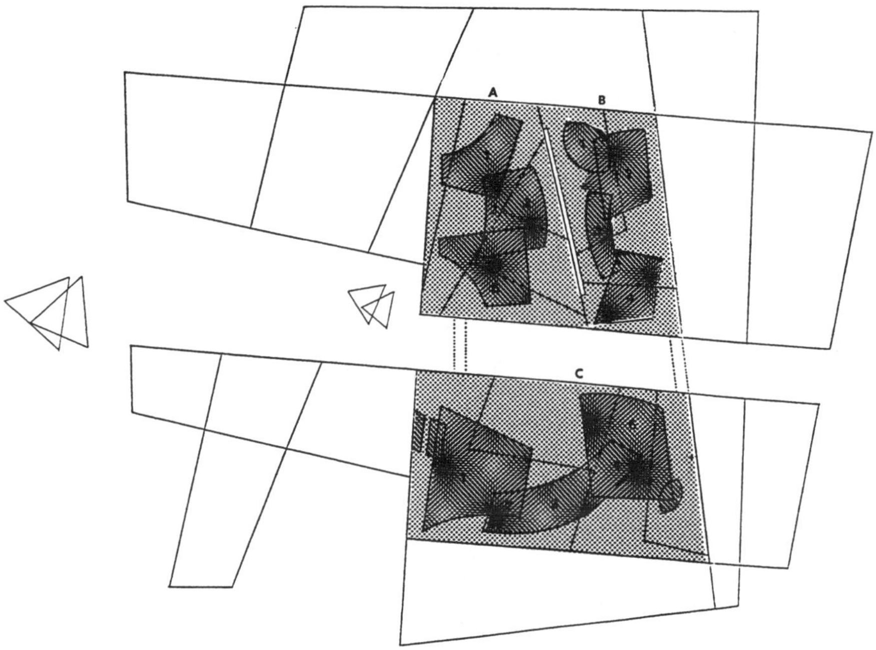

Inter-Centre. Global meeting point of the Eco-region
Red central plan

A. Cultural zone
1. World theatre
2. Meditatorium
3. House of the Arts
4. Intersport
5. Cultural contact

B. Social zone
1. Future Museum
2. World Inform
3. Mall
4. Big Bang Club

C. Economic zone
1. Global Commodity
2. Enterprise companies
3. Inter Congress
4. Technology Centre
5. Museum of Ecology

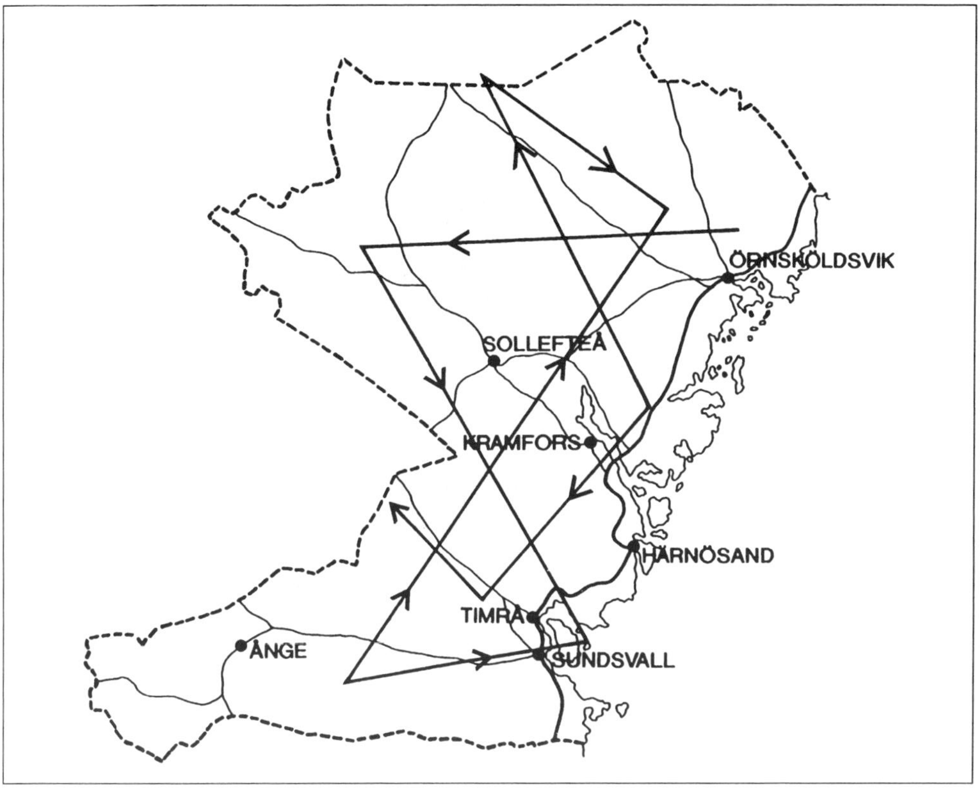

RENEWAL PLAN FOR THE LOCAL LEVEL OF THE ECO-REGION

In the local regional plan, Västernorrland is regarded as a life frame for locally oriented activities in production, social life and culture. The local regional plan safeguards the special values of the intimate community and life qualities associated with the needs of sensory life. The local level also serves as an important buffer and a safety net in the event of failures of inter-regional relations. Retaining control over the local basic structure creates greater regional security but also provides guarantees that power over living conditions will not leave the region. In many cases the local activities can be expanded to an intra-regional exchange of goods and services.

Here we confine our attention to the spatial structure. In our draft version of a local renewal plan, the focus of interest is on a particular locality in the region, the Härnösand conurbation. Härnösand is especially interesting because the city already includes a number of qualities which we like to associate with an ecological urban renewal.

Contents:

Härnösand conurbation: Green urban plan
" " Blue urban plan
" " Red urban plan

HÄRNÖSAND CONURBATION. GREEN URBAN PLAN

The green renewal plan for Härnösand shows, firstly, the conurbation as an organic life-space for the local population.

The natural areas of the place are accentuated and linked together so that a green holistic plan emerges. It is important that the green stretches form a distinguishable outer zone which, through its character of park, is experienced as part of the city. It is also important that this natural stretch be allowed to penetrate the urban centre without breaking it up altogether.

Through these measures, the green urban plan takes the shape of a continuous and enclosed natural-urban space. This natural-urban space forms the existential base level of the three-layer ecological plan; the space harmonizes with the structure of ecological identity as described in the ecological theory of interaction.

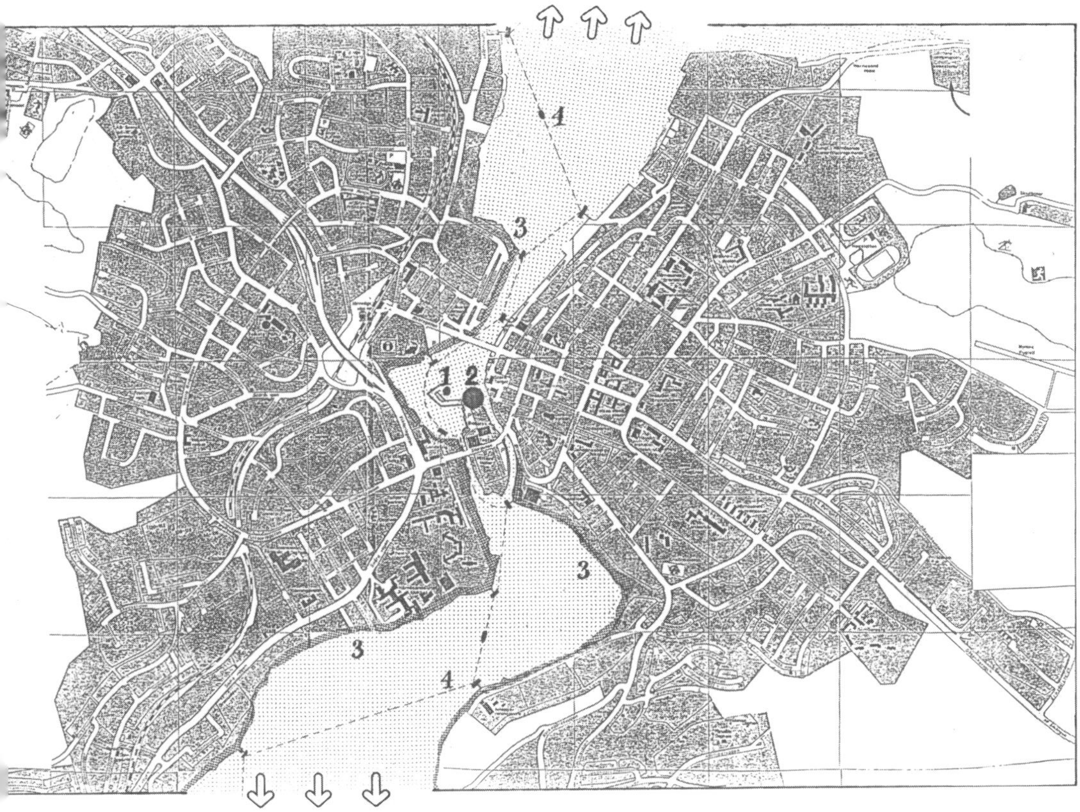

HÄRNÖSAND CITY CENTRE.
GREEN URBAN PLAN,

The green urban plan forms the very foundation of the city's ecological life. This plan contains, mainly in the centre, a number of functions which attract attention and provide impulses for visits in the zone. The most important functional accents are represented by a couple of facilities which we have dubbed "Ecotopos" and "Ecoforum" (see below). During the summer months, a regular ferry link across the fiord in the city-centre acts as a unifying element.

Härnösand. Ecological urban renewal

Green urban plan

- **1. Ecotopos**
- **2. Ecoforum**
- **3. Green stretch**
- **4. Vaporetto**

THE GREEN URBAN PLAN.
VIEW FROM THE CENTRAL BRIDGE

This picture shows how the natural elements of the green urban plan penetrate the central parts of the city and at the same time form a distinct outer background to the urban space. In terms of natural conditions, Härnösand is a privileged city, with high natural-aesthetic qualities in its townscape.

Ecotopos, Härnösand
Green urban plan

THE GREEN URBAN PLAN. ECOTOPOS

The placing of a dynamic accent in the centre of the green urban plan - and thus in the centre of the city - underlines the concentration of the natural plan and emphasizes its sovereignty in the life of the local community. In the ecological vision, nature automatically assumes a place of honour in the life-space, instead of being regarded as a contrast to it. The dynamic accent, however, must be designed in a manner congenial with the green plan. In the actual case, a geyser-type fountain has been placed in the centre of the Nattviken fiord.

Ecoforum, Härnösand

Green urban plan

1. Eco-inform
2. Gallery
3. Auditorium
4. Library

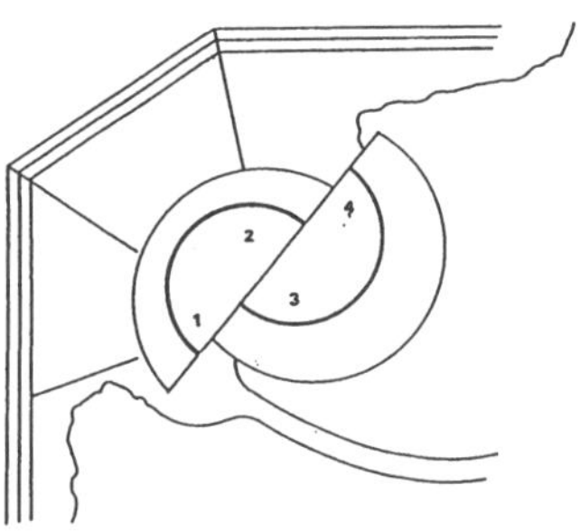

THE GREEN URBAN PLAN. ECOFORUM

The “Ecoforum” is another accent in the green plan which underscores its dignity in the life of the local community. The Ecoforum is a meeting point for information and debate concerning ecological aspects of life and the environment. A well-stocked library highlights this thematic context in local debate. The Ecoforum has been placed eye-to-eye with Ecotopos. We can describe the Ecoforum as the reflective facet of a green suite in which Ecotopos represents the meditative facet. The Ecoforum is also visible and easily accessible from the surrounding city districts.

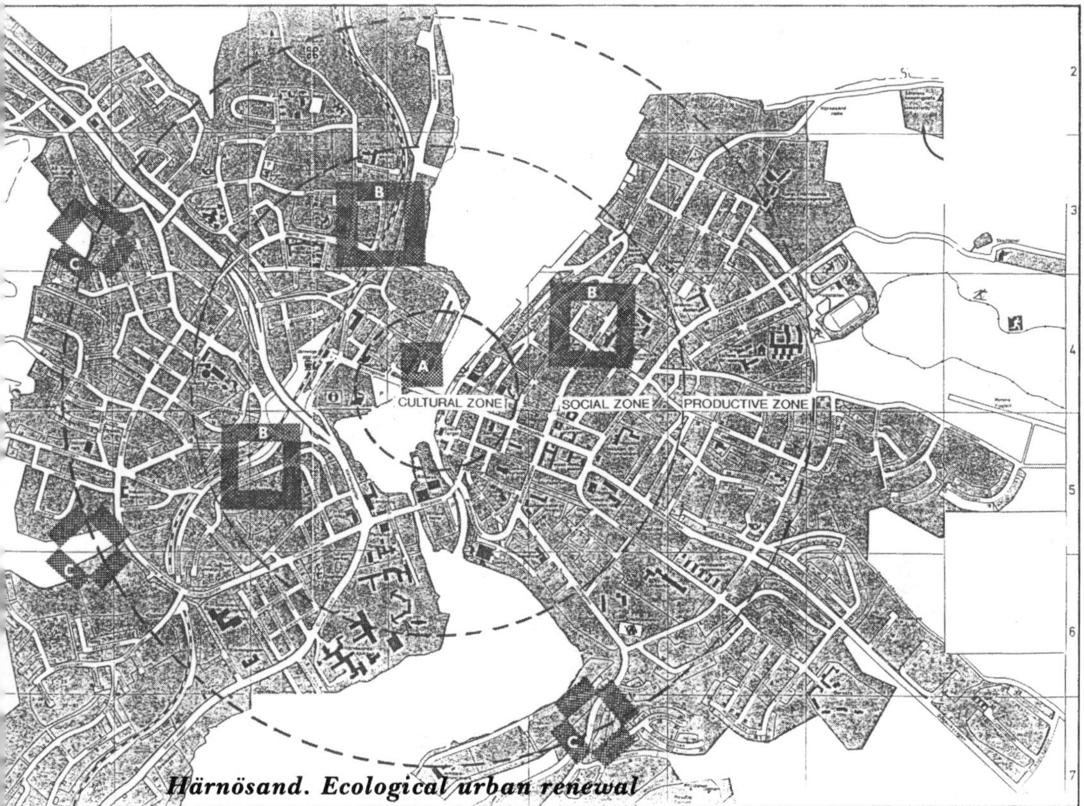

Härnösand. Ecological urban renewal
Blue urban plan

A. Entrance to the city
B. Boundary canal
C. Public avenue
D. Human Academy

HÄRNÖSAND CONURBATION. BLUE URBAN PLAN

The task of the blue urban plan is to lay the public foundations of the city and to turn these into a framework for the institutional life of the community. This plan represents a second level of ecological urban life and is an overlay to the green, organic plan. The most important formal characteristic of the blue plan is the two-dimensional urban floor with its streets and squares, prospects and outer boundary zones. From the floor of the city there arise, with special authority, the buildings housing the community's permanent institutions.

In the renewal plan for Härnösand, the prescriptions of the blue urban plan have been met by means of several complementary measures. The wholeness and self-containment of the plan are emphasized by symbolically marking the city boundary with architectural city entrances and boundary canals. The main strategy for the inner parts of the city is a couple of "public avenues" cutting through the city centre (see below).

BLUE URBAN PLAN.
CITY ENTRANCE TO HÄRNÖSAND

At the north exit from the E4 highway for Härnösand, an architectural entrance has been implemented on a downhill slope where the view of the city opens up for the first time. In the centre of the frame formed by the two gateway fragments one glimpses Härnösand Cathedral. (Note that this drawing, like many others of the sequence, is just an explanatory sketch.)

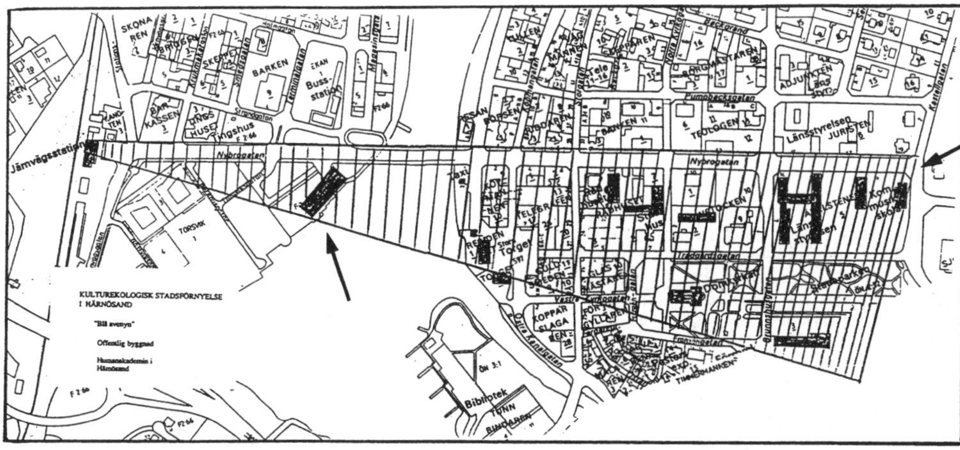

BLUE URBAN PLAN. PUBLIC AVENUE IN THE CENTRE

An architectural accentuation of the blue urban plan has been tried by linking the sequence of public buildings in the centre together to form a spatially continuous enfilade. This intention is achieved by opening the ground level round and between the buildings into a wide avenue expanding from the centre towards the periphery; the result is a distinct urban townscape. In terms of expression, the blue ensemble can be held together by means of boundary markings towards the surrounding areas and through careful treatment of the interlinking urban floor.

The photograph shows a section of Nygatan in Härnösand, which is included in the renewal area. The character of this street as a traffic route has been toned down by means of a ground articulation emphasizing its character of city floor (not to say parquet). In the distance a city entrance serves as an outer boundary mark.

BLUE URBAN PLAN.
SECTION OF THE PUBLIC AVENUE (MAIN SQUARE)

The Main Square in the centre of Härnösand is already an important part of the city's public space, by virtue of the moderately monumental buildings which surround it. This character can be further underlined by aesthetic articulation of the square's pavement, and by linking it more closely with other public environments in the neighbourhood, especially the Cathedral and Courthouse.

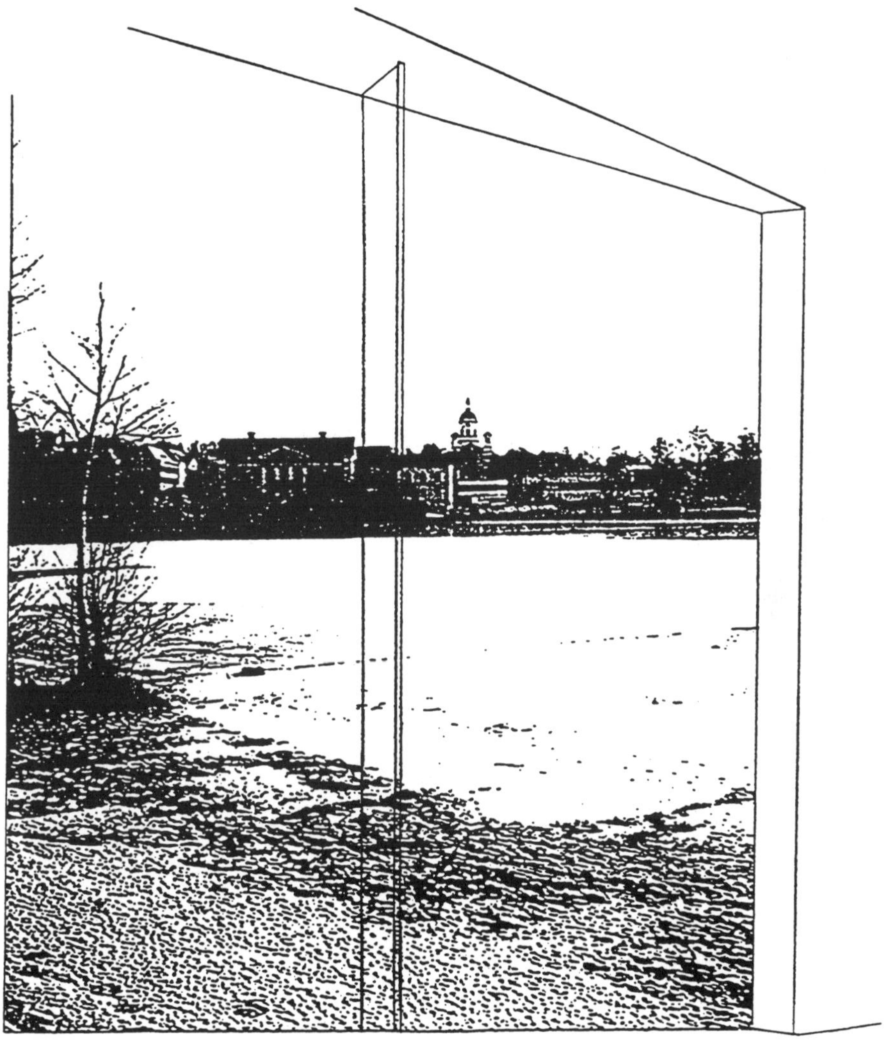

Human Academy in Härnösand
View from the Civic Forum

BLUE URBAN PLAN.
THE HÄRNÖSAND HUMAN ACADEMY

The objectives of the blue plan are also accomplished in the public buildings included in the plan. The establishment of a local civic forum responsible for practical aspects of democracy in various living sectors is one such example. Here we have dubbed this forum "The Härnösand Human Academy". The academy has been given a prestigious location overlooking the Nattviken fiord and facing the city.

The building is characterized by a modest classicism harmonizing well with the group of historic city centre buildings from the 18th and 19th centuries. This serves to underline Härnösand's old epithet of "Athens of the North".

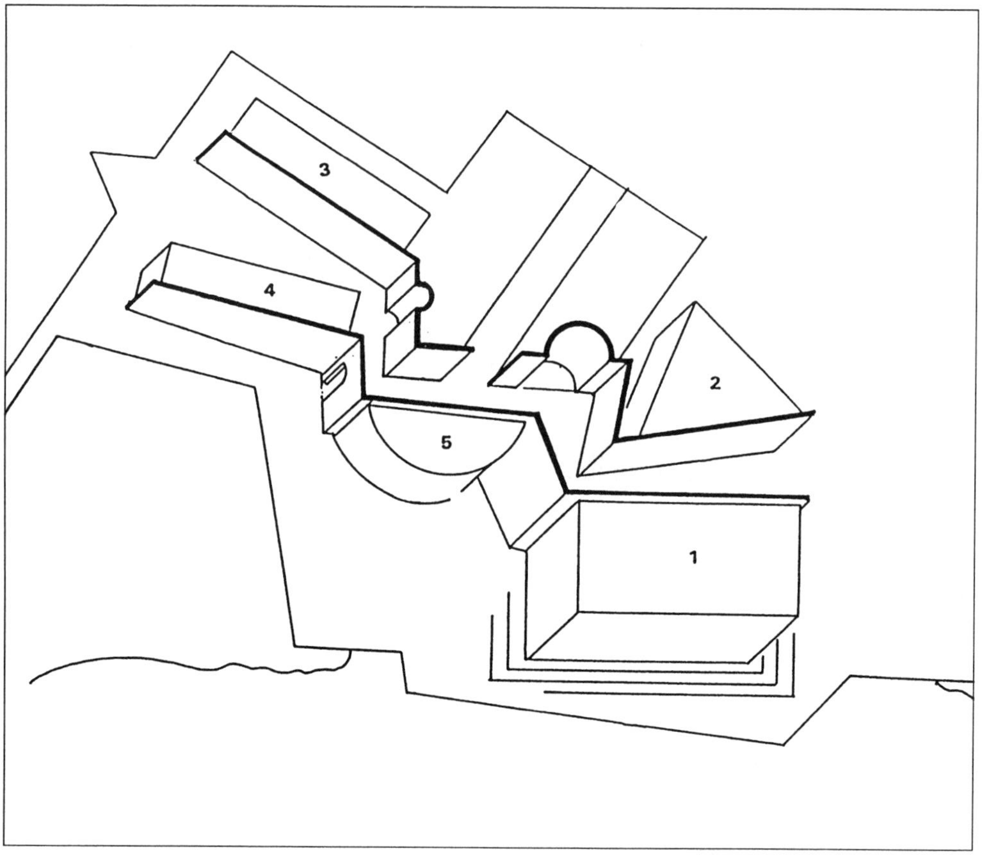

Human academy

Blue urban plan

1. Civic Forum
2. Local economy, secretariat
3. Local democracy, "
4. Local culture, "
5. Local planning, "

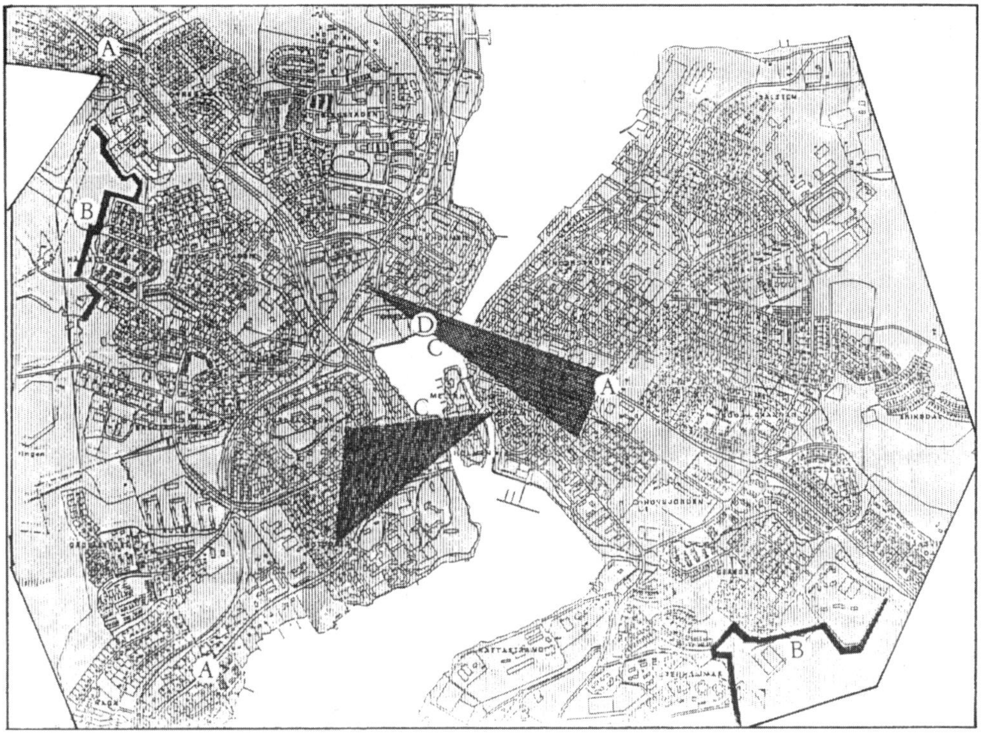

Härnösand. Ecological urban renewal

Red urban plan

A. Cultural Forum
B. City neighbourhood
C. Resource co-operative

HÄRNÖSAND CONURBATION. RED URBAN PLAN

The task of the red urban plan is to frame and articulate the present-day operative structure of the local community. The ecological model distributes the operative functions to three zones: the central cultural zone, the intermediate social zone and the peripheral productive zone. A spatial spread of this kind facilitates the dynamic circulation between the various life functions which we regard as characteristic of ecological behaviour.

In the physical urban stage, the operative level is symbolized mainly by the three-dimensional structure of settlement. It is articulated with the means of architectural form. We have made a sketchy attempt to inscribe the three operative eco-zones in the existing plan for Härnösand. We imagine that such ideas can serve, in practical planning, as general guidance for the localization of new activities when the opportunity arises.

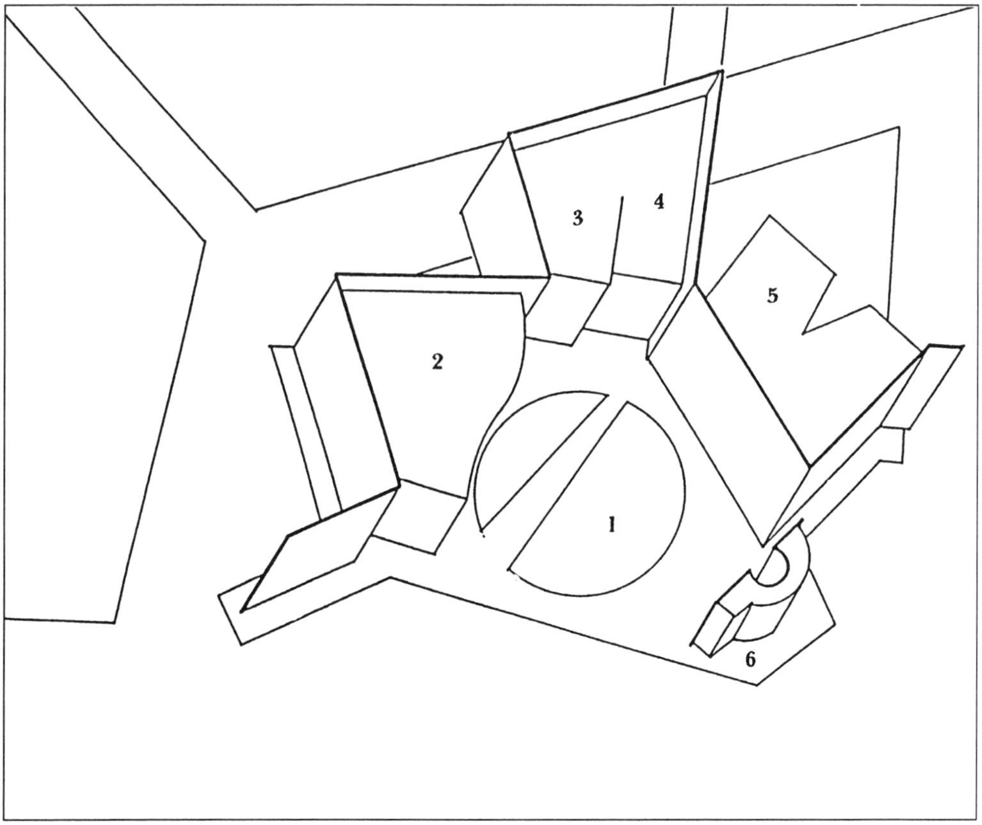

Cultural forum, Härnösand

Red urban plan

1. Urban spring
2. Aestheticum
3. Colloqium
4. Refectorium
5. Market place
6. Information

RED URBAN PLAN, CENTRAL ZONE. CULTURAL FORUM

The Cultural Forum is intended as a central meeting point for civic activities with an aesthetic-cultural focus. These activities cover the entire spectrum of sensory manifestations of life. The Cultural Forum is an autonomous unit which also serves as a starting point for the aesthetic mobilization of the entire local community.

The positioning of a cultural action forum in the centre of the city underlines the role of qualitative needs in the life of the local community. It will be an advantage if the facility also harmonizes with the socio-ecological vision architecturally (without offending the existing urban characters).

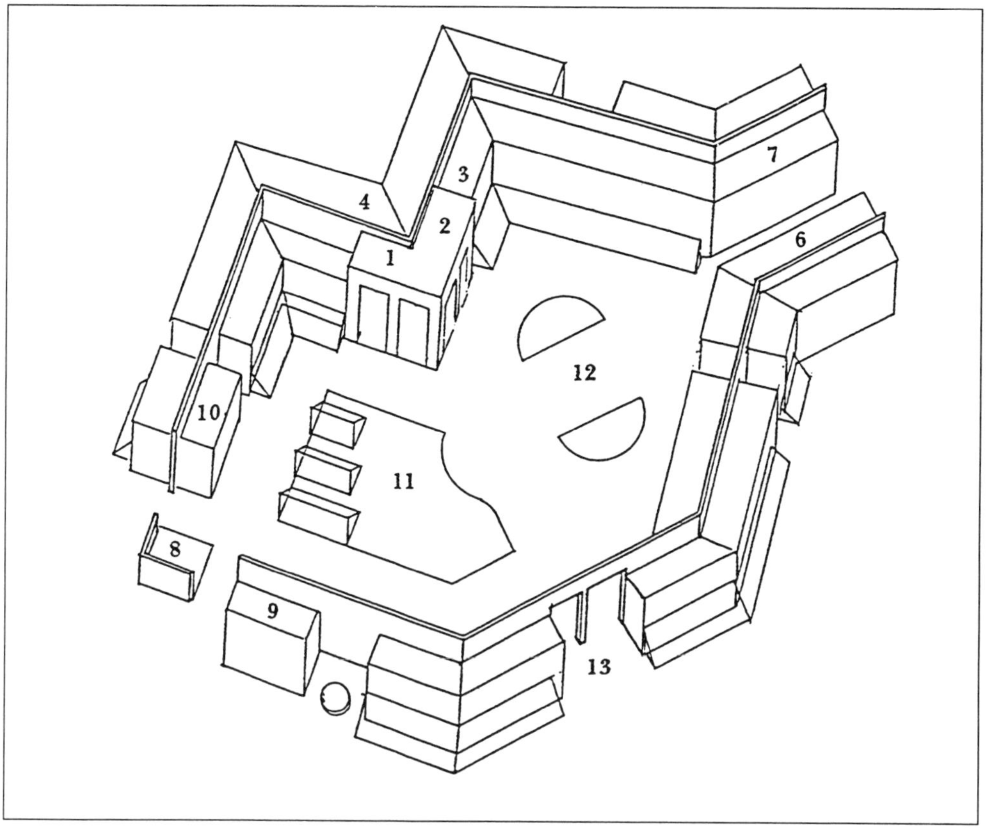

Urban neighbourhood, Härnösand

Red urban plan

1. Neighbourhood forum
2. Lobby
3. Restaurant
4. Local office
5. Catering
6. Workshops
7. Service shop
8. Memorial park
9. Local archive
10. Day nursery
11. Kitchen garden
12. Outdoor amusements
13. District gate

RED URBAN PLAN, INTERMEDIATE ZONE. URBAN NEIGHBOURHOOD

The city's operative intermediate zone we have dubbed the "social life zone". One planning target is to articulate within the zone collective neighbourhoods of such scope that the collective, parallel to its residential function, can assume direct responsibility for a large proportion of public community services. Arrangements of this kind are regarded, not as a burden but as a qualitative increment. In this way fresh new energy will be infused into the currently languishing social life sphere.

The urban neighbourhood as depicted here is an architectural prototype incorporating all these functions; at the same time it stands out as an independent unit in the townscape. The task is to shape an environment which arouses a desire for identification and feeling at home. (The prototype shown is a compressed model. The neighbourhoods normally have a larger population base than can be accommodated within this format.)

Urban neighbourhood on Kronholmen, Härnösand

A. Housing area
B. Common services
C. Handicraft quarter
D. Cultural forum

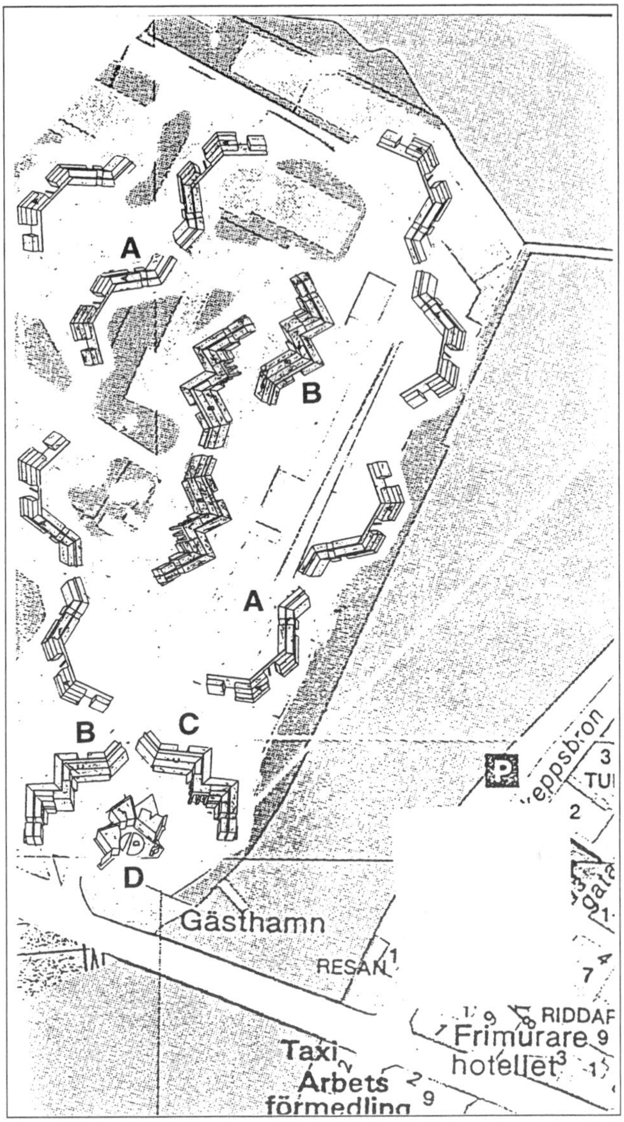

RED URBAN PLAN.
URBAN NEIGHBOURHOOD ON KRONHOLMEN

This picture shows, very schematically, a free implementation of the ideas of the urban neighbourhood in a development plan for the central parts of Härnösand (Kronholmen). In this we should notice the scale of the settlement unit, which is larger than that of the neighbourhood co-operatives which are normally discussed (social co-operatives, eco-villages etc.). The reason, as already mentioned, is a desire to elevate the scheme above the level of purely housing and to regard it as a genuine social alternative in the local community. The term "urban neighbourhood", therefore is quite an illuminating designation.

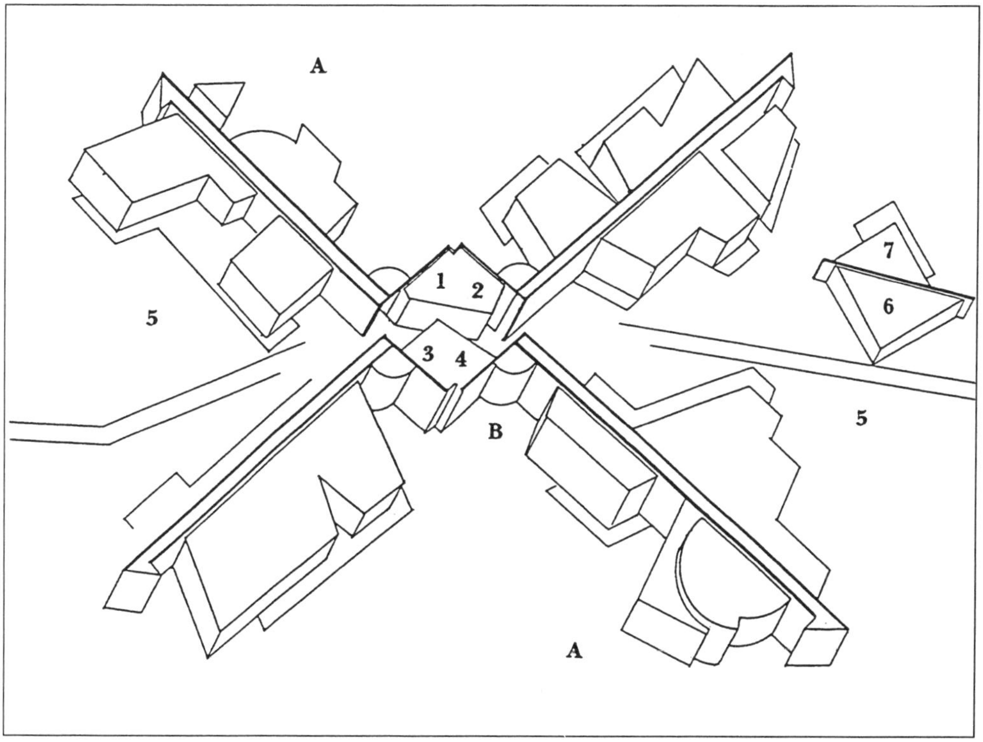

Resource co-operative, Härnösand

Red urban plan

A. ENTERPRISES

B. COMMUNAL FACILITIES

1. Administration
2. Personnel
3. Expo and sales
4. Restaurant
5. Parking
6. Raw materials stocks
7. Stocks of finished products

RED URBAN PLAN, PERIPHERAL ZONE. RESOURCE CO-OPERATIVE

The peripheral productive zone of the operative plan is placed close to the social zone. The creation of opportunities for convenient circulation between the functions of everyday living is an important part of ecological planning. In the ecological renewal plan for Härnösand, the aim has been to find, on the periphery of the city, niches which will accommodate locally oriented production on a moderate scale. The facilities ought preferably to have an "ecological" profile and to be of good quality architecture.

This sketch of a resource co-operative for Härnösand attempts to satisfy both requirements. In the resource co-operative a number of small local businesses have moved into a complex having a number of communal facilities. One common denominator is their shared need of certain raw materials and semi-manufactured products. With the aid of architectural ideas, the local community's productive unities are given the appearance of attractive, forward-looking branches of enterprise.

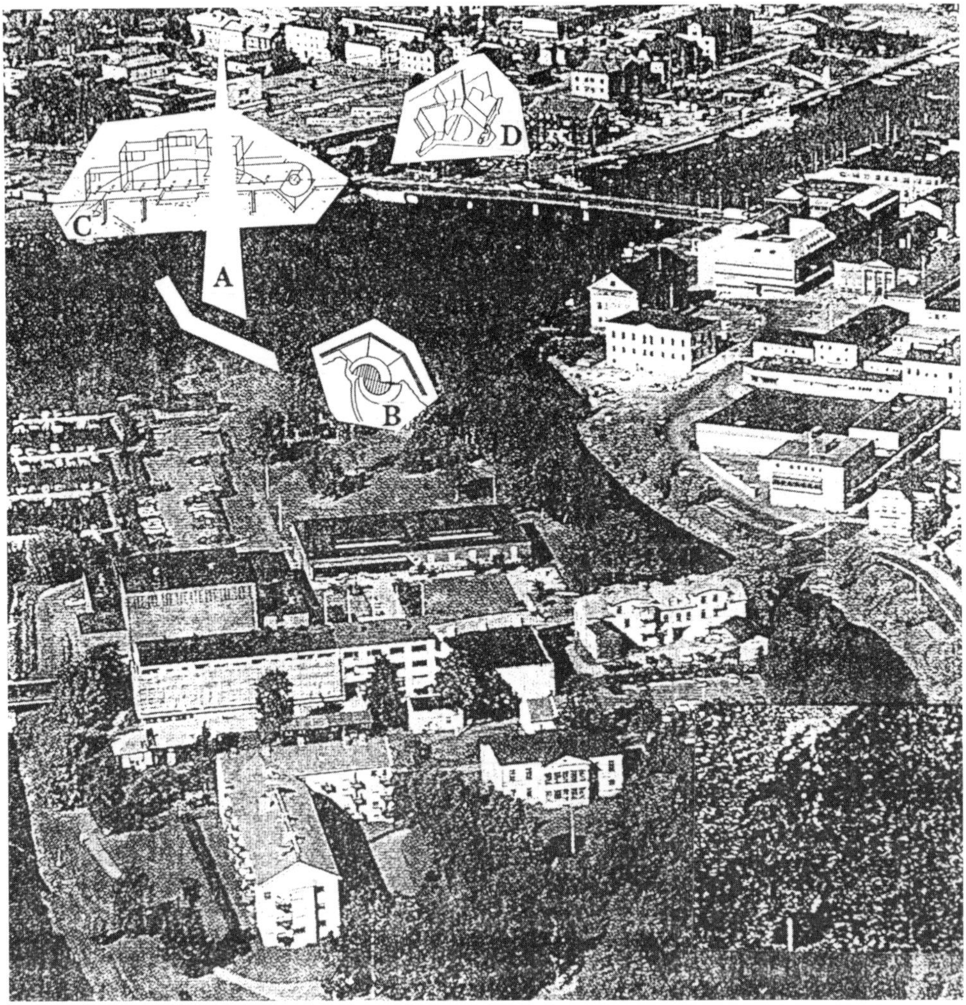

City centre of Härnösand.
Ecological urban renewal

A. Ecotopos
B. Ecoforum
C. Human Academy
D. Cultural Forum

CITY CENTRE OF HÄRNÖSAND. ECOLOGICAL ADJUNCTS

This air photograph conveys a good picture of the city centre of Härnösand and the location of its historic buildings. The montage shows how an articulation of the city centre in keeping with the intentions of the red, blue and green eco-plans, would create an interesting new constellation of buildings. Through their form and content, the new buildings help to impart a radical socio-ecological profile to the townscape. But even without ideological preferences, they can add new qualities to the centre.

III REGION AND CULTURE

"Window Vision". Photo by Hans Marklund

Culture and Regional Identity

Michel Bassand

INTRODUCTORY NOTE

The concept of regional culture can be understood in two main senses. It can be applied in the broad anthropological sense meaning the totality of the socio-cultural conventions inherent in a specific society. But it can also be read in a narrower perspective, designating the aesthetic and artistic sector of the society.

Very often in praxis the two perspectives overlap and influence each other in creative ways. Sometimes, though, it is advantageous to clear the terrain by looking at the concepts one by one. On this special occasion the focus is directed towards the aesthetic definition of the regional-cultural concept.

Although I highly sympathize with the chosen path, I would like to start my paper by applying the broader approach. By looking at the regional question from the wider angle of recent geopolitical development in the Paneuropean society, I believe it will be easier to appreciate the values and possibilities inherent in artistic culture for the sake of local and regional mobilization.

THE REGIONAL QUESTION IN EUROPE

This question appears in at least three overlapping perspectives: 1) urbanization or uneven regional development, 2) regionalization, and 3) regionalism.

1. Urbanization or unequal regional development concerns the transformations tied to industrialization which concentrate jobs, services and population to the benefit of certain places (cities, urban agglomerations, metropolitan areas) at the expense of rural areas. In 20th-century European history, the process has augmented itself considerably: cities have been continually drawing more to themselves, becoming larger, wealthier, more

powerful and extended, while the less attractive rural areas were losing population and becoming more dependent, marginal, and impoverished. After World War II, these changes accelerated considerably. The spatial organization of Europe came to look like a hierarchy of central and peripheral regions. The regional disparities and inequities became profound. The socio-economic regional map of Europe is changing perceptibly because of the overlapping between regionalization and regionalism which we distinguish for analytic purposes (Quévit, 1986; Aydalot; Greffe; Bassand, 1990; Kuklinski; Brunet).

2. Regionalization has a rather socio-political character; it is strongly linked to urbanization, to the crisis of the protectionist state, and to the globalization of exchanges and transactions. It is tied equally to those demands of local and regional democracy which are developing across the whole of Europe.

Let us remember that etymologically the term region means to govern. Therefore the regionalizations now evolving correspond to state efforts at decentralization and deconcentration. And so, in order better to regulate social change, respond to democratic demands, and correct uneven regional development and its harmful effects, countries are proceeding regionally to redivide their territories. It is this that we call regionalization. Sometimes it implies a more or less generous redistribution of jurisdictions and fiscal resources, sometimes the granting of new forms of regional sovereignty of autonomy, and sometimes both.

3. Regionalism cannot be understood independently from urbanization and regionalization. Regionalism differs from the other two in the sense that it possesses socio-cultural foundations and claims to identity, and further is being characterized by a rising social dynamic. The regionalism that has made its appearance since the 1970s has nothing to do with the highly reactionary regionalism at the beginning of the century. Contemporary regionalism results from a menace to the regional socio-economic structure and to the socio-cultural heritage. Such regional community is based on a history, language, religion, habitat, and on ways of life, art, and popular traditions. Urbanization and sometimes regionalization when their spatial partitioning fail to match those of the oldest regional communities, erode and even destroy this heritage and so place in question the very existence of regional community. It loses its autonomy, identity, and ability to act. Thence come the regionalist reactions, occasionally violent; these shake up the national states; they no longer wish simply to reconstitute some more or less mythical past, but rather to generate a new society that is more open, dynamic, and democratic (Gerdes, Ricq).

Regionalism is sometimes simply autonomist - that is, the regional actors ask for more autonomy for their region within their constituent national framework; this demand can lead to federalist structures. Sometimes regionalism turns nationalist or separatist, showing the region's wish to become its own sovereign national state. And then sometimes regionalism is 'nation-building': here this term means that the regionalist movement connects a social revolution to its demand for identity (Petrella, Quévit, Guindani and Bassand).

Let us add, internal to the regions of Europe, there frequently rise questions of uneven micro-regional development, of micro-regionalization, of micro-regionalism.

In ways that can at times be violent, nearly all the countries of Western Europe, and recently also those of Eastern Europe, are confronted by the regional question, whether it be under one aspect or another.

At the heart of the regional question throbs a cultural problem, in spite of its socio-economic dimension.

The socio-economic typologies of the regions of Western Europe are numerous. Most writers recognize the center-periphery dimension (Aydalot, Greffe, Bassand, 1982). This does not mean that the regions of a country, or of Europe as a whole, divide themselves between these two categories. They divide themselves into many nuanced categories along the scale between these two poles, proceeding from center through semi-periphery to periphery. Therefore the more central a region is, the more urbanized, the more important, dense, rich, and diverse its economic, cultural, and demographic potential, the more it will take part in social dynamism and project itself into the world. Conversely, the more peripheral a region, the less urbanized it is and the more its economic, cultural, and demographic potential is impoverished, scattered, and homogeneous. Also, it is more isolated, participating less in world social dynamics (Bassand, 1982).

This typology may seem to give a static view of things, but it is nothing of the kind. A region is in constant change - certain specialists rightly maintain that a region is a process. Analysis shows that over more or less extended periods, regions can change considerably in this center-periphery dimension. Central regions become more or less peripheral, and the converse can also be observed. In the matter of regional development, the processes of reproduction, decay, rebirth, and bifurcation are frequent. In the long run regional development is no fated thing.

Another idea is involved in the center-periphery dimension. Central and peripheral regions maintain power relations, whether on the scale of a society, a continent, or the world. Very often - but not always - the central regions dominate the peripheral ones. This relationship sometimes sets in motion within peripheral regions movements of resistance and even of struggle. This phenomenon also rises within an urban region, between its urban and peripheral collectives.

As we suggested above, the socio-economic diversity of Europe's regions is fundamental but it is far from being the only dominant, for socio-cultural diversity also counts for a great deal. European countries with cultural homogeneity are rare indeed. Quite often the cultural diversity tallies with administrative regional divisions, but not always. This cultural heterogeneity shows itself in regions whose population speaks a language other than the official one. This minority tongue, within a given nation, may be 'universal', as for example with German in France, French in Switzerland, and Swedish in Finland. It can also be typically regional and unique in Europe, as with Breton, Rheto-Romansh, Basque, and Lapp. Sometimes the regional linguistic diversity takes the expression of a specific dialect. Cultural regional diversity also shows in religious terms.

Regional cultural diversity also expresses itself through a more or less diffuse identity which has its roots in a history variously mythical and glamorous. Regional identity also corresponds to a certain geographic specificity.

In most cases these linguistic, religious, historical and geographical aspects of culture overlap and mutually reinforce each other; they often interact with central and peripheral socio-economic aspects. These cultural characteristics give birth to movements and political parties which frequently are as structuring as their cultural foundations. Let us pass on, then, to the question of regional identity.

THE COMPLEXITY OF IDENTITY-FORMING PROCESSES

Identity is a collective cultural creation, forever coming into being. Identity is permanent creativity, tireless exploration. In this process, "self and other aim at a common future. The self does not find a separate existence by cutting itself from the other, but in establishing a relation with him. (...) Therefore the question lies not in cutting oneself from the other, but in building a relationship with him, while remaining true to oneself" (in: V.A., 1984, p. 226). It is therefore in interpersonal and intergroup relationships that the actor's identity forges itself; it is the transformation of these relationships that makes an identity outdated and obsolete and requires the elaboration of a new one.

In our time, identity can no longer be founded exclusively on the quest for and the cult of roots and traditions. This tendency would only bear the seeds of suffocation. It follows that identity has meaning only if it is confronted and associated with the dissimilarities of both present and future.

These various ideas hold for regions. It is in confrontation with other regions that a region fashions its identity according to numerous modalities.

Even when a region lacks great cultural specificity, it builds an identity that becomes a highly significant element in its development. Frequently, regional protagonists employ terms other than 'identity': trademark, emblem, symbol, etc. Each of these of course possesses a specific quality which for our purposes we simplify with the term 'regional identity'. It is that image which the individuals and groups of a region fashion in their relation with other regions. This self-image may be more or less complex and grounded, whether on a cultural heritage, a natural environment, a history, an aim for the future, a specific economic activity or some combination of these different factors. If identity is a cultural process, it does not only have foundations that are cultural. Lastly, let us emphasize once more that this image-building is more or less negotiated with actors who live outside the region.

Regional identity is often stimulating for its carriers, arousing a pride in belonging, a wellspring of regional cohesion, a will to act on the region's behalf. It is quite evident that this identity is only rarely unanimously accepted, for whatever stands as emblematic for some will seem a mark of shame to others. Then too, regional identity is often criticized because, as some say, it threatens to lead the region to fold upon itself, at the very time when people are widening their horizons. Thus, it is preferable to build cosmopolitan attitudes. This debate reasserts itself in different contexts, yet there is no incompatibility between regional identity and openness to the world. Quite to the contrary: the greater and more generous such openness is, the more the regional identity is forced to be strong and differentiated. Only when a region becomes a dynamic and authentic partner with other regions in Europe and in the world will its own identity be a living one. This being said, regional identity is no universal panacea and it needs not become one, but it remains one important aspect of regional development.

Practically all regional cultural policies include the idea of the construction, defense, revitalization, and advancement of a regional identity, whether within a national framework or on the European and world scenes. In every region, cultural policies of the following types also aim at identity-building:

- the organization of music, theatre, and film festivals, etc.;
- the mounting of art or history exhibitions in cooperation with local and regional museums;
- the formation and support of artists that project themselves not only throughout the region but also the country and abroad;
- the rehabilitation of the urban and rural architectural heritage;
- the safeguarding of natural areas;
- the publication and promotion of literary, artistic, and scientific work, both regional and cosmopolitan.

Even when these events take on a cosmopolitan, avant-garde, elitist tone and are therefore criticized by citizens of the region, even (and perhaps especially) in these cases they contribute to the region becoming known and recognized, its reputation and prestige enhanced, and its identity strengthened.

For these reasons, regional authorities assert more or less exclusive claims over cultural policies. No one knows and can stimulate the cultural reality better than they do, no one can contribute more to its enrichment and promotion. This point of view is often objected to by national authorities because, for one thing, cultural policy is equally important to a nation in its relations with other countries, and for another, only a national culture - so they say - can be avant-garde and truly innovative.

Also let us face that more and more local and regional powers are throwing themselves into voluntarist actions in order to create or protect an identity; generally, these actions do not rely on the term 'identity', but on more sophisticated labels, such as "strategies of communication in regional marketing" (Benoit and Benoit). These activities illustrate the importance of identity in regional development. We shall take up this idea again later.

Now for still further details concerning the concept of identity.

We believe, in accordance with numerous sociological and psychological studies, that each group and, it follows, each region exists only insofar as it possesses an identity. This claim does not mean to say that every group and so every region actually has one. Some project quite forceful images of themselves, while for others identity may rest near degree-zero. That is, a region without identity "is the pawn of others", and the probability is very high that it will suffer domination. Conversely, the existence of a regional identity moves residents to behave in accordance with that image, or even to transform it.

The absence of regional identity, on the other hand, does not mean that the inhabitants have no identity: a person's identity may be local, social, functional, and not necessarily regional. In the same way, all of a region's inhabitants do not necessarily identify themselves with their region, even if it has a strong identity.

Now for more evidence that there are several kinds of identity. We distinguish between at least three types:

- the historical and patrimonial identity: this is constructed around past events crucial for the whole population and/or around a socio-cultural and socio-economic heritage;
- the projective identity: this is founded on a plan for the region; in other words, the identity is a more or less worked-out scheme for the region's future which may or may not take its past into account;
- the experienced identity: it reflects everyday life and the current existential style in the region; sometimes historical, projective, and heritage-based elements combine with the everyday dimension.

These three types often interact in defining and making up regional identity; but sometimes they can be taken up in isolated fashion by the regional actors.

This being said, identity of any type can be positive or negative. In the first case, we would speak of 'emblems', that is to say of signs or symbols that have been chosen by the actors for presenting themselves to others and setting themselves apart. It is by means of regional emblems that regional differentiation operates.

Negative identity is described as 'stereotype' or 'stigma' (Goffmann). It implies the negation and devaluation of the region and its members. The stereotype or stigma most of the time is manufactured by persons outside the region and is often inherent to a relationship of domination. Quite frequently, regional struggles have as their goal the removal of a stigma. Simultaneously, they shake loose the socio-economic and socio-political domination which had imposed the burden of stigma, and they enable an emblem to be worked out.

Regional identity is a process, "it cannot cease being a process without running the risk of congealing into a history that is over or into a territory which is purely abstract. Collective identity must constantly produce new forms; otherwise it will become a folkloristic construct or a sectarian image" (Centlivres, 1981).

In brief, the practice and the attitudes generated by identity - of any type - may be either offensive or defensive.

THE VARIATIONS OF REGIONAL IDENTITY

Taking into account the definitions set forth above as well as the results of recent studies and experiences made within the framework of the Culture and Regions project (Bassand, 1990), we are able to attempt the construction of a typology of regional actors and their identity.

This typology is still a research hypothesis and a provisory basis for action. Let us add that, obviously, we are passing no value judgments upon individuals, groups, or their region.

1. *The apathetic and the resigned actors:* they are characterized by the fact that they identify themselves with neither their local community nor their region. For them, neither one exists. Their identity and personal plans are vague, even nonexistent; their participation in local and regional bodies amounts to little or nothing, and they adapt rather poorly to circumstances. They are uncritical and isolated consumers of mass culture. This type often includes migrants, persons lacking a profession, unqualified wage earners and the aged.

2. *The potential emigrants:* they likewise do not identify themselves with the region in which they live. However, they formulate plans and elaborate personal identities which cannot be realized in their current place of residence. Their present environment means very little to them, and so they are on the look-out for the right moment to leave. They belong mostly to the middle levels and are thus critical and filled with contemps for the social, cultural, and political life of the region in which they live because they only see its stigmas. Once they have left, these actors often develop nostalgic feelings for their region and sometimes elaborate rather idyllic fantasies about it.

3. *The modernizers:* they are well integrated socially, economically and politically. They buy into all aspects of modernity, and strive for the systematic modernization of public facilities, businesses, living spaces, household goods, etc. They are open to anything that is new and approve of it uncritically. Although they are thoroughly integrated into their region, they identify with neither its history nor its heritage; to them such things mean only obsolete traditions, parochialism, and even backwardness. While they are not numerous, modernizers are influential, recruiting among every social group, and having 'external relations' that are both multiple and positive. They generally take up official regional posts in economic, political, social and cultural affairs. Their style is usually technocratic, but of course not always.

4. *The traditionalists:* they possess a very strong identity in terms of emblems, history, and heritage, as well as a plan for the region that consists in freezing it at its current stage of development or, still better, refashioning it along the lines of an older mythical model. They actively resist every kind of change and take a militant role in actions that are inherently conservative.

5. *The regionalists:* they have as their chief preoccupation the development of their region, but not regardless of the price of the method. In this they differ from the modernizers. To the regionalists, the natural, cultural, and historical specifics of the region are positive values in which they take pride. These values stand as emblems and they include them to their elaboration of a regional project. In other respects, regionalists are open to modernity, but on condition that it harmonizes with their plan for the region. This type of actor is in the minority, poorly organized or simply, on principle, unorganized. He is frequently, but not exclusively, young. He often takes part in new social movements arising within the contemporary society.

Three remarks seem necessary pursuant to this typology:

-Individuals and groups do not belong to one or another of these five types of actors once and for all. Passage from one type to another is frequent.

- The relative importance of these actors, their power and alliance vary considerably according to the position each region occupies on the center-periphery scale. Problems related to identity-forming vary according to the center-periphery scale.
- Regional identity is plural, comparable to a kaleidoscope. A certain 'stock' of identities exists in a region, some positive, others negative. Depending on the power of the actors in charge, a certain configuration of identity prevails; it varies with combinations of circumstances, and does not prevent other actors from affirming their own identity. This fluidity takes nothing away from the importance of the identity-forming process; it is one of the stakes in regional development.

THE CONCEPT OF REGIONAL IDENTITY

Collective Memory

We have just seen that one facet of regional identity rests in its being rooted in the history and the heritage of the region. How can patrimony and historical inheritance generate regional identity?

The concept of 'collective memory' connects heritage and identity. Let us go back to M. Halbwachs, the creator of this concept. While history tries to be objective and stresses temporal discontinuities, collective memory, on the other hand, "is the group seen from inside". It presents a tableau, or a system of images, that persuades members of a group that it has remained the same in spite of the changes that have taken place. Thus, collective memory is an image of the past constructed by the community; it becomes its identity.

Therefore the concept of collective memory helps in understanding how history and heritage intervene in social dynamics: there is collective memory whenever the group appropriates its heritage and history.

It follows from these points that the architectural heritage, its restauration as well as the museums that often carry such actions further, are the setting by means of which a territorial community makes itself visible to its own members as well as to outsiders. The same thing happens with festivals: these become the occasion on which a community acts out some more or less mythical phase of its history, thereby putting its identity on stage. Let us emphasize that these processes are never mechanical.

The Identity-Makers

The identity-forming processes that we have described up to this point are the product of a large number of actors. With the emergence of new techniques of communication and information, under the rubric of 'mass marketing', a specialized category of professionals has sprung up that specializes in fashioning trademarks not only for products and corporations but also for public collectivies: local communities, counties, regions and countries. The functions of this form of communication are of at least three types:

- to notify the inhabitants of a territorial group about the manifold services at their disposal,
- to inform citizens about the leadership actions of the authorities, and
- "to facilitate, specify, and even create a comprehensive discourse about the community" (Décentralisation à l'affiche; Benoit et Benoit 1989); in other words, to fashion a collective identity.

Let us take the Nord/Pas-de-Calais region as an example to illustrate the process by which it is possible to create an identity. This region has a problematic image, as the expression "The Inferno of the North" shows. This negative image places a curb on regional development which it shares with every region having an industrial past. But until the mid-19th century, the image of the Nord/Pas-de-Calais region was powerful, attractive, and filled with contrast and diversity. At the end of the industrial period, it only applied to the economic and industrial domain. During the 60s, this image became worse, with connotations of regression and decline. The inhabitants of the region themselves absorbed this negative image, and the stereotype grew stronger. For change to come, it is therefore necessary that the inhabitants take a different view of their region (op. cit. p. 137).

Several studies clearly showed that the regional stigma kept companies out of the area because their staff refused to settle there.

A campaign was started in five domains: youth training, lines of communication with Europe, cultural affairs, the quality of life, and the natural environment. Public-relations people singled out the fact that the Nord/Pas-de-Calais is part of a northern European civilization. Following the analysis, the identity makers retained key words that clearly post the way to the Nord/Pas-de-Calais's northern identity: "proud as Vikings", "dogged as the Flemish", "conquerors like the Dutch", "cultured like the British", "disciplined like the Germans".

AESTHETIC CULTURE AND THE REGION

Let me finally attach a postscript which has been specifically designed for the actual context: the aesthetic culture as a carrier of regional identity. Although strong reasons point towards applying a broad cultural scheme in the debate on regional development, it is necessary also to distinguish the existence of an aesthetic concept of culture and to consider its genuine and intrinsic values. Otherwise the risque is considerable that dilution and shallowness will blur the emotive content of the regional cultural idea and that it is reduced to an empty cliché. Conversely, recognition of the creators and artefacts of aesthetic culture might entail strong synergetic effects in the entire region and contribute to a general heightening of the quality of life in it. Ultimately also economy will profit from a high level of ambition in the aestetic culture. A high aesthetic standard not only entails a well motivated staff in the corporations, it also shapes socio-cultural stability in the region, thereby attracting enterprises to locate their plants there.

The regional aesthetic culture covers a broad spectrum of activities. Many of them are contained in the list of cultural initiatives presented earlier in this paper. The main cultural themes in conventional public opinion are the topical pictorial arts, music, theatre, cinema, crafts and literature. The sphere of the topical manifestations however is dramatically widened by adding the cultural heritage maintained in our museums, libraries and musical collections. Often the inventory of aesthetic culture comes to a halt with the immense treasury of artefacts which surrounds us in the private and public environment. A great part of that treasury belongs to the fine arts sector, another has its roots in popular culture.

However, our longing for aesthetic experience in our lives looks out for prey in a much broader domain than the field of cultural artefacts mentioned. The architectural environment of our private sphere of life or in the public townscape represents a ubiquitous aesthetic frame for our living with an extremely strong emotive impact, irrespective of how we wish to classify it. And to an even higher degree the space of the natural landscape in its unlimited undulations and nuances represents an experiential background to our lives which we consider indispensible. Especially if we are interested in shaping a regional and territorial concept of culture, these two categories advance to a central position as creaters of value in society.

Creating a regional aesthetic culture and then maintaining it through the years and centuries is a large and prestigeous task. Fixed norms for handling it hardly exist and no holistic approach on the part of the state can be found, despite the fact that society praises itself of possessing a cultural policy. The various sectors develop side by side according to established traditions and laws of organic growth. They interact with each other and society's functions and spontaneously form a carrying structure. However, because of a lack of systhematic care the cultural web might easily be damaged or even break into peaces. Better knowledge of the elements of the cultural process in combination with more conscious care probably would result in considerable aesthetic rationalization, if the expression is allowed.

At the political and public administration level but also in many academic circles the active interest and the idealist ambitions are often limited to the organizational tasks of the cultural sector. This covers the whole spectrum of communal, national and privat cultural institutions and their well established staffs. The output of their efforts are exhibitions, concerts, performances, festivals and fairs on various themes. It is at the same time a prestigeous and money consuming sector which normally has close connections with the other sectors of official community life. This is also the most conspicuous part of the cultural life in the town and the region and therefore, unconsciously, we often regard the institutional cultural sector as synonymous with the whole of cultural life.

Well functioning cultural institutions however imply active and alert visitors. Only the dense flow of visitors and the benevolent references and applauses from the audience inserts life into the institutions and gives legitamacy to their efforts. The cultural and art critics of the daily newspapers represent the fastidious cultural audience in a condensed and sublimated form. Quite often though, the interest from the side of cultural policy towards the visitors remains with their role as a quantitative phenomenon: what counts is the number of visitors, listeners or readers. Their number stands forth as a meausure of the success of the event and is used as a justifying factor in the institutions annual budget requests. The subjective content of the visitors cultural experiences and the deeper existential needs that culture supplies in man are seldom asked for. But even the individual cultural consumer himself often remains unaware of the motivating forces activating his cultural contacts, and remains uncapable to elucidate if they are founded in status-hunting, bodily or mental uneasiness or deep inner needs. Hereby a false hierarchy is easily established in our cultural consciousness, in which the public is regarded as a servant to the museums and institutions rather than the opposite, which ought to be the normal case.

The third group of actors in aesthetic cultural life consists of the creators, the inventing artists in music, theatre, pictorial art and architecture. In spite of being a minority this is the most indispensable of the three categories mentioned. The reason is not only the artist's role as a producer of that culture which we are supposed to consume, or the aesthetic experience which we share. At its deepest it is due to the aesthetic inspiration which the experience of art is able to awake within all men and women and the qualitative heightening of our existential demands which is evoked hereby. The arts help reinforcing the often hidden and suppressed aesthetic nature of us all, and makes us capable of taking a creative or at least constructive stand in our social and humane relations. It is a depressing fact in our time that the social status of the artist remains far behind the status of the audience and above all the institutions. This may indicate how little our age values the deepest creative driving forces in man. In a fully developed regional society the creative artist, by way of his genius, is a model for our unbending love for liberty as well as for our deepest demand for personal sovereignty. This is the only stable ground on which a democratic society can ever grow.

REFERENCES:

Allardt, E., Implications of the Ethnic Revival in Modern Industrial Society, Societas Scientiarum Fennica, Helskinki, 1979.

Aydalot, Ph., Economie régionale et urbaine, Paris, Ed. Economica, 1985.

Bassand, M., Villes, régions et sociétés, Lausanne, Presses Polytechniques Romandes, 1982.

Bassand, M., Brugger, E.-A., Bryden, J., Friedmann, J., Stuckey, B., Self-Reliant Development in Europe, Hampshire, Engl., Gower, 1986.

Bassand, M., Hainard, F., Dynamique socio-culturelle régionale, Lausanne, Presses Polytechniques Romandes, 1986.

Bassand, M., Hainard, F., Pedrazzini, Y., Perrinjaquet, R., Innovation et changement social, Lausanne, Presses Polytechniques Romandes, 1986.

Bassand M., Identité et développement régional, éd. Lang, Berne, 1991.

Bassand, M., Culture et régions d'Europe, Presses Polytechniques et Universitaires Romandes, Lausanne, 1990.

Benoit, J.-M., Benoit, Ph., Décentralisation à l'affiche, Nathan, Paris, 1989.

Bourdieu, P., Questions de sociologie, Paris, Ed. de Minuit, 1980.

Brunet, R., Les villes "européennes", La Documentation française, Paris, 1989.

Camartin, I., Rien que des mots? Ed. Zoé, Genève, 1989.

Caussat, P., De l'identité culturelle, Desclée de Brouwer, Paris, 1989.

Centlivres, P., Identité régionale. Approche ethnologique, Suisse romande et Tessin, Rapport de synthèse. Document ronéo, p. 52, Neuchâtel, Institut d'ethnologie, 1981.

Furter, P., Les espaces de la formation, Lausanne, Presses Polytechniques Romandes, 1983.

Gerdes, D., Regionalismus als soziale Bewegung, Frankfurt/Main, Campus, 1985.

Greffe, X., Territoires en France, Paris, Ed. Economica, 1984.

Le Goff, J., Guieysse, L., Crise de l'urbain. Futur de la ville, Paris, éd. Economica, 1985.

Goffmann, E., Stigmate, Paris, Ed. de Minuit, 1975.

Guindani, S., Bassand, M., Maldévelopement régionale et identité, Lausanne, Presses Polytechniques Romandes, 1982.

Halbwachs, M., La mémoire collective, Paris, Presses. Universitaires de France, 1950.

Kuklinski, A. (éd.), Social Issues in Regional Policy and Regional Planning, La Haye, Mouton, 1977.

Meier-Dallach, H.P., Hohermuth, S., Nef, R., Anliker, R., Zwischen Zentren und Hinterland, Diessenhofen, Verlag Rüegger, 1982.

Meister, A., L'inflation créatrice, Paris, PUF, 1975.

Petrella, R., La renaissance des cultures régionales en Europe, Paris, Ed. Ententes, 1978.

Quere, L., Des miroirs équivoques. Aux origines de la communication moderne, Paris, Aubier-Montaigne, 1982.

Quevit, M., La Wallonie, l'indispensable autonomie, Paris, éd. Ententes, 1982.

Quevit, M., Le pari de l'industrialisation rurale, Genève, Ed. Régionales Européennes SA, 1986.

Ricq, Ch., "Région, régionalisation et régionalisme" in: Genève et l'Europe, Genève, I.U.E.E., 1979.

Rossel, P., Hainard, F., Bassand, M., Cultures et réseaux en périphérie, Réalités sociales, Lausanne, 1990.

Touraine, A., Le retour de l'acteur, Paris, Fayard, 1984.

Touraine, A., Dubet, F., Le pays contre l'Etat, Paris, Seuil, 1981.

V.A., Le complexe de Léonard, Paris, J.C. Lattés, 1984.

The Avant-Garde: A Door to the Cosmos

Jean Fischer

1960 was the year in which western avant-garde art completely changed direction. 1960 appears to be a "farewell" to western art and a "well met" to the art of other cultures, oriental culture not least.

MODERN WESTERN ART

At the end of the 18th century, western art was segregated from religion, dance, from things collective and social. It became art. First and foremost it was a matter of individual and soul. The solitary artist expressed his mental feelings to an audience of anonymous individuals. Kant gave theoretical expression to this segregation and made aesthetics the theory of the specifically artistic.

Art originates simultaneously with democratic society. Detachment from the collective is at the same time emancipation from servitude to Church and prince. But, paradoxically enough, art appears to become more and more undemocratic and nonpopular as democratization advances! Art becomes more and more inaccessible. This is also connected with another idiosyncrasy of western art, namely its attitude of social criticism. Starting with Romanticism in about 1800, it was critically condescending towards the middle class, the Philistines, and towards trade and material things. It looked down on everyday things and banality. True, it was most often on the side of the people and considered itself an advocate of the people - peasants, the child, the woman, the unspoiled. But in reality, paradoxically enough, this was the very reason for its non-popular nature, in that provocation and scandal often became the criterion of artistic quality, freedom from opportunism and a truly revolutionary attitude.

This was the case, not least, with avant-gardism, which developed during the second half of the 19th century (Baudelaire). Avant-gardism was an exploration and glorification of very special, unconventional experiences which the rest of society found negative and repellant. Political radicalism coupled with artistic avant-gardism has acquired high status in a society which puts a value on development, progress and innovation. The collapse of society is looked on by radicals and avant-gardistes as a good thing, but it is undeniably peculiar to have an art which sees its task as being to break down the culture and society by which and in which it lives.

The beginning of this century saw the rise of Modernism, which, often inspired by ethnic culture and traditional culture, looks for a less spiritually expressive, less romantic art. Modernism in music can be summarized in the slogan: Back to the pre-romantics. This can mean a return to the Baroque and Bach, when music was still collective and religious. But this only applied to the outward forms that were used: the counterpoint and rhythm of the Baroque. Not its collective and religious frame and content. And so the mental expressiveness of romanticism and individualism was still retained, most conspicuously by Schönberg and twelve-tone music, which, in spite of its reaction against the late Romantic period, perpetuates the supreme mental expressiveness of that period. In the conventional progressive way of looking at things, Modernism is regarded as a big new beginning. From a conservative viewpoint it can rather be considered a desperate attempt at blowing new life into forms which modern (Romantic) music itself had broken down, and at infusing new content into a burnt-out mental fire. In the 1950s this non-popular development was brought to a head, when development and novelty became the height of dogma. A good musical composition had to betoken something completely new and, preferably, incorporate "sounds never heard before". An absurd aesthetic which soon died by its own hand. Anti-art and anti-novels were a necessary consequence of this suicide. Perhaps, quite simply, this was necessary in order for the western avant-garde to escape from its blind alley and venture into a more universal and less western attitude.

FROM GEOMETRY IN SPACE TO THE DYNAMIC OF TIME.

Western Culture 1800-1960

Before turning to consider the new developments post-1960, let us look at a decisive change in culture and society generally which occurred at the genesis of modern western society about 1800, simultaneously with the appearance of western art. This was the transition from space to time and the emergence of dynamics, enterprise, tension and a rapid escalation of time. Before 1800 people had on the whole lived in space, and geometry was the preferred discipline of the Baroque. Physical culture, architecture, music and science were conceived of as geometry. Concepts like evolution and revolution were spatial and denoted choreographic movements in space and changes in those movements, be it the movements of the horse in dressage or the motions of the planet in the heavens.

But all this changed in about 1770. Suddenly it was time, not space that people experienced and lived in. Evolution became a development in time, indeed it became the development. The slow choreographic, spatial

"Window Vision". Photo by Hans Marklund

movements of the Baroque were straightened out, becoming an escalating movement in time. Time passed more and more rapidly, and tempo and tension became vital phenomena. The dynamic of time would be a fitting superscription for the modern epoch. The dynamic of time permeated the whole of society and culture, thus embracing the growth of the economy (capital accumulation), the expansion of politics (imperialism), the development of art (constant renewal and transcending of limits) and our innermost feelings, thoughts and motives (the craving for novelty, crossing of boundaries and tension).

FROM TIME TO SPACE. THE NEW AVANT-GARDE

It is, not least, in this perspective that we can view the significance of developments in art after 1960. This can be generally described as the reconquest of concrete space, as slow time, i.e. as a rejection of the modern dynamic of time which reduced concrete space to points, numbers and abstract space. Concrete space is visible in such forms of pictorial art as installations and land art. Slow time is distinctly expressed in many musical forms which are slow and monotonous music and also more like an experience of sound in space - space music - than tones in time.

Whereas, about the turn of the century, Impressionism and Art Nouveau were, it is fair to say, influenced by Japan and Modernism by African sculpture (Picasso), the breach in about 1960 was a great deal more radical. People reached all the way back to the Pyramids, Stonehenge and the civilization of the Incas. And instead of just grasping the outward forms, they were completely absorbed by the fundamental attitude behind them.

ISMS AFTER 1960

But before trying to delineate these new features, I would like to sketch briefly the various renewing isms and art forms post-1960. Pop Art, fluxus, happenings and various but related new forms of artistic expression appeared all at once in about 1960. Also in the course of the 1960s came minimalism, concept art and land art. By 1970 virtually all the new forms which we, in 1992, may consider decisive in the last three decades had come into being. Changes have occurred since then, including for example a certain aestheticization. For example, happenings have been succeeded by more streamlined performance. During the 1970s there appeared a new tendency, pattern and decoration. During the 1980s there appeared the much-discussed wild painting, Neo-Expressionism, which, however, today is generally to be regarded as a species of concept art. Today, by and large, concept art appears to be the prevailing idiom.

When the post-modernists claimed that 1968 is now over and done with and that sculpture has now returned, instead of the insufferable installations of the sixties, this is, to say the least, a doubtful contention. Perhaps 1968 is over, but the sixties are very much alive and have defined the frames for the art of ensuing decades. Installations and concept art, which in the 1960s were small avant-garde groups, are now wide, dominant phenomena. Installations today are the most-utilized form of art, even if more traditional sculpture does occur. In a word, the closer we look at our own time, the more obvious the artistic victory of the 1960s becomes.

But what do all these new forms and isms contain? I shall try to summarize this under a number of descriptive headings.

THE AVANT-GARDE WORLD PICTURE AFTER 1960

The substances and processes of nature (process art and arte povera)

Process art and arte povera were typical art forms of the sixties. Often they consist of wholly elementary investigations of the properties of materials and are made to include humble or "poor" materials: grass, ice, felt, cement, stone, rubber, grease, coal. There are a number of varied but related intentions behind this. Often there is a feminist underlining of the organic and biological, with emphasis on the visceral and original. One can mention a prominent woman artist like Eva Hesse. But also male artists devote themselves to material studies and ask elementary questions about the phenomenological properties (the despised secondary sense qualities) of materials: hardness, softness, smoothness, roughness, dryness. Here we can mention Bjørn Nørgård's material experiments in the 1960s. He often uses plaster and sits, for example, with his bare feet in the wet material. Hans Haacke exhibited meteorological or biological systems concerned with how water turns to ice or how water evaporates. This is almost concept art, and concept art and minimalism often drift into process art or arte povera. The latter variant is almost metaphysical or alchemistic: the relative size of things is investigated with the magic numbers and inherited numerical formulae such as, for example, Fibonacci's sequence of numbers. This applies, among others, to the Italian Mario Merz.

Reconquest of concrete space (installation)

Whereas sculpture is traditionally neutral in its relation to space, one of the points of installation is in fact relationed to the concrete space in which it appears and for which it is often created. Often that space takes the form of dilapidated old factory buildings, where the premises can be more alive than in many rooms of more recent vintage. Installation tries to sense and use the properties, the aura, of the concrete space. Installation is often more than a temporary underlining of the character of the room than an autonomous, "eternal" work of art. Rather, the purpose of the installation is to be a catalyst for a genuine experience of this room, a revelation of its distinctive character.

The Observatory: place, time, earth and space (land art, site works)

Land art and site works are works in the landscape in much the same way as installation are works in a concrete indoor space. The best-known work of land art without any doubt is Robert Smithson's Spiral Mole in Salt Lake, Utah. Like so much land art, that work is only expected to exist for a limited period of time (from a couple of months to a couple of years) and is meant to capture and demonstrate natural processes or man's relation to nature. Gradually these works then enter the natural cycle, as for example in the case of the Spiral Mole, which was broken down by the lake.

Several works of this genre are a kind of observatory, but more akin to Stonehenge than to the observatories of modern science. Because of their primitive character they facilitate a perfectly elemental experience of man put in this given place at this point of time in this infinite space. Man here acquires a rare opportunity of experiencing what we often rest content with commanding as theoretical knowledge. Our embedding in time and the universe is made physical and visible.

Nancy Holt's famous Sun Tunnels can be used as an example. The place is a desert in Utah (Nancy Holt was married to Robert Smithson). Four cement pipes are placed opposite each other, two by two, so that viewed from above they form a cross. The Sun Tunnels show the annual extreme positions of the sun on the horizon, an idea inspired by Stonehenge. In addition, there are holes, of various sizes and variously positioned, in the walls. During the day the light of the sun, shining through the holes, creates a variety of patterns on the lower half of the tunnels. At night the moon projects its paler light. Robert Morris has also composed a famous primitive cult work, Observatory, which is a good deal more archaic than Holt's.

Body art and sound meditation

Body art is a new and distinctive art form which also originated in the 1960s. There are at least two variants. One of them is American and left-wing Freudian and is concerned with the happy sexual bodies. The other is Austro-German and sado-masochistic and is concerned with orgies, passion, sacrifice, death, purification and resurrection. The first variant can be instanced with the Ann Halprin-inspired Judson Dance Theatre and by Carolee Schneemann with her famous performance Meat Joy.

The Austro-German variant includes the Vienna Group and Joseph Beuys. Hermann Nitsch and his Vienna Group evoked the strongest reactions with their Orgy and Mystery Theatre, which performed ritual and orgiastic scenes. Powerful sacrificial scenes in which blood and entrails were smeared or poured over naked human bodies. By a positive interpretation, OMT

Robert Smithson.
Spiral Jetty, 1970.

Hans Haacke.
Floating Ice-circle, 1971.

Eva Hesse.
7 Poles, 1970.

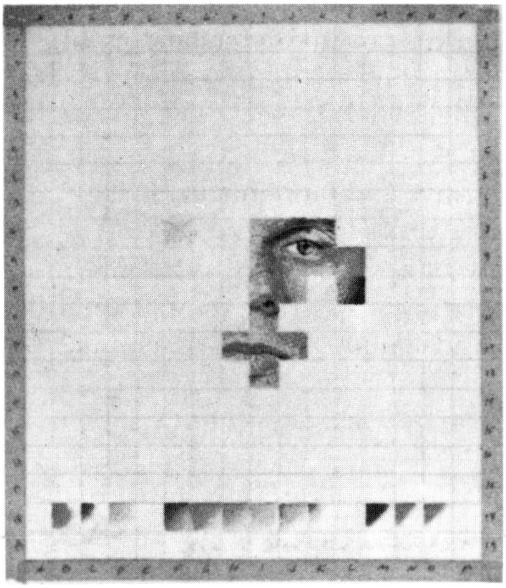

Chuck Close.
Kent, 1971.

Nina Sten-Knudsen.
Ladyhawk, 1985.

Nancy Holt.
Sun Tunnels, 1984

personifies the philosopher Merleau-Ponty's body phenomenology, which regards the body as the point of departure of the ego and understanding of the world. Negative critics speak of sado-masochistic exhibitionism. To Nitsch himself, the question was one of a kind of catharsis, a living-through of the suppressed.

Female body artists are often less provocative and more poetic. Joan Jonas and Tina Girouard use, respectively, Zuni and Hopi Indian ceremonies as their models. Other, more restrained forms of body art merge with music and come close to therapy, meditation and healing, exploring, for example, breathing and its importance for body, soul and sound. The composer and singer Meredith Monk often makes breath the centre of her vocal compositions, just as Joan La Barbara explores the voice during both ex- and inhalation.

During the 1980s, some of the main pioneers from the avant-garde of the sixties steadily moved towards an avant-garde form of living with meditation, healing and interest in other cultures. The composer Robert Erickson has studied Balinese music and is absorbed by the inherent power of the individual sound. The composer Pauline Oliveros is known above all for her Sonic Meditations and often uses the mandala. She has made a recording in an American cave where the character and acoustics of the cave itself are important, which rather puts one in mind of installations. Her group during the recording is called Deep Listening Band, and comprises the composer and trombonist Stuart Dempster. He is interested in the therapeutic and healing potentialities of music and has a sound massage institute. Dancer and choreographer Ann Halprin conducted great ceremonies at Mount Tamalpais during the 1980s to promote a more coherent and peaceful life.

Religiosity and new spirituality

A lot of the things which have now been mentioned have an unmistakably eastern spiritual atmosphere. To take the two greatest gurus of the avant-garde since 1960, John Cage and Joseph Beuys, they are both obviously "spiritual". Through the decades, John Cage has transmitted an oriental, Zen Buddhist inspiration. He relates that it came as a revelation to him, hearing an Indian musician proclaim that the object of music was to make God visible. Cage is not to be looked on as part of the western avant-garde, but as part of a strong American tradition which goes back all the way to the 1920s (Cowell, Lou Harrison) and looks eastwards, being inspired not least by Balinese gamelan music. It is much more pragmatic and experimental than Western European theoretically founded avant-gardism and, to distinguish it from the latter, was often called experimental music. Today's music bears the profound imprint of this American tradition, as a part of which Edgar Varese should also be mentioned: music is sound rather than tones. Two characteristics distinguish this music from conventional western music. Firstly, its character of sound, noise, silence rather than a succession of notes. Secondly, its deliberate ritual monotony, which to a great extent is inspired by gamelan music. These two characteristics combined give the music the character of space and slow time. And they are connected with Cage's endeavour over the years to detach music from the soul and the intentions of the individual composer. As with the Italian Giacinto Scelsi, this music is not communication by the individual but the cosmos transmitted through the composer. Recent rhythmic music itself uses sound, noise, monotony, e.g. Sonic Youth, Loop and Swans and, most recently of all, the Seattle sounds of Nirvana, which is the dernier cri of rhythmic music. Earlier popular forms of music of this kind were minimalism, New Age music and World music.

Joseph Beuys has been the great guru of the Western European avant-garde since 1960. His basic inspiration is clearly oriental, transmitted through Rudolf Steiner and anthroposophy. Not surprisingly, then, during the last years of his life, Beuys often organized actions together with the Greens in West Germany. Given Beuys' attitude, the Green movement must be one of the most essential things in existence. Several of Beuys' works are concerned with the amalgamation of oriental and occidental, or spiritual and material.

EXCURSUS AND MODIFICATION

Needless to say my picture of the avant-garde since 1960 is selective, but I venture to say that I have identified central tendencies, though they have seldom been described in this perspective. I have dwelt least on the very well-known phenomenon of Pop Art, even though I admit that a neo-Pop Art has arrived which, to some extent, characterizes the painting of the 1990s, especially the more kitsch-like productions like those of a Jeff Koons. I can see one positive thing about Pop Art, its reaction against the western hostility between art and life. But in Pop Art, it seems to me, art and life merge on false premisses. Otherwise one should not become too preoccupied with Warhol's more spectacular advertising pictures. Other important Pop Art figures like Oldenburg and Rauschenberg often drifted into fluxus/happening. Moreover, one can object that the latest changes in art partly constitute a modification of the above. There has clearly come into being a new professionalism, aestheticism and market-awareness which is at variance with the avant-garde of the sixties. Some critics today reject fluxus, because it neglected artistic expression or aesthetic. Finally, some maintain that the Neo-Expressionism of the eighties constitutes a breach with the tendencies of the sixties and seventies. But there is widespread talk today of Neo-Expressionism as a kind of concept art, that is, a sixties phenomenon. And besides, the painting of the 1980s includes many ritual and cult themes: Indian rituals, totems and scenes at the fire.

Certain post-modern writers (e.g. Carsten Juhl, on the subject of Morten Stræde's sculpture) maintained that sculpture has returned, succeeding the flipped installations of the sixties. This claim, as stated earlier, is untenable.

Installations have never been more widespread than today. Otherwise, there is an interesting and ambiguous development to be mentioned. In recent years, installation and land artists have collaborated with landscape architects on several projects, e.g. landscape reinstatement. This is ambiguous compared with the traditional western aesthetic, which tells us that art ought preferably to be segregated from more profane, useful activities. But in a more holistic perspective it is exciting that art should enter a wider context. In doing so it loses some of its absurd liberty. On the other hand it was able to move forward to social and physical significance and influence.

Lastly, I would like to add a few words about the avant-garde as a phenomenon. In this presentation I have taken issue with western avant-garde and argued its partly negative significance. And yet here I am saying that after 1960 the avant-garde suddenly acquired great positive significance? I have two answers to this. Firstly, talking about American and oriental-inspired avant-garde and not about the current western variety. This avant-garde, as I said earlier, ought to be called the experimental tradition rather than avant-garde. Secondly, the negative of the current avant-garde is that it has, so to speak, spearheaded an unfortunate development, it has taken the lead in terms of progress, the breaking of bounds and disintegration. Whereas the avant-garde since 1960 has also tried to build up new positive values and world pictures. These values and world pictures come very close to the Green movement and to a wide and many-faceted movement like New Age.

AN ATTEMPTED SUMMARY

All things considered, I regard the avant-garde since 1960 as an expression of a wider change or a paradigm shift from time to space or from dynamic time to slow time. Concrete space and slow time of the new paradigm. Secondly, we are moving towards a new image of nature, a new understanding of nature. Whereas mostly in modern times nature has meant nature in the wild state (the Alps meant a great deal to early Romanticism) or, more recently, the landscape of the Danish Golden Age, the historic, man-made landscape. Nature has been a meditative landscape, relaxation, enjoyment, passivity, the antithesis of the practical activity and concerns of the city. But the new image of nature in process art and land art is more of an active nature, nature as processes, in which man participates.

Nature in modernity is passive beauty (art) and also active control (natural science/technology). Nature in the new paradigm is active, is processes, and man forms part of those processes. Art here is a kind of science which is not reduced to control but is openness and curiosity. It is the very observation of and participation in natural processes which is the objective. "Knowledge" here is more like art or religion.

Religion here is no longer belief and dogmas but actions, rituals, activity. Religion is dance, music, meditation, land art, astronomy. In other words, religion incorporates elements of art and science. In addition, religion is active and collective. Whereas religion in the Christian Protestant tradition has mostly been passive and individual, thereby unleashing an explosion of worldly activity, religion here is collective and connected with art and science, leaving less room for purely worldly activities - that is, for "consumption" and "production".

THE AVANT-GARDE WORLD PICTURE AS A STARTING POINT TO UTOPIA

In about 1800, art became separated from the other social spheres - from the corporeal, the collective and everyday life. The post-1960 avant-garde shows a distinct tendency towards reunification.

In my personal dreams I hope for an ecological society in which we live in local communities and in a slower time, that is, on a lower material level. Where the artistic is included in people's association with one another, with nature and with themselves. Where meditative associations with individual things replaces the multiplicity of things. The total liberty of the individual will be restricted not only by a lower level of material consumption but also by the collective. Economic growth, consumption and centrifugal self-realization will be superseded by artistic-scientific-meditative association and centripetal self-realization. Expansion will be superseded by implosion. The crossing of boundaries will mean, not going round the world and as far as possible, but something ever closer. There are infinite revelations and experiences within ourselves and our relation to nature and other people, when they are viewed both artistically, meditatively, scientifically and religiously.

And there are also obligations and responsibilities in our relations to each other. Artists cannot just follow their own individual impulses, they have a duty to contribute towards the wealth of the local community. And the rest of us have a duty of taking an interest in art, of seeing new aspects, of receiving and using them. Not the total liberty of the artists and the total irresponsibility of the public. But the duty of the artists and the public one to another. Over-arching these things we will always have a state and a market. But it is the third instance, the community, which must be the basis. The community as a local society and region.

In the centre of the community and in the very centre of ourselves, a door opens towards a new cosmic experience. Revealing this experience might become the supreme task of the new utopian art.

Glossary of terms:

The following is a brief explanation of terms which are not explained at all in the main text.

Fluxus: A Dadaist-inspired movement from about 1960 which also included Zen Buddhist inspiration. Unfortunately the Dadaist inspiration is often more powerful, with the result that provocation became an end in itself. Through its more humble Zen inspiration, however, the movement has played an important part in the following development. Principal names: John Cage, La Monte Young, Nam June Paik.

Minimalism: This movement appeared in the mid-1960s. It attached importance to the minimal, but assumed a variety of forms - both a highly schematic and geometrical and puritanically sterile form (Donald Judd), and a more humble form, related to process art and arte povera. In music, an expression denoting a rhythmic and repetitive, monotonous form of composition (Glass, Reich).

Concept Art: This also appeared in the mid-1960s. It emphasizes the linguistic, meaning-creative. It attaches more importance to ideas and model than to the realization of the work. Often the ideas are more exciting than the accomplished work. Principal names: Kosuth, Weiner, Art and Language.

Otherwise the various trends and forms merge with one another, making it impossible for them to be adequately described and defined one by one.

Mass Culture Meets Man

Sidsel Mörck

"Freedom - we moaned.
Freedom from disease and starvation and death in childbed.

Freedom - we said.
Freedom from wooden plough and yoke.

Freedom - we cried.
Freedom from poor relief and drudgery and a 12-hour day.

Freedom! Freedom!
The shouts rose to a sky whose name we did not know and exploded there.
While the dangerous precipitation leaves us at liberty for
headline sensations, celebrity features and fictionalized murder."

The title of my talk is "Mass Culture Meets Man", and I would like to start by saying that mass culture does not meet man. Mass culture addresses itself to an average person who, originally, does not exist but is gradually created by the same mass culture.

Political, technological and economic developments in the western world have to a great extent taken physical exertion away from us and given us mental paralysis - as I began by saying in the poem.

In 1836, G. Bennet wrote in his newspaper, The Herald: "Books have had their day, the theatres have had their day, the temples of religion have had their day. Now we can put the newspaper at the head of all the great movements in human thought and civilization. A newspaper can send more souls to Heaven and save more from Hell than all the churches in New York put together - and make a profit out of doing so."

Those words, then, were uttered more than 100 years ago, and before such things as radio and television existed. Even though books, theatres and temples are still with us, there is little doubt as to where the power over hearts and minds is located: in the audiovisual mass media.

The following criticism applies, in principle, to most of the cultural industry, but to keep my subject within bounds on this present occasion, I propose taking audiovisual mass media, and especially television, as my point

"Window Vision". Photo by Hans Marklund

of departure.

I venture to claim that mass media of this kind reinforce the structure already existing in society. They are implements for various types of power groupings which consolidate their position. It is the wielders of power in industry and commerce, politics and bureaucracy who dominate news reporting in Norwegian broadcasting. The same people take part in radio and television debates of other kinds, and they are the people who are talked about and interviewed on the news. And, yet again, it is the same people who write books about their merits and hobby horses and achieve good sales because they are already marketed through the media.

There is a far-reaching coalition of the various kinds of power and media abuse. A former director of entertainment on Norwegian television has said: "The fact is that most of the people one is always seeing on television are also one's friends in private life." By saying this he confirmed two things. Firstly, that there are people one is constantly seeing on television. And secondly that he chooses his friends to take part in the programmes, or else he eventually becomes friends with those who do take part. (His use of the word "one" instead of "I" in this connection I take to be no more than a circumlocution.) But these participants in programmes are actors and artists who become well-known and popular because they appear on television and, consequently, find it easier to obtain parts at theatres, attract audiences to concerts, sell more records, cassettes and so on. And are interviewed by the glossies, become still more well-known and popular, and consequently appear still more often on television - and so the circle closes. To a certain extent this also applies to politicians and celebrities of other kinds who keep appearing, more and more frequently, in entertainment programmes.

Entertainment programmes - can we perhaps say that everything is a kind of entertainment? There is something imperialist about entertainment, it is taking up more and more transmission time and increasing numbers of subject fields. In this consumer age we are all consumers of products and media. Expressions like "consumerism" and "the throwaway age" apply equally to the commodity and media markets. Quantity prevails over quality, over-production and high turnover rates make themselves felt in all media: the volume of books is growing, with light-weight, instant publication ousting books which demand empathy, concentration and depth of understanding. The amount of material in our newspapers is growing, with light-weight, instant items ousting material which demands room for structure and argument. There is no mistaking the infectious influence of radio and television. The consumption process - or perhaps we should call it the process of digestion - is being made easier and simpler all the time, to enable us to consume - or digest - as much as possible.

A culture can fade and wither, and there are two ways in which this can happen: on the lines of either Orwell or Huxley. What we are experiencing today is not Orwell's prison culture but the variety culture of Huxley's "Brave

New World". Orwell prophesied that truth would be controlled and kept under lock and key. Huxley prophesied that truth would become just as inaccessible because it would vanish in a sea of indifference. And Huxley was proved right by events. Today we have to deal with a smiling enemy with a mask concealing his power-hungry face. And in this society, no guards or prison gates are needed, because there is nothing to guard. Cultural life becomes a society of children, public discourse becomes a babble of voices and wise-cracks, and a proud and active concept of "the people" becomes the servile, passive "public".

This entertainment syndrome is an ideology, even though strictly speaking it does not belong to that category. But the characteristics of an ideology are there: a new lifestyle and new relations between people and ideas are forced upon us. But we have not fully realized that technology is ideology, even though technology is transforming the whole of our existence - literally before our very eyes.

When the alphabet is introduced into a culture, this also changes its way of achieving recognition, its social relations and its notions of its own society and the world at large. The same happens with the introduction of printing. The same happens too with the introduction of image transmission at lightning speed. And this is a cultural revolution! But it has taken place with neither discussion nor polemic, without war or struggle. And its total success was due to our having developed a blind trust in the blessings of progress.

Now that the revolution is a fact, has it given the people what the people want? And what did the people want? Happiness - but more than the fleeting smile. Involvement - but more than a slight irritation. Meaning - but more than a superficial thought.

"This Saturday too
There is blue light in the windows.
This Saturday too
We sit with our long drinks, smiling at catch phrases,
While the satire flirts with pop problems
And titivates the social animal within us.

Smile - a little wanly
Because our only participation is smiling
- And a little faintly
Because we do not have the strength
To think the thought behind the thought.

Our own strength, of course, is long since mortgaged
To the receiver
To the drink
To having such a cosy Saturday evening."

A philosophical digression. It is a bitter fact that generations of toil have got us where we are today - that is, to "having such a cosy Saturday". Or to hours of stupefying time-wasting. Was it worth the price? Was this the objective?

Historically speaking, we are faced with a completely new situation. Large groups of the population, for the first time ever, now have leisure - free time - to do what they want. But do we know what we want?

We have no tradition, no knowledge for coping with this new freedom. So we ramble about looking for our personal links for someone to take responsibility for us, for someone to decide. That, after all, confers a kind of security. And so we open our doors wide open to the new mass media and submit to their smiling reign of terror.

The mass media are not only gentle, smiling despots, they are also strict, and television especially: television goes a long way towards deciding the agenda of public debate and talking points in the home and at work. It decides what is in and, consequently, what is out. It creates its own language: Monday film, late-night cinema, soap opera. ...... It writes a timetable for all of us, with programmes and serials pre-empting hours and days. It decides when people are to have meetings, when they are to meet privately, when they are to have their meals .... And in extreme cases - or perhaps not so extreme? - where they are going to lay the table, so that they can watch while they eat!

Of course, not everything on television is equally bad, neither so-called serious programmes nor the entertainment sector. But here we have a paradox. The aim, needless to say, is the ideal: improve the quality! But if one party - television - succeeds in creating quality products, the other party - viewers - must succeed in choosing actively. For quantity, no matter how high its quality, has the negative effect of new impressions erasing previous ones. As recipients we have a limited capacity for absorbing impressions and converting them into recognition.

This was by the way. I don't think there is any danger of the engaging, activating programme style gaining ground.

On the contrary. In the future visions of Norwegian Broadcasting, competition with foreign transmissions is taken for granted. This means, quite simply, accepting commercialization as an ideology. Commercial media policy also means blatant contempt for the public: the public is only of interest as a market!

The Danish researcher Ole Thyssen has said much the same thing: "Today it is called freedom when the cynical market regulates things. It is called coercion when the state tries to intervene and reduce cynicism. And the media adjust to this market, willy-nilly. Will the adjustment become so powerful as to make an illusion of diversity?"

Those words were uttered four years ago, and many would say that diversity has been an illusion for a long time now.

Now, as everybody knows, there is nothing about the actual paper which causes us to receive the books and newspapers we do. And there is nothing about the radio or television receivers which causes us to get the programmes we do. It is people who fill the media with material, according to certain criteria and laws. The strongest law is economic. But the laws of economics cannot be automatically applied to all activities. The media give the impression of also wishing to convey artistic experiences, but efficiency and profitability as yardsticks spell the death of all art. For example, it takes just as long today as in the 18th century to produce a theatrical performance, but a shirt can be made in only one-twentieth of the time it used to take.

While the unreality of the media gives us escape and absence, the real world quite often requires us to have a standpoint and to be present. The real world with its interpersonal demands, problems and conflicts calls for a debate on ideas and ideologies, calls for new visions and views. The real man - as opposed to the average man created by the media - needs to be taken seriously with his needs of empathy, sympathy and involvement, to find a use for his resources and creativity.

Instead the invisible hand of Adam Smith is steering us into Huxley's "Brave New World", which is not brave at all. It is the reaching-out of commodification, the commodity market steadily gaining ground. And with advertising on radio and television, economics and the media are entering into a powerful alliance.

And what is advertised? Individual happiness through individual purchases of goods and services. Neither market nor media care that it may not be happiness for ourselves or for coming generations.

"Perhaps it is all the sunbeds
That make us so pale.
Perhaps it is all the jogging overalls
That make us so lazy.
Perhaps it is all the love stories
That make us so loveless.

Ideologies are sold cheap
While we count our royalties.

Perhaps it is the future
That roams the streets
Begging a coin for hope."

The collision course between the speculative interests of the market and the genuine interests of the individual is most obvious in the relationship between the car and the child. The motor industry has the economic means to advertise its products, while the child has no economic means of advertising its right to a non-toxic future.

The arguments about advertising revenue for radio and television making us more competitive internationally can be dismissed.For we have not decided the premisses, we have not had a hand in shaping the terms of competition. The game is rigged, we are issued with our counter and the starter fires his pistol.

Why do we have to compete about entertainment serials with more and more wealthy, dumb women, dubious men with fast cars, jealousy at the pool-side and bodies in the basement? We are a small country with a limited economy. We will never be capable of producing so much glitter and opulence, so many fantastic settings, such large casts of famous actors as the big television producers in the big countries. If it were a world language we were helping to construct, if it were common values we were contributing to, then we would gladly participate. But we do not create genuine community as a result of many million people knowing the names of the hero and the rogue in the soap opera Dynasty. On the contrary, we create an artificial proximity to artificial people. In the world of reality, this means detachment from our own inward life and, consequently, from a meaningful existence of interaction with others.

I have been talking about mass culture as televised entertainment. I would also like to say a little about mass information in the same medium.

We are said to be living in a society with an unprecedented abundance of information. It is certainly true that more people today are living by producing and selling information than were doing so a hundred years ago, but this only means that we have more information in the "commodity" sense. By contrast, we have a good deal less of another type of information. Now, for example, we get our weather forecasts from the radio and television, with somebody selling and other people buying. In past times we got our weather forecasts from indications given by nature itself and from knowledge handed down to us.

The fact is that in many ways we can be said to have acquired a society which is poorly off for information. An office employee sitting by a computer screen today receives necessary working information that way. Earlier the information came in the form of hand-written notes or by word of mouth. What has been lost on the way is the human touch, all the information which could be sensed from handwriting or from somebody's tone of voice.

Of course we receive a lot of useful information through the modern media of today, but there we have the problem - it is such a lot. There are too many impressions, they edge one another out, they are too fragmentary, they do not reach our innermost feelings. And so they also fail to generate understanding. Besides, the information supplied by mass media is very often reduced: we see nature programmes on the screen, but we do not smell the flowers. We see drought and people dying of starvation, but we do not feel the burning heat. On the contrary, we continue eating our dinner while a smiling face in the studio announces the next programme.

Daily news broadcasts usually have a fixed allocation of time, which reveals how casually the so-called news is selected. On Norwegian Television, the Dagsrevyen programme is the same length every day. It goes without saying that the degree of importance varies a good deal, depending on what has actually happened, and it takes really extraordinary events to disrupt this predetermined time allocation.

The media also create their own media events, not by making things up but by choice of emphasis, dramatic phrases, pictorial coverage and so on. News programmes are a result of technology. A "primitive" culture using smoke signals for communication purposes obviously could not have a daily news programme. Smoke signals are quite inadequate as vehicles of news. And not all information can be conveyed by word of mouth - take the flavour of an apple, for example. Not everything can be communicated through books - the strength of a gust of wind, for example. And not everything can be conveyed by television - an intellectual analysis, for example.

Modern media usage is a topic of research. But what is the position of the researchers? Are they implicated in the system and thus indirect spokespersons for media culture?

The Norwegian historian and media researcher Hans Fredrik Dahl declares: "All media research tells us that a media-rich society provides more opportunities for political innovations than one which is poorly off for information. The political process becomes more democratic, and democracy itself more accessible, in a society with a heavy flow of information. Individual activity rises, artistic manifestations are distributed stock. Even children's play is said to become more abundant in a society with many mass media than in a society with only few."

I profoundly disagree with Hans Fredrik Dahl. And the media research he refers to must be very limited indeed. For if these claims were true, then the USA would be an example of extended and accessible democracy with a high level of individual activity. For there, of course, the information flows of the mass media burst their banks. But in fact it is in the USA that the presidential election goes off like a carnival. It is in the USA that children's individual activity consists largely in watching MTV! And it is the people of the USA who, on average, spend more than 5 hours a day watching television.

The idea of "news" as a selection of the events of the age is turning reality into stage scenery and people into extras. And this picture of the world, again, is helping to turn our minds away from fundamental human issues. The result can easily be a paradox: the more we know, the less we know.

This is putting things drastically, for of course we need information. The closed society can easily become full of prejudice and self-satisfied. What I am criticizing here is the quantity and, not least, the methods: the chit-chat and the platitudes, information as a commodity and information as entertainment.

The media are always hard at work to create the average person they want. They have been highly successful in doing so, but there is still hope, because all of us carry a different type of information with which we can face up to the media: our supremely personal experience of life, experience through people we know and rely on, and, not least, we bear within us the accumulated experience of the human race. That experience speaks to us - often without our hearing or seeing it - through our history and geography. It speaks to us through our bodies, feelings and senses.

It is profoundly aggravating when the mass media discourage and make fools of their public. Ideally speaking, modern audiovisual media should elevate the level of awareness and knowledge. Turning people into idiots is a totally misconceived populism. In a modern media world, the words of Goethe are more relevant than he ever imagined: "Talk to a person as he is and he will go on being as he is. But if you talk to a person as he can be, you will help him to become what he can become. There are fantastic possibilities in all people."

Modern technology, mass media included, is its own objective, is concerned with its own diffusion and growth. Human growth is not subjective.

I have drawn a rough picture of the influence of mass media on our culture, and I have taken television as an example when focusing on negative consequences. But indication of the negative always bears within itself the vision of the positive, because what is healthy becomes healthier when the sick is cut away.

And so we have the thorny paths towards a radical popular culture. This culture, needless to say, must be controlled from the bottom up, and not as at present, top-down - or, more concretely in the case of the mass media: literally from the outside inwards.

Summing up, the mass media supply premisses for public and private discourse, pre-empt our time and our resources and in this way actively intervene to control our lives. That sort of thing gives the mass media power and authority. And what are our experiences of authorities? Not many. What training have we had in asserting our own view of things against that of the authorities? Not much.

On the way from childhood to adulthood we have met a succession of authorities to whom or which we all had to adapt. We were born into a family where father and mother know best, and have the power to punish or reward. For them, of course, obedient and, as the expression goes, "good" children are the most convenient option, and this "goodness" is rewarded with hugs and kisses. School is the next stage. Here it is the teacher and, indirectly, the text book that know best and have the power to punish or reward. Obedience and diligence are rewarded with good reports and proud parents. Many young people go to be confirmed, and here they meet the authority above all authorities: God and the priest as God's spokesman. God not only knows best, He can also read your thoughts, hear what you say and see what you do. God also wants good, obedient children and has the power to punish or reward - with Hell or Heaven. Most young boys perform military service, and the armed forces are essentially based on strict discipline under authorities. Obedience is a precondition and, consequently, rebellion is punished with the glass house, black marks in the pay book and, later on, trouble in finding a job. A job is the final stage on this pilgrimage towards adulthood. A minority become superiors or bosses, the majority become subordinates and have somebody who orders them about. At work it is the boss who knows best and has the power to punish or reward. Obedience and diligence are rewarded with promotion and pay increases, disobedience is punished with non-promotion or, at worst, with the sack.

After this pilgrimage of subservience we are expected, as adults, to straighten our backs and display independence, critical minds and independent opinions. Obviously, this is not easy, and many people never embark on such a painful process. Unfortunately, the banal truth is that authorities are only authorities as long as we do not actively rebel against them. Power is only power as long as it does not come up against an actively opposing power.

Up till now this has been a wretched description of mindless cultural imperialism, and perhaps I have given the impression of being totally pessimistic about culture. But the picture is not all dark. I can also see cause for optimism.

What signs of the time give cause for optimism? What chance has culture of operating as a living communication between human beings? And how strong is the culture of the local community in the region in relation to the massive pressure exerted by the great centres? First and foremost I can see something positive in this, right now, being one of the "open" moments of history, a changing of the times. We are the last generation of an old culture and the first of a new.

There are three interconnected factors which provide unique opportunities for creating a new and different future. Firstly, changes in the sex roles and the break-up of the patriarchal cultural heritage. Secondly, changes in industrial society and the move towards post-industrial society. And thirdly, the new ecological awareness of large sectors of the population. On the strength of all this, we are more entitled than previously to say that "the future is now".

This is a good basis for hoping that intellectual life, art and culture can become powerful steering mechanisms on the way ahead, because the centrally controlled mass culture has passed a boundary: it is gradually being suffocated by its own tremendous growth, it has become too imperialistic and too suppressive. And so today we can see the first signs of a cultural revolution. We are surfeited with observing and we feel the lack of participation. We are surfeited with information and feel the lack of perception. We are surfeited with amusement and feel the lack of happiness. And gradually we are realizing that the traducers of culture are just that. And this mobilizes the profound need of the people to live as what we are by origin: people with the will to choose.

A quotation from Dostoevsky exactly describes our situation today and outlines solutions for tomorrow: "The way out of the desert of unimaginativeness, apathy and amusement - and consequently away from social apathy as well - passes through one point only: the capacity of the individual for involvement! Involvement is the opposite of apathy. The opposite of amusement as a pastime is not seriousness but also involvement! This capacity, however, does not exist automatically, it has to be awoken, developed and used. Its development occupies the remainder of our lives."

Indeed, but when and how is this capacity awoken, and where and how can it be developed?

I have had a lot of unkind things to say about the new information technology as it has functioned hitherto on the whole. But that technology also presents possibilities of decentralization, electronics opens the way to communication over great distances. This technology also presents possibilities for creating our own local and regional networks for the transmission of knowledge and experience. But this, needless to say, depends on how much cultural resistance we can display, how much countervailing power we can create against present-day power in cultural reality - or rather: irreality!

The fact is that community radio and local television have become popular media, and here we have horizontal communication between equals, instead of the vertical, top-down communication of the traditional mass media. It is also a fact, at least in Norway, that dialects have acquired a completely different status compared with just a few years ago. And this brings us to a central point: language in itself is a power of resistance. Language is created through history and geography and transmits knowledge of life and landscape. Language tells us something important about who we are and where we came from. And even though the mass media have tried to create an average person with an average language, Oslo will never be the same as Dallas, and Umeå will never be the same as Los Angeles. Because a society's culture is so closely bound up with the landscape and the natural foundations.

In the USA, this realization has resulted in the "bio-region movement". A bio-region is a region with a fairly well-defined ecosystem: soil, climate, flora, fauna. It can be a coastal strip, a planeland, a mountain region ... Gradually almost the whole of the USA has come to be divided into bio-regions of this kind - regions strikingly similar to the way in which the Indians once divided up the country between their various tribes.

Post-industrial society can perhaps become a modern version of the old rural society, in a kind of electronic edition. But if this happy future is to come true, then as I said earlier, there will have to be active resistance to the main features of the cultural scene today. And this resistance struggle will have to be waged at all levels. Obviously, not all hopes can be pinned on community radio and local television, dialects and bio-regions in the USA.

Schools are an effective starting point. As I mentioned earlier, schools are one of the many authorities we encounter on our way to adulthood.

"It is always the pupil
Who has adjustment difficulties
The class
That has behavioural problems
The teacher
Who has trouble keeping discipline.
It is never school.
School
Which fails its own examination
Because it has forgotten the curriculum:
The right questions
Are more important
Than the right answers."

As an active participant in a radical popular culture, our present school must be completely re-structured and set itself new objectives. As I say in the poem: It is the right questions which must be rewarded, more than the right answers. It is the right questions which advance the world - and advance the pupil's own development. In the form of greater independence, insight and involvement.

Of course the school of the future must provide basic skills. But as the foundation of a popular culture it must also make clear that knowledge is an implement in the service of the community. Knowledge is not private property for the purpose of supplying the individual with material privileges. Knowledge is a common heritage stored, administered and upgraded by succeeding generations, to be inherited by new generations as common property. In this way too, a happy union will develop between information and communication.

And here, obviously, clearly defined regions, small communities and small units will be in a good starting position, because networks and paths of a kind already exist between people. Even if the networks are fragile and the paths almost overgrown, because they are so little used under the tyranny of mass media culture.

Intellectual life, art and culture can never be detached from their landscape. Norwegian painters like Tidemend and Gude, Norwegian authors like Bjørnson and Ibsen, derived their material from their bio-region - long before the term existed. Ibsen's Peer Gynt is rooted in the nature and culture of Norway, but in the particular we also recognize the universal - as in all true art.

To respect regional culture means reinstating the individual, bringing him out of the oblivion of the masses and making him visible. This means releasing all the creativity inherent in the people. It means creating communication between body, senses and intellect within the individual, between people and between people and nature.

It does not mean that the great wide world is to be shut out. On the contrary, it means opening doors. But on our terms! We are to be the hosts and the world at large our guest. So far the opposite has applied. We have sat at the table of the great, but we are tired and hungry, because we were given no spiritual sustenance.

The stealing of people's time and vitality by the mass media and the robbing of nature and the environment by large-scale industry are two sides of the same thing. Solution of one crisis demands the solution of the other. If we conserve and respect our resources as consciously acting human beings, then we will also conserve and respect our landscapes, our small places on earth. And none of these places is called Disneyland: they are called home!

The Culture of Nature

Per Råberg

THE RHETORIC OF CULTURAL POLICY

In the neo-liberal nineties, aesthetic culture has come to occupy the focal point of market interest. Cultural life is being highlighted by a new direction in science, cultural economics, as if it were a newly discovered raw material resource whose exploitive value society, for some inscrutable reason, has hitherto overlooked. From their studios and workshops artists, musicians and all the other aborigines of aesthetic culture look at the newcomers with amazement.

Which, then, are the new arguments? And which is the new stage of cultural policy? In the first flush of enthusiasm, the cultural sector is described by converted economists as a source of motivation and creativity in the market society. But this mild expression of utilitarianism is only a starter. Culture is also regarded as an economic sector in its own right, with a turnover equalling that of many industries and manning on the scale of entire branches of ordinary enterprise. Investments in culture, we are told, are not a waste of resources but a paying proposition, both in various indirect ways and from a productive perspective. Progressive regions in Europe have rapidly assimilated the neo-liberal message. Whereas formerly culture was looked on as a drain on the local economy, today it is seen as a nurturing factor, a resource of hard-hitting entrepreneurial strategies.

The economists' take-over is seemingly radical and total. It is now recognized everywhere that public cultural promotion is a key factor in persuading companies and entrepreneurs to locate their branches in the region. Communities boasting a generous cultural infrastructure are thought to be the winners of the stiffening competition for power in the new united Europe. The managers of liberal cultural policy invest large sums in museums, music festivals, art fairs, cultural tourism and other projects. The success of the new strategy is confirmed by cultural-economic statistics of earnings and numbers of visitors generously presented to the public.

Primarily this new trend is an urban phenomenon, modelled on the expansive metropolitan cities of Central Europe. The cultural image is being cunningly marketed as part of the propaganda for the metropolitan regions as the locomotives of development. Surely the cultural style also bestows a coveted aura upon the new economic and political élites of the European continent. Bold comparisons are frequently made with Renaissance Florence and the flourishing princely courts of the North Italian Quattrocento.

In this neo-liberal euphoria there is really but one group whose voice is hardly ever heard in public. These are the makers of culture themselves, the unique creators of art, music, dance, theatre, film, poetry, media and communication. It takes a lot of searching among all the ins and outs of cultural policy to find viewpoints from the professional artists concerning the cultural future of the new Europe.

Why is that so? One can only guess: poets and artists are seldom requested to contribute towards the vision of Europe's cultural future. Culture, in the neo-liberal vision, seems more than anything else to betoken a representative scope, preferably adorned with a cachet of historical authenticity. Most typically it may comprise a mixture of French Renaissance and German Biedermeier, of Mahler symphonies and rock operas, possibly completed with a few works by Christo and Beuys.

But is this cleverly composed mix really the true culture of today, and of the true new Europe? Before we answer this question we shall take a look at the precursor of the neo-liberal vision in cultural politics. Hereby we can better judge what is new and what is merely common thought in the winning approach.

In many municipalities the neo-liberal ideology overlays a classic cultural policy which, a little drastically, can be termed social democratic. Social democratic cultural policy took shape as part of the welfare politics of the inter-war years in the Nordic area and the western industrialized countries. The so-called Nordic model, it will be remembered, was characterized by the alliance between industrial and democratic forces and by endorsement of the nation state as the upholder of society. This ideal scheme also included the launching of cultural policy programmes with great social ambitions. In Sweden their cardinal manifestation was the 1974 Government Bill on Cultural Policy, the core of which was made up of the eight cultural goals. Those goals have subsequently provided a benchmark for local as well as central cultural policies in this country.

No doubt there is a decisive difference between democratic and neo-liberal cultural policies. The tone of the classical model was set by the strong notion of community responsibility. This meant, firstly, that the cultural sector was looked on as an action field of the municipality, but also that closer attention was paid to the role of the sector in overall social politics. Secondly, the model was permeated by a democratic pathos, by the desire to make all citizens and classes cultural participants, in whatever part of the country they

"Window Vision". Photo by Hans Marklund

were living. The principal democratic targets were to be achieved by relocating public cultural institutions to different parts of the country, but also through self-activation in the local community.

The practical implementation of cultural policy, however, did not become the subject of heavy debate. Activation was expected to be achieved through administrative measures and financial support for various purposes. This is a policy which, at least in official assessments, has achieved well-expected results, as was recently confirmed, for Sweden's part, by an examination undertaken by the Council of Europe.

A PSEUDO-CULTURAL POLICY?

Now, how do the traditional policy descripts relate to the new liberal trend? Both directions coexist in the "mixtum compositum" of the postmodern society. The ideological differences between them are there for all to be seen and do not require any further elucidation. In practice, however, the two models are not quite as incompatible as the political rhetoric may suggest. Both of them are designed by highly competent administrative experts, although originating from different offices. They are characterized by a good deal of bureaucratic pragmatism, as well as great amenability to corporative solutions.

On one important point, though, democrats and liberals stand not simply on parallell tracks, but fall directly into a clinch with each other. That is in their astonishingly reserved and bashful attitude towards the creative culture itself. The same diffuse, but unshakeable, alienation from the living art characterizes - by and large - the neo-liberal and the social-democratic vision. Both programmes are, in fact, equally devoid of inspired viewpoints concerning the aesthetic experience which is the indispensable core of culture and of a true cultural policy. Nowhere an uneasy thought is delivered of the jeopardy of the aesthetical sphere in a civilization becoming more and more shot through by rationalist and technological valuations.

To the seriously involved public it is quite evident that cultural policy is regarded by both persuasions from a rather instrumental vantage point, be it as distributive activity or remunerative patronage. In both cases, culture is treated like an inexhaustible stock-in-trade, a river gushing from never-failing sources. It hardly ever occurs to the policy writers that, on the contrary, art and society's cultural awareness - like many other natural assets - are a finite resource, a dispersing commodity. It does not seem to have dawned on the policy makers that the aesthetic sphere of society, like many of our lakes and seas, is threatened with serious pollution and that really good catches are beginning to be few and far between.

Confronted by the massive rhetoric pouring from well-funded tribunes, an indigent aesthete is sorely tempted to play the devil's advocate. Does not the snorting optimism of cultural policy - irrespective of political origins - testify to an underlying ignorance of the real world of art and cultural creativity? Isn't the depressing fact of the matter that official cultural policy - possibly unintentionally - serves as a repressive force against true aesthetic and cultural creativity in society, and this so effectively that the latter remains almost invisible? That cultural policy, by way of its official status, justifies a social order which is fundamentally hostile towards aesthetics and the experiential values of life?

These are sharp words, and they should be underpinned with good arguments. Below I shall put forward the thesis that aesthetic culture in our times is in a deepening state of crisis, and point to the logical connection between this crisis and the prevailing development paradigm of today. Further, I would like to call attention to the fact that this paradigm, by way of its undisputed exercise of power, has become a serious threat to the ecological and socio-cultural balance of contemporary society, and that a re-establishment of homeostasis demands major achievements also in the aesthetic and cultural fields.

Finally, I will focus on the need for an ecological cultural policy founded in man's need for existential wholeness and quality of life, and assert that a policy of this kind can come to play a leading part in a longed for shift of paradigm from a quantitative and mechanistic towards a qualitative and humanistic course of development.

THE CRISIS OF CARTESIANISM

Let me start with a glance at the prevailing paradigm. The goals and paths of human progress in the ageing modern society are still decided by the almost total official support for the Cartesian Weltanschauung. The central idea of the Cartesian paradigm is the strong belief in the power of rational thinking alone to solve all the mysteries of the universe, and to steer social development towards higher and more dignified goals. The flagship of the Cartesian view are the exact sciences, including subjects like mathematics, physics, chemistry and astronomy. Their cousins in the living community are technological progress, economic expansion and bureaucracy. The good technocrat is the human ideal of this ideology, cultivated by an educational machine whose indoctrination is applied when the individual is in his most ductile age. In technocratic high society, culture is ever the poor relation. The cultural creator himself, for better or worse, is an outsider.

Cartesian rationalism has led to indispensable advances in many areas. Owing to its formidable success, however, one-sided rationality has become established as a hyper-ideology, claiming to represent the supreme reason of the whole community. In many fields where techno-rationality has gained power, it has without a doubt introduced one-eyed, inhumane attitudes. In the scientific community this has been done by writing up natural science and writing down human science, and in politics by giving rigorous priority to economic and technological development at the expense of the social and humanist goals of society. In the cultural field Cartesianism has avoided displaying its attitudes openly. Its deep anti-aesthetic stance is realized first and foremost by marginalization of the cultural sector, and by using conservative definitions of culture. Cultural life has thus become a kind of luxury ghetto at the periphery of society, but without any influence whatsoever upon mainstream social development.

One of the general consequences of the strong breakthrough of the Cartesian paradigm over the last century, has been a dramatic reduction in the role that aesthetic culture plays both as a human stimulus and as a developing force in society. The basis for such a categorical statement rests primarily on an appreciation of the relative position of the cultural sector vis-à-vis sectors like economics, technology and politics. Aesthetic dismantling can however also be traced in the development that the various sectors themselves have undergone. The aesthetic retreat is revealed in the gradual intellectualization of avant-garde art, and in its transformation into a museum culture, distanced even from the wealthy public. It can also be seen in the degradation of popular culture from a tradition of civic creativity to a passive consumption of the assortment offered by the mass media.

The decline of aesthetic quality in the social environment is another evidence. In modern times, environmental culture has gradually stepped out of its anonymity and into the hands of experts. As far as the aesthetic aspect is concerned, this change has not necessarily constituted an advance. The aesthetic quality of the interior domestic environment has undoubtedly improved for large groups, and architectural artistry has long been producing unique creations which arouse our admiration. However, at the same time there has been an aesthetic regression in large-scale planning, in rural as well as urban areas, which has eclipsed all the unique efforts. And all this has happened despite the fact that there is no lack of committed architects.

Aesthetic deconstruction has not only affected the various branches of art, it also affects the aesthetic cultural awareness in politics and amongst the citizens. Ignorance and lack of interest in aesthetic culture is spreading through politics, schools, higher education and research in a disturbing manner. The lack of critical debate about the structural hostility of modern techno-society towards aesthetics, and the social consequences thereof, is in itself an alarming sign. Equally strange is the lack, within planning and art education, of comprehensive ideas concerning the means of expression, role and responsibility of aesthetic culture in future society.

Most disturbing of all is perhaps the growing theoretical petrifaction that we meet within the aesthetic disciplines of the university, for example in Art History and Philosophical Aesthetics. These are disciplines that should long ago have initiated a debate about the internal and external crises of artistic culture in the age of techno-culture.

What none of the expert groups within the cultural sector have been willing to acknowledge, is the serious threat to civilization which one-sided rationalization and the neglect of aesthetic culture entail. A deep-seated myth is used to support this stand - the myth that aesthetic culture is a superstructure perched on top of the material and technical culture, with the banal purpose of *decorating* the human environment. This myth is also a part of the process of deconstruction, and there is reason to anticipate with anxiety the final verdict that the arts lie outside the responsibility of society and can, without too much hesitation, be dispatched into history. The culture of the senses is thus reduced to a mere cosmetic on the entertainment market.

AESTHETICS AS A DRIVING FORCE

Aesthetics is a complex field, but behind the myths still great challenges are waiting for a radical human science with the courage to advance new hypotheses. One alternative hypothesis with high credibility states that the view of aesthetic culture as a superimposition on the material culture is in fact false. From an evolutionary viewpoint, technical society may be seen as a superstructure, while aesthetic culture is the more fundamental and indispensable sector. It seems, rather, to represent a deep level of our human constitution which is close to our central value system. It is through the system of values that our qualitative requirements of life are met, and that we obtain motivation for renewed vital efforts. The aesthetic sphere is perhaps linked to the sensory core of our consciousness, which lies far deeper down than the technical and rational level with which evolution has rewarded us.

It may be within this sphere that the whole creative cosmos of sensory impressions, emotional expressions and aesthetic nuances, has its beginning. It may also be from here that we draw the zest for life and the inspiration which allow us to submit to the strain of everyday life, and find its routines worth the trouble. A culture which cuts off this vital nerve rapidly may find itself sunk in barbarity. Its life-force becomes significantly reduced and soon it is caught in a downward spiral of development.

The hypothesis which has been presented here was taken from the theory of social ecology, developed by the University Group for Humanist Futurology in Umeå. The socio-ecological theory as a whole is a contribution to the current debate on sustainable development, which has so far focused mainly on the natural ecological environment. The University Group in Umeå belongs to those who wish to widen this debate to include the corresponding crisis at the socio-cultural core of society, and to attempt to apply a systematic ecological approach to this crisis as well. Both crises are logical expressions of neo-Cartesian philosophy and are signs of its increasing insufficiency. The researchers in the Umeå Group have formulated a theory of existential needs, based on an ecological approach to man. All this has been discussed in detail in preceding essays in this book.

What I wish to draw attention to here, is the place of the aesthetic dimension within this theory of human needs and within a strategy to remedy the present crisis in civilization. In the model of the Umeå Group, the aesthetic dimension occupies a key position in the discussion concerning a shift in paradigms. The aesthetic deconstruction of society is regarded as a threat to the very living core of our civilization and one of the most alarming symptoms of the crisis. One of the main projects within the approaching shift in paradigms, therefore, should be the recreation of a dynamic aesthetic culture. The reason for this is not only to safeguard the riches of aesthetics. One significant motive, considering the threat hanging over us, is the potential for aesthetic culture to bring back human dignity and vital force to the whole of society.

A meaningful realization of these intentions presupposes an extensive awareness of the possibilities which the aesthetic field offers. The attempt to map out an aesthetic eco-identity, made by the University Group, can give a certain amount of guidance. Aesthetic consciousness includes both individual and general characteristics, and contains both overtones and undertones. It is the overtones and the individual characteristics that have interested traditional cultural politics, because they have never entailed any threats to the established order. Artistic avant-garde culture is the best example of this. In the project of the Umeå Group, it is the general characteristics and undertones of the aesthetic spectrum which have attracted attention. By undertones we mean the wide and profound thoroughfare of qualitative experiences in every day life, which can be named *the life culture.* This aesthetic culture is of immeasurable importance to an individual's experience of quality and value in society.

If the general features of this sphere can be described and put on paper, we will have created the basis for a cultural politics, capable of defending not only the unique high culture, but also the aesthetic life culture in society as a whole. Enormous socio-economic gains could be achieved if technological and economic efforts were, more directly than at present, aimed at socio-aesthetic targets. Here we actually have the foundation for a totally new welfare politics which simultaneously bears humanist and ecological characteristics. Not economic growth, but the individual and collective aesthetic quality of life, is the prime objective for an ecological humanist welfare policy.

From what has been said here, it is easy to imagine the central role that aesthetic cultural policy will play in overall politics in a coming ecological society. But already at the very breakpoint where a shift in paradigms is accomplished cultural policy assumes a strategic position. Correctly handled it can become a battering-ram in the struggle to establish an efficient value-based parameter within politics and planning. - It is primarily in its ability to realize high-level aesthetic demands, that a humanist vision of society presents its intelligent face, and can become a competitive alternative to the solutions, brilliant within their own category, offered by technical society.

AN ECO-CULTURAL POLICY

If, however, we are serious in our intentions, we must take the step from rhetoric to practical cultural policy. Necessity then forces us into a new discussion, where the saying "politics is the art of the possible", comes into play. Our starting point is the current practice and established conventions within the cultural field. What can be done to radicalize these conventions in the desired direction? What complementary efforts should be made to broaden the ambitions of cultural policy? In this way, alternative ecological cultural policy can be more easily assimilated by established bodies, which would facilitate the launch of the new overall model. Let us see which consequences speculative thinking within these quarters will entail. What path and direction should a cultural policy with eco-systemic foundings follow, if applied in the contemporay society?

The primary object of such a policy ought probably to be the caring for the life-related aesthetic values which are located in the daily surroundings of the individual, all the way from the environmental qualities down to the empathic stimuli of neighbourhood life. But an important aspect of an ecological cultural policy must also be to overcome the passive voyeur attitude so typical of cultural behavior in the ruling paradigm, and to promote an active attitude, which can inspire the individual to shoulder the role of a co-creator in his social setting. But an ecological cultural policy should not be restricted only to serve the needs of ordinary people's aesthetic self-realization. It also has to confront the challenges of frontline artistic culture and the exclusive demands of the leading representatives of world art. Otherwise cultural policy would make the serious mistake of cutting itself off from its own carrying branch in societal evolution. The end result would be a general decline in cultural awareness, and a loss of power vis-a-vis other sectors of development. However, an ecological cultural policy at the same time must take as its task to integrate professional culture in the living society in a more efficient manner than our epoque has managed to do.

There exist still other domains where an ecological cultural policy should play a major role in an ecologically profiled society. One of the most important is the field of environmental planning. If there is one major blind spot in the cultural awareness of our time, that is the lack of response to the cultural content of our living environment, with its multivariate forms and meanings. For the new policy it should stand out as a major challenge to awaken people's consciousness for the deep experiential values embedded in the panorama of our physical living space, be it an urban center, a provincial village or a regional landscape.

More important than catalogizing all the potential areas of application of an ecological culture, however, is to indicate its overall societal object. Radical criticism ought to be directed towards the contemporary conception of aesthetic culture, for its tendency to alienate culture from its human and sensory roots and to put it on high stilts above the routines of community life. An ecological cultural policy, instead, should have its main attention directed towards the citizen in his social setting. Its primary objective ought not to be to fill individual homes with aesthetic luxury, but to strengthen the aesthetic quality of community life in all conceivable ways.

However, let us try and turn this discussion somewhat more concrete. The University Group for Humanist Futurology has made an experimental contribution to the actual debate. It has drawn up a cultural policy model which places aesthetic needs and the quality of life at the centre of arguments. The contribution has been designated a socio-ecological approach. Below I shall present a brief sketch of the model. It should be seen as a follow up to the discussion in earlier chapters of man's ecological identity. It must be remembered that, at the present stage, the model represents a hypothetical attempt. As such it can still serve well as a source of inspiration in the task of uncovering useful knowledge in the humanist soft-data sphere. In the programme we distinguish between general and sectorial guidelines. I will start by enumerating the general lines.

GUIDELINES OF AN ECO-CULTURAL POLICY

The Experiential Values.

The objective of an ecological cultural policy is to defend the experiential values of human life and assert their legitimate place in the community.

Eco-cultural policy designs its strategies in full awareness of the serious threats which are hidden in the actual course of development towards the domain of sensory values. It therefore introduces a broad definition of culture, whose field of activity extends from the primary body-senses and expressions of social empathy, over the area of cosmopolitan aesthetics to transcendental experience. The culture of sports and games with its many branches is also contained within its responsibility.

At the same time ecological culture wishes to refute the ongoing exhaustion of cultural concepts by formulating more elucidating aesthetic definitions.

The Quality of Life

The intent of an ecological cultural policy is not primarily to support unique top achievements on the margins of society. At the centre of interest is a socio-aesthetic purpose: to achieve an improvement in the quality of life of all citizens and in the whole community.

In overall politics this implies that the quantitative schemes of contemporary welfare politics are replaced by qualitative and humanist objectives. Creating an aesthetic life-culture is the overriding object of the alternative policy, not just making sectoral economic contributions. The many tasks implied in this contain important commissions for the professional artists. But a major object is also to promote the creative forces among the rank and file of citizens.

A Holistic Perspective

Ecological culture carries strong holistic ambitions. Its intention is to create syntheses: of life situations, of living environments and finally of human existence as a totality.

But what is intended are humanistic syntheses, which take the scale of man and of civil society as a measure for their proposals. Realizing in all their dealings the human and holictic aims of an eco-cultural policy is probably a utopian wish today. But a demonstration of the holistic intent may already have considerable effects upon cultural development.

There are tasks enough for practical reform work. Forming aesthetic environments for the social community functions is one object. The neighbourhood, the local community and the home-region are all examples of fields where qualitative contributions are much needed.

Universal Knowledge.

Ecological cultural policy wishes to support the manifoldness of the life-culture. But within the various contributions it discerns not only the unique but also the general traits, which possess aesthetic universality.

The radical culture sees it as an especially vital task to draw attention to the lavish circumstances which surround man's qualitative experience. If aesthetic culture is to survive in the future at a level appropriate to its existential import, it will have to put forward general policy and planning goals and defend these in competition with the other sectors of politics.

Cultural Pedagogics.

The opening of minds to the field of cultural experience and creativity is a prime task of an ecological cultural policy.

We are living in an era which is far from estimating the aesthetic and experiential culture to its full value, and the implementation of an ecological culture, therefore, demands extraordinary efforts. Without persistent preparations, genuine cultural awareness will never take hold of minds and be capable of flourishing in municipalities and regions. Implementing a deepened aesthetic consciousness is an important task which can possibly be delegated to the school and the educationalists.

However, a serious cultural-pedagogical initiative certainly also demands a reassessment of the educational institutions and their training programmes.

SECTORS OF AN ECO-CULTURAL POLICY

An ecological cultural policy according to the actual model contains four main sectors. We have named them *the local culture, the cosmopolitan culture, the cultivated environment and the natural aesthetics.* All spheres are equally important in satisfying basic aesthetic needs, and are interacting with each other in a complex manner. In social reality they also often overlap in imaginative patterns. Basically, though, they represent separated attitudes and levels in man's ecological constitution.

Local culture

The ecological local culture has many traits in common with genuine popular culture, as anthropologists have decribed it in different epochs and parts of the world, i.e. in the history of the industrialized countries. This is an aesthetic culture which emerges from a local tradition of crafts and skills and from the wish of the population to enrich its daily living environment. Local culture is often closely related to the forms of sensory expression in an elementary bodily or communal guise. By its intervention an important part of our striving for sensory self-realization is satisfied. The aesthetic local culture can quite correctly be defined as *a culture of expression.*

However, ecological popular culture does not wish to copy historical patterns. It shows perfectly modern traits, not least in its broad range of activities. It achieves its genuine impact in different kinds of play, sport and outdoor life in the neighbourhood, in the culinary and sartorial habits of the region, in popular participation in dance, theatre and musical events in the municipality, and in the entire spectrum of collective sensory activities in the commune. Local cultural life partly remains an unpredictable mixture of reception and creation within every single individual and within the population.

For a long time modern community planning, in symbiosis with politics and educational ideals, has inadvertently frustrated the prospering of local culture in the popular sense. Due to the systematic dismantling of communitarian values in the highly urbanized society, the informal aesthetics of civilian life have been dealt a knock-out blow. A reconstruction of the organic local community and its aesthetic culture therefore constitute a central point of the programme for an ecological cultural policy.

A cultural policy for the local sector may be profiled round the following points:

1. ***Socio-cultural activation.***
Here we have in mind civic initiatives for local popular culture in the neighbourhood in cooperation with children, elderly persons and minority groups. Initiatives for social cultural activation may further include: the insertion of meeting points and walkways in the public townscape and filling them with experiential events; establishing a local festival tradition in connection with annual celebrations, carnivals and public meetings; aesthetic mobilization at all levels of the school system; caring for niche cultures such as ethnic restaurants, handicraft tradition, sporting specialities etc.

2. ***Protection of the aesthetic environment.***
Aesthetic environment protection may consist of civic initiatives for the beautification of the dwelling environment as a whole and in details. Other suggestions for creative initiatives are: aesthetic concern for the urban environment, its public amenities and greenery through local environment groups; protection of the natural surroundings through civic initiative; the adornment of the urban environment by temporary artistic stage sets.

3. ***Local cultural centre.***
A meeting place for local civic culture can be created amid the urban bustle. Autonomy and freedom from age boundaries are a condition for its well-functioning. The facility has accomodations for a variety of activities, e.g. workshops, music and theatrical studios, craft shops, children's cultural centre, exhibition hall, hothouse, canteen and meditation room.

Cosmopolitan Culture

The cosmopolitan culture derives its inspiration from other sources than the local culture. It is closely related to the reflective and inter-regional levels of our ecological identity, which belong to modern rational man. The aesthetic culture which is realized here emerges originally from the sphere of elementary sensory experience, but has been cultivated in an abstract and esoterical direction. It can best be described as an aesthetic sublimation of the modern consciousness.

Its proper mode of expression is a highly creative avant-garde culture, the practice of which demands an extraordinary talent and the presence of a body of aesthetic experts. In the eyes of the ordinary citizen it represents - contrary to the case of local culture - *a culture of imagination*. But the imaginative part does not reduce the role of the citizen to pure aestheticism. It possesses an important existential meaning to him in that it confirms his participatory role in a world culture at the level of the modern identity. The cosmopolitan culture, therefore, is an unforfeitable element of the contemporary society.

The participation of the local community in a cosmopolitan culture puts heavy demands upon the single municipality. The demands consist first of all in the provision of arenas of various kinds for the large output of cultural events, extending from the exclusive culture of exhibitions, stage and meditation all the way to the huge events of sports and entertainment (if the overlaps are accepted which were mentioned earlier). Today, only the metropolises dispose of an economy which can handle such a task. But the provincial regions would also be able to tackle the cosmopolitan mission, if only they coordinated their efforts in regional projects.

However, achieving full membership in the global culture also requires an original contribution on the part of the local county and municipality. This means that the creative talents in the region should be given the opportunity to develop their arts in order to enter, in due time, the cosmopolitan stage. The world artists must also be offered reasonable living conditions in the community in the form of economic reward, schools and studios. Hereby the community pays its natural tribute to a living global culture.

Many provincial regions probably hesitate in front of the task of realizing cosmopolitan cultural ambitions. But progressive regions will quickly discover the opportunities to launch consummate alternatives. One means to create regional cultural arenas with a cosmopolitan stamp may be to gather the events of world culture into a cosmopolitan meeting point centrally located in the communications network of the region. Arguments in favour of a regional world centre are also readily available from other parts of society, e.g. from business enterprise and the public sector.

I will end this presentation by giving a few examples of cultural facilities which could be located in such a cosmopolitan centre:

- Advanced stage art, tying in with the theatre, dance, music and multimedia applications.
- Front-line exhibition routines in pictorial and sculptural arts, installations and virtual performance.
- A poetry and literary culture forum.
- Arenas for cosmopolitan spectator sports like football,athletics, gymnastics.
- Spiritual and meditative meeting-points.

The Cultivated Environment

The record of the ecological aesthetic culture is not finished with the action programmes of local and cosmopolitan culture. It has anchorage even at the deep level of ecoidentity where the community is regarded, not as a field for short time activities, but for generational development from past times into the future. We have designated this as the level of species-identity, and its field of responsibility is the management and renewal of the institutional and public structure of society. The name given here to this level of eco-cultural life is *the cultivated environment.* Inserting the regional institutions within an aesthetic sphere of experience is the purpose of cultural policy at the environmental level. One of the most important objects here is to articulate the architectonic frames of social life in the setting.

The beautification of public institutions has been, for a long time, a favoured object of those in municipal power. However, ecological aesthetics widens the field of artistic interest to embrace the total community space. The actual task is to articulate the local habitat into an aesthetic living environment, including not only the historic buildings but also the streets and the places, the memorial fund and, in certain cases, the inventory of the buildings. The task is not getting less onerous by the addition of the regional cultural and cultivated landscape.

In order for the citizens to become a part of the aesthetic environment they have to participate actively in public life, they must make use of the existential environment. The striving for holistic identification plays a key role in the cultural experience at the existential level. We describe its formal language as *an aesthetics of wholeness.*

In our day the environmental cultural sphere is defended by several groups of agents. Often, however, this is without full consideration of the aesthetic values which the public environment involves, especially on the large-scale level. Similar shortcomings are noticeable when we examine the contributions of cultural and social sciences to cultural analysis. Without hesitation, one can contend that the environmental space is a neglected field of contemporary aesthetic awareness. Measures in that sector, therefore, are an outstandingly important task for an ecological cultural policy.

The aesthetics of cultivated environment in the region comprises two main levels: the infrastructure of modern townscape and the historical cultivated landscape. The classification is functionally somewhat arbitrary, but quite important from a symbolical point of view.

1. *The modern townscape.*

The modern townscape includes the buildings and tectonic structures which frame the social activities of our own epoch. The cultural importance of a well-articulated community environment expressing the values of the contemporary era can hardly be overstated. A municipal or regional cultural policy not having the architectural environment as a main point on its programme can not be said to be taking its mission quite seriously.

In spite of qualified architects taking part in micro-planning and despite the existence of a profession of landscape architects, it is extremely uncommon with aesthetic syntheses which include larger sectors of the living-space or the life environment as a totality. Similarly, one looks in vain for an aesthetic critique which pleads for the socio-aesthetic values in the public community space. Great efforts are needed to create professional and political awareness of the importance of the local and regional infrastructure as a cultural factor.

2. ***The historical landscape.***

This term inscribes the regional cultivated landscape as it has evolved through the centuries, with arable plots, road systems and settlement. But it also comprises the historic buildings inherited by urban and local communities from various epochs. Both the urban environment as a whole and the individual blocks and their buildings are included. If we go into details even representative interiors and artefacts should be covered by the aesthetic interest. The historical cultivated landscape and its artefacts represent an indispensable aesthetic memorial fund in the regional community.

Heritage conservation today is an important part of the safeguarding of the cultivated landscape. This guard duty is performed mainly with historical and nostalgic arguments, but with a great deal of diffidence about references to aesthetics. As a contribution towards an aesthetic heritage conservation in the ecological sense, this is not enough. We can see this now that we are faced with the threat of set-aside for the open cultivated landscape in the Nordic countries. This is a question of such dimensions that it can only be dealt with if we acquire a regional policy which realizes that the cultivated landscape is also an irreplaceable aesthetic element of a living community setting.

The Natural Aesthetics.

At the very nuclear level of our cultural awareness we discern the sphere of natural aesthetics. It is closely connected with man's basic natural ecoidentity, which we have described as the ecological habitus' most original domain. This is, in short, the place for man's identification with the natural space as a resource-base, but above all as a source of life. An aesthetic consciousness plays a key part in the events taking place here.

It ought not to be necessary to argument in favour of the significance of natural experience to the vital energies in man, but in technological society aesthetic man has good reason to feel uneasy. We have better to say it plainly: The natural environment with its organic forms of life, its topography and phenomena of the sky, represents the primary stimulus for man's obtaining confirmation of his participatory share of the organic environment - hereof he is a genuine descendent and herein he is irrevocably enclosed. Refering to the instinct towards emotional transcendens provoked in us by natural experience we designate the environmental culture as *an aesthetics of transcendence.*

The natural landscape has never been regarded as a domain for cultural policy. It has been seen as an easily accessible resource of sports and recreational life, certainly posessing a primitive aesthetical value, which, however, is negligeable because of its abundant occurance. First through representation in the arts or pruning by a landscape gardener does the natural landscape achieve the status of art and, after a while, of historical culture.

In recent decades there has been an increased interest in the natural landscape, but the focus is on its bio-ecological properties. At the same time the modern life-style has continued to develop in the adverse direction, into the technical labyrinth of the man-made infrastructure, with the townscape of the metropolis as a backward ideal. With the help of parks and green areas town planners try to transplant lungs into the stone dessert, and simultaneously to implant aesthetic values therein. But the effect is only strong enough to awake a diffuse yearning within the individual. This is a poor substitute for the genuine aesthetic experience of nature.

Ecological cultural policy, on the contrary, makes the safeguarding of the natural space as an aesthetic field to a primary task. It points out that modern development has brought us so far away from the basic foundations of living that the most self-evident values of life are threatened and must be made the object of politics in order to be maintained. The natural environment is one of the most important among these threatened values. It is an affair of the heart for an eco-cultural policy to place the natural environment at the centre of the aesthetic discussion, and to formulate viable strategies to defend it.

Hereby, cultural policy is confronted with tasks of a new kind, and of huge dimensions. But the conditions for a successful contribution vary considerably between places and countries. In countries with a geographical structure like the Nordic, the natural landscape foundation is not lacking. Culturally it can, if anything, be termed a latent resource. New understanding of the aesthetical value of natural space causes our interest to turn away from the stone desert of the metropolitan cities to focus on the qualities of the small-scale local community. Here the natural space still forms the perfectly discernible outward frame of society. To these local communities we address an appeal to safeguard their natural cultural values and to rejoice in their qualitative lead on the big cities.

DEVELOPMENT OF ECOLOGICAL CULTURE

These, then, were all the good ideas - the aesthetic utopia of the socio-ecological strategy, if you will. Let me finish by taking a look at grim reality. What should be done to establish an ecological cultural policy in municipal and regional planning?

Merely delivering draft programmes like this one to the actors and hoping for implementation is probably too optimistic. Invisible powers, academic lethargy and well-worn routines will see to that. Nor should we expect the Nordic or continental metropolitan areas to be seriously impressed by the ideas propagated in an ecological cultural policy. They are far too occupied with launching themselves in the role as winners in the ongoing international power-game. We do not expect them to be in a hurry to advocate a cultural policy which questions several of their well established cultural convictions - no matter how socially and ecologically well-motivated the former might appear.

It is an altogether different case with the many provincial regions with intermediary or peripheral localization that make up the predominant part of the national territories. To many of them the future is uncertain, and they realize more and more clearly that the leading political strategies of the day are hardly designed with their future prosperity in the foreground. Especially the peripheral regions - in Europe and all over the world - have much to gain by furnishing themselves with alternative development strategies which focus on values like self-reliance and an aesthetic quality of life.

Aesthetic quality of life is, in reality, a remarkably good substitute for economic welfare, where exterior conditions have reached an adequate level. Naturally, this is to be seen as an understatement. Knowledge about what constitutes the central aesthetic qualities of life brings with it the ability to direct public welfare efforts directly at qualitative requirements and to satisfy them. In many cases, the most important effort may well be the revaluation - by those in power and people in general - of the assets at hand, that is the sleeping aesthetic capital. I have already implied that large savings can be made in this way.

In reality, the provincial regions make up a central arena for ecological cultural politics. This is because they own, in their natural landscape and cultural historical heritage, and also in the fact that their scale is better adapted to man, a long line of aesthetic qualities, held in high esteem by ecological cultural politics. Carrying out an aesthetic renewal and reconstruction of the many metropolises, run down both culturally and humanly, is of course also a deserving task in a short-term perspective. In a general and long-term perspective, however, this seems an almost insurmountable task, and may even be an act of supererogation.

A sensible strategy ought instead to entail creating out of the regions sound aesthetic living spaces, taking the present cultural ecological resources as a starting point and encouraging people, mainly from the younger generations, to establish themselves there. The ecological life region would, within such a project, also achieve the realization of a dream which has been growing in the inner sanctum of the author of this essay, and doubtless within that of many others: a vision of the aesthetic society.

REFERENCES:

Adler-Karlsson, G., 1981: Humanism och konsumism. En pamflett om kulturpolitiken i det snabbt föränderliga samhället (Humanism and Consumerism. A Pamphlet on Cultural Politics in the Rapidly Changing Society). Statens kulturråd, Stockholm.
Arnheim, R., 1969: Art and Visual Perception (6. ed.). Berkeley.
Bassand, M., 1990: Culture et Regions d'Europe. Presses polytechniques et universitaires romandes, Lausanne.
Bassand, M. et Hainard, F., 1985: Dynamique socio-culturelle régionale. Presses polytechniques romandes, Lausanne.
Berlyne, D.E., 1971: Aesthetics and Psychobiology. New York.
Brain, R., 1980: Art and Society in Africa. Longman Group, London.
Brothwell, D. (ed.), 1976: Beyond Aesthetics. Investigations into the Nature of Visual Art. London.
Duelund, P. (ed.), 1992: Kulturens spaendetrøje. Om kultur og kulturforskning i Norden (The Strait-Jacket of Culture. On Culture and Cultural Research in the Nordic Countries). Nordrefo 1992:1. Akademisk forlag, København.
Duelund, P., 1995: Den danske kulturmodel. En idépolitisk redegørelse (The Danish Model of Culture). Forlaget Klim, Århus.
Gerritsen, F., 1988: Evolution in Color. Schiffer Publishing, West Chester, Penns.
Gustafsson, K.E. (ed.), 1990: Lokal mediestruktur och lokal medieutveckling (Local Media Structure and Local Media Development). Handelshögskolan, Göteborg.
Henriksson, J., 1988: Kulturkraft. Kultur som regional utvecklingsfaktor (Cultural Power. Culture as a Regional Development Factor). Länsstyrelsen, Luleå.
Jakle, J.A., 1987: The Visual Elements of Landscape. University of Massachusetts Press, Amherst.
Kultur, politik och planering (Culture, Politics and Planning), 1988. Svenska kommunförbundet, Stockholm.
Kulturen och kommundelsnämnderna (Culture and the Municipal District Committees), 1988. Svenska kommunförbundet, Stockholm.
Kulturen som livgivare (Culture as Vitalizer), 1987. Statens råd för byggnadsforskning, Stockholm.
Kulturpolitik i praktiken (Cultural Policy in Practice) (7. ed.), 1991. Statens kulturråd, Stockholm.
Lenars, C., 1980: Ritual and Seduction: The Human Body as Art. London.
Levin, K., 1988: Beyond Modernism. Harper and Row, New York.
Lindeborg, L., 1991: Kulturen som lokaliseringsfaktor (Culture as a Factor of Localization), Industriförbundet, Stockholm.
Lynton, N., 1989: The Story of Modern Art. Phaidon Press, Oxford.
Länsvisa program för bevarande av odlingslandskapets natur- och kulturvärden (County Programs for Conserving Natural and Cultural Values of the Cultivated Landscape). Riksantikvarieämbetet, Underrättelser 1991:l. Stockholm.
Monaco, J., 1978: Media Culture. Delta Book, New York.
Myerscough, J. 1990: National Cultural Policy in Sweden. Report of a European Group of Experts. Stockholm.
Nilsson, S. 1981: Vägen till kulturpolitiken (The Road to Cultural Politics). Lund.
Nørregaard Frandsen, J., Andersen, M., Kayser Nielsen, N. (eds.), 1990: Naer og Fjern. TV-Fyn som regionalt medie (Near and Far. Fyn-TV as a Regional Medium). Universitetsforlaget, Odense.

Offensiv regional kulturpolitik (Offensive Regional Cultural Politics). 1991, Landstingsförbundet, Stockholm.

Proposition 1974:28. Den statliga kulturpolitiken I (National Cultural Politics I). Stockholm.

Rees, A.L. and Borzello, F., 1986: The New Art History. London.

Ross, M., 1982: The Development of Aesthetic Experience. Oxford.

Skogerby, E. og Lundby, K., 1988: Medieutvikling, kultur og regional identitet (Media Development, Culture and Regional Identity). Institutt for medier og kommunikation, Univers. Oslo.

Skot-Hansen, D. 1984: Kulturpolitik og folkekultur (Cultural Politics and Popular Culture). Akademisk forlag, København.

Skot-Hansen, D., og Duelund, P. 1990: Lokalkultur, livsformer og ny informationsteknologi (Local Culture, Life Forms and Information Technology). Danmarks Biblioteksskole, København.

SOU 1972:66. Ny kulturpolitik. Nuläge och förslag (A New Cultural Politics. Present State and Proposals). Kulturrådet, Stockholm.

SOU 1995:84. Kulturpolitikens inriktning (The Direction of Cultural Politics). Kulturdepartementet, Stockholm.

Spilling, O.R. og van der Ros, J., 1988: Kultur og regional utvikling (Culture and Regional Development). Östlandsforskning, Lillehammer.

Swedish State Cultural Policy. A National Report, 1990. Stockholm.

Sydhoff, B. och Nilsson, S. (eds.), 1992: Folkkonsten - All tradition är förändring (Popular Art - All Tradition is Change). Carlsson Bokförlag, Stockholm.

Tafuri, M., Dal Co, F., 1986: Modern Architecture 1-2. Rizzoli Int., New York.

The Futures of Culture. 1994, UNESCO Publishing, Paris.

Mobilizing Regional Culture

The Case of Västernorrland

As the introductory chapter makes clear, the essays in this book are contributions to the R&D project "The Ecological Life-Region", which has been conducted by the Umeå Futurology Group together with agents in the County of Västernorrland. The purpose of the project was to outline and test ideas for endogenous, i.e. self-reliance oriented, regional development. Many crisis-ridden regions in the world and on the periphery of Europe today are in great need of future models based on regional resources (both material and spiritual) and safeguarding the qualities and competence of the home region. These essays were originally presented as papers at three symposia in Härnösand and Örnsköldsvik, both of which are leading municipalities in the county.

Conducting a research project into endogenous development without consulting the people truly affected, i.e. the regional population groups, however, is making bricks without straw. The Umeå Group realized from the very outset that the project would have to proceed in dialogue form. For this reason, activities were concentrated within thematic conferences which were offered to the municipalities in the county and required their active participation. The conferences were also addressed primarily to a regional audience of local habitants and experts.

But the research group was keen to establish the dialogue in still greater depth. Therefore, prior to the Härnösand symposium, which dealt with territorial planning, a regional planning study was carried out for the county, taking as its starting point the socio-ecological theories of the Umeå Group.

At the regional cultural symposium in Örnsköldsvik, a different part was chosen. In order to concretize some of the project ideas about cultural development, but also to elicit a critical reaction, a regional reference group was set up, consisting of cultural actors from various fields (about 30 persons altogether). Prior to the symposium, a series of seminars were held, hosted and co-ordinated by the Örnsköldsvik cultural affairs office.

At the seminars the ideas launched by the project were vigorously discussed and confronted with regional experience. The talks resulted in a number of local study groups being formed to scrutinize the new perspectives and to try to make suggestions for a mobilization of regional culture in the county. The intention of the working groups was to take part in a special session at the symposium and to compile a written summary to be published as part of the symposium proceedings.

The following themes were chosen and formed the basis of the group discussions:

- Local cultural mobilization in a civic perspective.
- Global cultural activation, with special reference to professional agents.
- The regional natural and cultivated environment, urban settlement included.

The contributions by the three study groups are presented in the following chapters in their original state, i.e. with no editing or criticism on the part of the project team. The group work was started with support from the Umeå project, but then took shape in accordance with each group's interpretation of its remit. The authors themselves are primarily responsible for the ideas expressed in the memoranda.

Concerning one specific topic, however, the ideas of the Umeå project have affected in a more decisive way the proposition of cultural mobilization presented below. This is the continuous marketing of the museum as an arena for the renewal efforts. This fact is not only a consequence of the project group having an attachment to the new department of museology at Umeå University. The main reason is the potential for socio-ecological renewal within society which the modern museum seems to contain. This potential lies in the museum's latent ability to contribute to a general conscious-raising concerning the role of aesthetic sensory culture in social existence.

I use the words "potential" and "ability" since museums have as yet hardly used or brought into operation these latent possibilities. The contemporary museum has, in its public role, distinguished itself as one information medium amongst others in the expanding knowledge society. It sees its task as presenting in a critical way pieces of knowledge culled from history and the flood of ideas which pour forth in our post-modern society. The employee of the museum regards himself as a mediator whose duty it is to present complicated messages to the public in a well formulated manner, and thereby to enrich them with knowledge or culture.

A discussion is taking place within museum circles about the best way in which to communicate museological information to visitors. The visualization of the message is to the fore, considering the spatial frame of reference that the museum offers. Written information is however also of great importance,

"Window Vision". Photo by Hans Marklund

both in exhibitions and catalogues. The interplay between text and image and between intellectual information and visual expression, are central issues in the discussion about communication techniques within museology. Sooner or later the discussion will also come to deal with the intricate question of the role of the aesthetic means of expression. This happens particularly when the museum's scenographic possibilities are brought to the fore. The museum room invites the construction of large-scale scenic arrangements around the exhibited objects. In this instance the museum employee can, more than usual, work with colour, lighting and sound effects, and thereby create suggestive environments; more or less explicitly he takes on the role of an artist. When the discussion of method touches upon this important topic, however, his thoughts are quickly redirected to terra firma, with a reminder of the importance of clearly subordinating the aesthetic expression to the cognitive message, to the factual information itself. The magic of aesthetics contains dangers and temptations that the professional museum employee must also learn to withstand.

Modern museum employees normally appear as cultural personalities, with an ear both for the dramatic events of our time and their nuances. In the exercise of their profession, however, and not least in their ethics, they prefer the role of social technician. This is the role of the neutral communicator of dramatic information from the world scene to an audience which is to be kept informed, but whose own task it is to take sides. It is a role interpretation which the museum employee shares with many other professional groups in post-modern society. The museum thus becomes a neutral arena for displaying the events of our time, no matter how bloody and shocking they may be, and the messages conveyed, often with strong effects, come to change at the same pace as the exhibitions passing before us. The museum remains as clinically clean as before and its cultural service technically just as irreproachable. It remains a disciplined cog in the abstract and spotless machinery of contemporary techno-society.

Seen from a socio-ecological point of view, this is a role interpretation, which does not accord well with the inherent possibilities within the museum concept. It can be claimed that the role leads to a deformation of the museum's basic development possibilities. The museum is at its core a spatial concept and *spatiality itself* contains creative possibilities and limitations. The museum space is a stationary and, in its activities, also a singular medium. It is located in the centre of a communal space in which citizens move about daily; it is in actual fact a part of the local life-world

The stationary museum through these attributes, also takes up a position in the proximal perceptional space which human scientists define as the sensory sphere in man's existential understanding. This is a sphere in which we interpret reality in terms of sensory perception, empathic experience and emotional context, far more than in terms of cognitive and calculated information. We perceive this life-world as an aesthetic whole, and see ourselves not only as spectators, but as active and creative participants.

All of these unique qualities of the museum institution stand in sharp contrast to the cognitive, mobile, terminal-fixated ideal of the modern information society. In the utopia of the technological society the local qualities are nothing more than limitations, which make the museum appear as an ageing medium, even an atavistic relic. The fact that the museum has succeeded in claiming a niche of its own in the epoch of mass-communication technologists explain (in the doctrines of information theory) by its duty to bridge the gulf that exists between front-line technology and trailing civil society. Its provisional duty is to pass citizens, by means of gentle persuasion, over to the "real" (i.e. virtual) information society. The museum employees' commitment to their duties gives an aura of credibility to this act of illusion.

The socio-ecological movement sees the principal human and ecological dangers in this trend. It has established a contrasting vision, where the renewal of the local living environment with its aesthetic and empathic qualities is a major objective. One important reason for this commitment is that this sphere constitutes a more fundamental level of man's existential consciousness, and is therefore indispensable. *The modern museum appears to be a highly suitable platform from which to launch such a socio-ecological vision of society.*

One basic condition, however, for the realization of this vision is the active co-operation between the museum and the discipline of social (or cultural) ecology. Museum employees should make themselves aware of the present problematic situation, and above all of the epistemological basis of the museum idea. Only then will they realize that the museum, under the present paradigm, is a threatened sphere, and draw the logical conclusions from this. Two alternatives seem to be available in the long term: either to abandon the museum for other information channels, more suited to the times, or to consciously develop the inner potential of the museum, in close connection with its eco-cultural abilities. The classic thesis of the media philosopher Marshall McLuhan "the medium is the message" is a good starting point for such an effort.

The natural domain of the museum medium is the primary sensory environment in history and in the present, and a central part of its natural language is aesthetic communication. Exploring the contemporary and historical world of environmental experience is the given theme of the new aesthetic museology. A panorama of historical drama and evolutionary perspectives open up, ready to enrich museum activity. A more probing socio-ecological criticism contributes to locating the reformed medium in a wider political context, and to revealing its importance in a project for humanist and social change.

The radical *cultural ecological* museology, for which I am propagating here, cannot be produced over night. It must grow organically from internal discussion and experimental testing of new domains, but with its starting point in an established tradition. A reformist, rather than a revolutionary strategy is the surest route to success. If we introduce an indigenous cultural ecological profile in the museum, three natural levels of application present themselves.

Firstly, deeper medial insight can lead to a strengthening of competence in exhibition technique. It would open up for aesthetic means of expression which are freer and at the same time more thoughtful. It would also mean that the museums could work, even more than today, using an holistic approach, where sensory experience is mixed with cognitive information.

Secondly, a deeper medial consciousness means that themes for exhibitions can be chosen with greater consideration to the inherent possibilities of the museum. Themes which are better suited to other forms of presentation because of their cognitive complication can be filtered out. Focus can be placed on the primary sensory environment of life and society, and this can be done with an emphasis that penetrates the patina of historicism that sometimes obscures reality in the museum environment.

Thirdly, the museum can gain the courage to formulate an ideological profile, which will afford it new opportunities to make a personal contribution to the debate about society's future. I imagine that this contribution would first and foremost be on behalf of the humanitarian, social, cultural and aesthetic perspectives in the local society. The efforts would be made in close collaboration with contemporary voluntary forces for peace keeping, democratic ideals, ecological coexistence and respect for the humanist life values. The radical museum would act on the front line of a shift in paradigms, where the dystopia a repressive technologism would finally be absolved by the utopia of an organic rationality. The new museum could become a microcosm for the development of ideas concerning all these new possibilities, and perhaps a cult site for schooling in the new empathic consciousness.

This may have been a long digression, causing the reader to forget the main purpose of this passage: the introduction of a series of cultural mobilization studies which took place in co-operation with local groups in the county of Västernorrland. Let me conclude by returning to this theme with a piece of information. A short session was set aside at the Umeå project's cultural conference in Örnsköldsvik, for the contributions of the local study groups. A captivating discussion followed between the members of the public that had gathered there, and many local voices were heard.

The session at which the mobilization ideas were scrutinized began with a briefing on the cultural policy of the county and the history of its cultural development. The speaker was Harry Ördell, Cultural Affairs Officer at the Västernorrland County Council. His contribution forms an introduction to the section.

A Cultural Retrospective of Västernorrland

Harry Ördell

INTRODUCTION

It has fallen to me to present a status report on Västernorrland and its cultural heritage from a historical perspective. My report does not presume to be scientific or comprehensive. However, with almost thirty years experience of living in the county, and my continuous contact with cultural workers, politicans, bureaucrats and the consumers of culture in various forms, I believe that I have special knowledge and an overview of the subject that could be worth sharing.

THE COUNTY OF VÄSTERNORRLAND

Västernorrland is in the middle of the kingdom of Sweden, seen from north to south. The county is of average size, being 22 000 square kilometers in total. The population is a little over 260 000, which represents a density of 12 inhabitants per square kilometer. In other words, the county is sparsely populated even measured by Swedish standards.

With minor adjustments Västernorrland has had the same borders since the land reforms of 1810. In name the county is much older, dating back to a time when all Swedish territory to the West of the Gulf of Bothnia was known as Västernorrland. That was also the name given, in the administrative reforms of 1634, to the county designated to administer the north of the country.

EARLY HISTORY

The ice cap that once covered the area that later was to become Västernorrland, pressed the land down 200 metres. About 7300 B.C. the ice nearest the coast began to melt, and 700 years later the region was completely ice free. As the ice retreated the land rose, initially at the rate of 10 to 15 centimetres a year. Even today the land rise continues, but now at the rate of only 8 millimetres a year.

The irregularly undulating topography, with its deeply incised valleys, which typifies the county today, was formed before the onset of the ice age. The landscape underwent changes due to the action of the ice, including the formation of boulder-ridges and marshlands. The ice deposited different sorts of sediments which were to have a decisive effect on cultivation, which began around the time of birth of Christ.

Thus we live in an area that was partially formed by the ice cap. We can "thank" the ice for many of its features - such as the unusually difficult conditions we encounter when building houses and roads.

The ice left ridges of boulders in its wake, and it was in one such ridge that the people of Ragunda began to dig. As a consequence, Ragunda lake broke through the ridge on the 6th of June 1796 - giving the Indal river a new course and with catastrophic results for the farmers downstream. This year we start to celebrate the 200-year anniversary of these events, and on Sunday last, the county theatre gave premiere of "Vild-Hussen". So one cannot deny that the ice cap is a part of our culture heritage.

AFTER THE ICE AGE

After the ice age the climate was warm and dry. About 6000 B.C. the temperature rose even higher, making the climate damper and Atlantic in character. This situation lasted for about 3000 years and was, according to the archaeologists, the warmest period in the history of the north of Sweden. The area was covered by deciduous hard-wood trees with some birch and pine.

However, the climate once more became drier and the winters colder. The deciduous woods retreated and the fir trees invaded from the north, gradually getting the upper hand. Since about 600 B.C we have had the same basic climate as we have today, that is a relatively cool and damp climate.

THE COUNTY IS POPULATED

The first people are believed to have arrived here around 6000 B.C. They mainly lived on hunting and fishing. It was not until more recent times that a permanent population with an agricultural base was established in the area. The settlements followed the river valleys further and further into the inland regions, eventually pushing out the nomadic hunting and fishing Sami peoples completely.

Mixed farming was to be the main economic activity in the county. As late as the mid-1800s almost 90 % of the population was involved in farming.

INDUSTRY IS DEVELOPED

Utilizing water-power and the rich natural forests a small scale timber industry was established at the beginning of the 16th century. In the 17th century iron production was started. This may seem strange considering that the area lacked iron ore. However, due to the easy availability of charcoal and the transportation of coal from the mines of central Sweden, an iron industry was sustained in the county until the end of the 1800s.

It was not until the mid-18th century, when the first commercial saw-mill opened, that we had a timber industry to speak of. The industry really took off about 100 years later with the arrival of steam-powered saws and an increased demand for timber.

The importance of the timber industry for the region can perhaps best be seen in the fact that Västernorrland's saw-mills produced almost 30% of the country's lumber at the turn of the century. At one time the Sundsvall area was the biggest lumber area in the world. But nothing lasts for ever and by the 1930's the death of the saw-mills had reached this area. By the middle of the century the export of lumber had fallen to 50% of the 1900 levels.

After a method of producing paper from wood was invented, paper production entered the picture. The first wood-grinding mill in the county opened in 1870. At the turn of the century the first chemical pulp-mills came on line, further strengthening the county's industrial standing.

THE POPULATION INCREASES

Västernorrland has never been densely populated and at the start of the 1800's had a mere 60 000 inhabitans. By 1850 the population had incresed to nearly 100 000 and 40 years later had reached 200 000.

The population increase was unevenly distributed throughout the county. It was primarily concentrated in milling districts. The increase was greatest in the Sundsvall area where the population rose from 6000 to 32 000 during the latter half of the 19th century - making Sundsvall the biggest industrial region in the country. A similar population explosion took place in several places in Sweden, for example in Ådalen and the Örnsköldsvik area.

The large increases in population were connected to improved public health, unbroken periods of peace and an extensive relocation of people from other parts of the country - mostly from central Sweden, neighbouring counties and from Finland. When prices on the world markets fell, the population increase fell too. Had the population expansion continued at the same rate until the present day, we would now have a population almost twice the present size.

A VÄSTERNORRLAND IDENTITY

The late 1800's was a time when the county was rather rapidly and thoroughly converted from a farming to an industrial society. It was Sweden's most industrialized county and the one most directed towards exports. As a result, the economic and social life of the county was gradually altered, and the foundations were laid for much of what we feel gives Västernorrland its special character.

In the decades around the turn of the century there was a series of dramatic events which sent ripples far beyond the county boundaries; for example, the Sundsvall strike of 1879, the hunger-march of 1917 and the Ådalen riots of 1931. These three events have all provided material for artistic interpretations in literature, theatre, painting and film. They have directed the public's interest towards both the history of the labour movement and the social realities of Västernorrland at that time.

These events have doubtless contributed to putting Västernorrland on the map, thereby giving the county an identity. Even if it was only a matter of isolated incidents, which in one way or another were connected in that particular working-situation at the turn of the century, it is this kind of event that gets into the mass-media and brings publicity. One can only wonder what television coverage would have made of the incident at Lunde, where five unarmed men were killed by army bullets - if it had happened this year.

INTEREST IN NATURE

However, it is hardly the big dramatic events that affect our daily lives. They belong with anniversary celebrations and public speeches. For the people of the county life goes on despite these happenings, and little time is spent in reflecting over whether one is a Västernorrlander or not. Västernorrland is otherwise a rather anonymous county - for instance the county's tourist board no longer markets the county of Västernorrland as such but as Medelpad-Ångermanland - the two regions that constitute the county proper.

Regardless of whether one comes from Ångermanland or Medelpad, one is still a Västernorrlander - and one can ask oneself the question, whether there is something that unites the people who live here? I have posed that question to a number of people and in most cases the answer has been "hunting and fishing". Perhaps we can broaden the answer to encompass outdoor activities in general.

Admittedly this interest is shared by the majority of Swedes, but despite that I still feel there is something that is particular about the Västernorrlander's interest. Perhaps it is because of the proximity to waters and forests, wherever one happens to be, or the inheritance from the farming era when fishing and hunting were a necessary part of survival. Or maybe it is because it was a way for the industrial workers to stretch out their earnings - and have a meaningful free-time activity at the same time.

The consequences of the present day pollution of the environment, whether it comes from our own industries or from our Baltic neighbours, have been hard to take for the inhabitants of Västernorrland. Not least is this true of the Tjernobyl disaster.

THE HUMANITIES

An interest in the natural environment, outdoor activities and sports does not, in it self, prevent the development of an interest in humanities or in what we more narrowly can describe as education. Västernorrland has a long tradition in this area. Northern Sweden's first upper secondary school was founded here, and the Bishop of the See of Northern Sweden had his residence in the county.

Upper secondary school teachers and the ministers of the church had an effect on the cultural atmosphere in the county that is difficult to appreciate today. These educated people often made their mark on commercial life and there are several examples of Härnösand teachers who became succesful industrialists.

WORKERS' EDUCATION

The law of 1842 about parish schools can be seen as the beginning of Swedish workers' education. Initially these schools were known as farmer's schools, eventually being followed by the people's highschools, or community colleges. Workers' education, in the true sense of the phrase, came to Västernorrland with their establishment in the county at the end of the 1800s.

In the beginning the people's high schools aimed at satisfying the educational needs of the country people, but in some cases they accepted youngsters from the working classes. One such case was the Hola school in Ådalen, whose motivation was that in an area where industry and farming went hand in hand, it was only natural to admit industrial workers in to the school. It was partly students from this school who led the trades union movement at the turn of the century - making it into an important force in society.

Workers' education was thus at an early stage influential in the radicalization of the workers' movement. "Red Ådalen" became a concept which was strengthened by the events that took place in the area due to the social and labor union points of conflict - the folk high school was not left on the sidelines. Even so, Ådalen was basically solid farming heartland, and eventually it was hardly as red as the name "Red Ådalen" suggests.

The fact that the labour movement and the workers' education went hand in hand has certainly been of importance for the development of an interest in, and a consciousness of, the value of the humanities with respect to the union and the ideological struggle in Västernorrland.

Thanks to education provided by the temperance movement, the farmers' movement, the workers' movement and the Nonconformist movement, many people in the county have come into contact with vital social questions - and many kinds of collective cultural activities.

LIBRARIES

With the expansion of the schools and of workers' education came an increasing thirst for knowledge. Parallel with it came the need for books. Thus came the era of the parish library, which in 1864 found a place in the history of Västernorrland as the first cultural activity to receive County Council support.

With the rise of industry the need for education shifted to the towns, and we saw the development of the popular libraries. They were not at first equipped to answer the special requirements of the expanding workers' educational movements. The movement took great care of its own studycircles and workers' libraries, this epoch lasting until the early 1970s.

MUSEUMS

At the turn of the century the awakening interest in the county's history and the preservation of its cultural heritage bore fruit. Courses were arranged in the recording of oral history, local dialect and folk music - as well as in the preservation of historical artifacts, buildings and surroundings.

Early on there was also the idea of instigating a northern equivalent to the outdoor museums Skansen in Stockholm, and Kulturen in Lund, possible as the embryo of a northern university. These plans were abandoned after the various northern counties agreed on the organization of local-area work.

THE FINE ARTS

Music was one of the cultural activities that the workers' movement quickly made use of. Music groups were started, as were choirs and orchestras, both religious and secular in character. These had a central role to play in the life of the associations - even in the present day the county boasts extensive choir and orchestra activities with traditions going back to the hey-day of the workers' movement.

The theatre has followed the same trajectory as the musical arts, even if it has not as firm a foundation in the county as music. The performing arts have primarily been fostered by the temperance movement, and there are present-day examples of a flourishing co-operation between them.

At the moment, amateur theatre is flourishing in the county. During the summer Västernorrland is the scene of chronicles and pageants - sometimes rivalling the commercial theatres, enjoying a kind of popular support that they cannot easily achieve.

During the 1800's the literature of the region was very nationalistic in tone - the nation being the same as Norrland. The style of the county's authors has since followed the same direction as the great proletarian writers.

As far as content goes we find ourselves mostly in the 20th century forestry workers' environment and the industrial environment. With personal reminiscences from a life spent among timber, rapids and forest huts - but also from the closed-down pulp factories, co-op stores, railway stations and missions - the writers have captured something of the soul of the county and given Västernorrland a literary image.

In contrast to music, theatre, the libraries and museums, painting is not able to exhibit the same workers' movement background. Admittedly there are a few church buildings in the area where there are relatively rich wall paintings and several altar decorations of a popular character. Still there are few examples of earlier secular painting in Västernorrland, if we exclude painted furniture and grandfather clocks etc, where we really can boast about popular Västernorrland culture. As far as painting for high society was concerned, the popular painters were compelled to give way to the academically trained artists.

These artists, whether they painted sacred or secular works, arrived in the county at the behest of the clergy, the middle classes, the industrial leaders and so on. What the effect of these connections has been on the development of artistic life in Västernorrland is difficult to assess. At the same time as it restricted the local artists' chances of earning a living, one cannot ignore the possibility that it awakened people's interest in natural surroundings, and perhaps encouraged the young to undertake an artistic training.

It is easy to understand that the ever-changing scenery of Västernorrland was like a magnet for the artists who were interested in nature and tradition. There is still a lively influx of artists to the county, both new-comers and returners.

OUR OWN ERA

As we approach the end of the 20th century we can confirm that many of the processes and phenomena which typify our century, have their origins, directly or indirectly, in the last century. As an example, one can mention the rapid technical development, the improvement in living conditions and general education, the rise of democracy - but also the totalitarian states, the two world wars, civil wars, revolutions and so on. In one way or another we have all reaped the harvest of the ideals of the 1800's.

What ever our personal memories may be we can surely all agree that this century has been an exciting time to be alive in. The cultural sphere should be no exception and during this century we have experienced a dynamism that makes this period very special from an artistic point of view. Artistic expression now has an intrinsic value without precedence in history.

Surely no one could disagree that we have experienced a very lively century where ideas and news are quickly spread across the world, but where national, regional and local cultural forms have come under pressure and have had to make way for a global culture.

There is a saying "All great literature is local", which we took as our motto for the 1988 Central Nordic Literature Fair in Vasa in Finland. It is probably a truism, valid for all cultural expressions, for the simple reason that there must be a local foundation for all culture - that is, good earth in which to grow. The question is, how big does the patch of earth have to be, and how small can it be?

Perhaps we need to have an answer to that question before we decide our position on the concept of local culture. If we go out into the big wide world we discover that the local community is seen as something considerably bigger than the average Swedish county. If we use that measuring stick, Västernorrland is, as far as I can see, a local community. To give it a regional stamp perhaps we have to extend the borders as far as Central Norrland, Norrland, Scandinavia or possibly the Nordic Countries as a whole.

Be that as it may, we pursue something in our county that we call a regional cultural policy. It is based on the county as the smallest administrative unit, with responsibility delegated to the respective County Councils - the only organization at county level with the right to tax its citizens. This division does not restrict co-operation over the county boundaries, nor indeed over the national borders as is the case with Central Nordic co-operation.

THE CULTURAL-POLITICAL DECISIONS OF 1968

The notion of regional cultural-politics was born in the 1960's, and officially baptized in the 1974's government ruling on the new state cultural policy. The County Council in Västernorrland was, however, one step ahead as it had decided in 1968, as the first county administration in the country, to adopt a cultural-political programme for the region.

The first aim of the programme was to clarify the County Council's role in the area of culture and to establish rules, and types, of regional support within the cultural sphere. That the question came up should be seen in the light of the government's signals regarding new initiatives for culture, which presupposed regional responsibility in one form or another.

There are several reasons for Västernorrland being so quick to act. One reason is certainly the general need for County Council support for various cultural activities. One has to bear in mind that there were 34 local councils in the county in the 1960's compared to present 7. Furthermore, there was nothing corresponding to The Board of Culture, nor was there any local council funding set aside for cultural activities. Even the larger councils saw a possible partner in the County Council, with regard to investments in cultural infrastructure that could be seen as being of interest to the county as a whole.

Another possible explanation is the attitude towards the importance of cultural values, typical of the popular movements of the turn of the century, which were still alive, if in a weakened condition. This notion is supported by the fact that the prime movers in the setting up of the programme were firmly connected to the popular movements and workers' education. One cannot shy away from the thought that these individuals were driven by a will to see the role of the popular movements and the workers' educational organizations strengthened. Culture itself was seen to be at risk if these cultural-political questions did not come into the County Council's sphere of influence - where the organizations were well represented.

It is worth repeating that the popular movements had already lost much of their original momentum by this stage. Many of the demands they had been fighting for had been met, and with the constant increase in consumerism in the post-war era, the popular movements were not given much priority. So it is not to surprising that there was growing consciousness during the 60's of the importance of putting cultural issues on the political agenda.

THE PRESENT SITUATION

It has been almost 25 years since the cultural-political programme was adopted by the County Council. During that time there have been great changes in the cultural life of the county. This is probably seen most clearly in the expansion of the county's cultural institutions, which the County Council and the local councils concerned have taken economic responsibility for. With this, one of the main political aims has also been partially achieved - namely the decentralization of activities and decision-making in the cultural arena.

The expansion has resulted, among other things, in the number of employees in the county's cultural institutions increasing from around 20 in 1970, to around 200 this year. It is a development that would have seemed utopian when the directives for the programme were written.

The number of independent cultural workers, for example authors, members of free theatre groups, composers, artists, individual musicians etc., are harder to assess as there are no complete statistics available. My personal impression is that the building up of the county instructions has generated job opportunities for the many independent cultural workers - and that numerous new cultural workers have been provided by the county during the last twenty years. A contributory factor is the activities at the county's five People's High Schools and the aesthetic courses that they offer.

Independent cultural workers choosing to establish themselves in the big cities, or nearby, exercise a personal choice. They have to make that choice for themselves and no outside authority has any real influence over their decision. We know that there is always more than one reason for choosing your place of residence. If one can solve the family's problems regarding accommodation, work, educational prospects etc., I believe that Västernorrland is an area worth investing in. There are plenty of examples of high standard of cultural activity undertaken in the county and this is without loosing touch with what is happening outside the area.

If I may take the liberty of having a dream it would be this, that the number of active cultural workers in the county would increase. This concerns the independent cultural workers most of all. Even if there are one or two gaps waiting to be filled in the county institution I think that they are on the whole well covered. Personally, I think it is a very good result after twenty years of development, and one which I have been able to follow closely.

Local Culture in Örnsköldsvik

Local Culture Study Group:
K.E. Axenström, A. Söderström, A. Wistedt

INTRODUCING ÖRNSKÖLDSVIK

Örnsköldsvik is a small town in the centre of a municipality with 60.000 inhabitants. Most of these live close to or in the town itself.

The landscape is undulating with a high, indented coast and archipelago. The population live primarily in the valleys and round the coastal bays.

An agricultural society with elements of commerce has developed into a strong industrial society with many small enterprises.

This primarily concerns the central locality, the town, in the municipality. In the municipality there are several villages and smaller communities with their own identity, where cultural activities of various types are carried out by a large number of people, groups as well as individuals, and which can be developed further. There are village days (mini-festivals), markets, auctions and village spectacles. There is i. e. the Gideågården, a small centre for local culture, which stands as a model for other villages.

In Örnsköldsvik many people live within walking distance of the centre, their work and their schools. The children can walk to the place where their parents work. Anyone going down to the centre will probably meet several people he knows.

THE TOWN CENTRE

The centre lies at the innermost edge of a bay and the houses are mostly on the surrounding slopes. It is easy to walk down to the centre.

Storgatan - the Main Street - is a communal and well-visited thoroughfare during the day. This thoroughfare is now extended from Stora torget - the Town Square - with a pedestrian street through a service centre with the main library. It reaches down to the harbour. The ambition is to develop the inner harbour into a meeting place for companies, education and culture.

A CITIZENS CULTURE

The municipality of Örnsköldsvik between 1986 and 1990 had a committee which produced several proposals on the contribution and involvement of the citizens in various connections. Citizen groups were supposed to take care of different common tasks by delegation from the municipality. Several examples are presented in the project report.

Coming into contact with other people, to see and be seen, is one important ingredient in cultural activation. The physical arrangement of the area either is or is not conducive to this. Are there meeting places, semi-official or official, in the form of thoroughfares, local buildings, meeting rooms etc? In our climate it is important that these places are protected from the weather and have plenty of sun. Slope locations with a view and panorama are good examples.

Örnsköldsvik is not expanding today, and the need for new housing is not great. It is therefore in the existing environments that improvements and additions can be made.

The influence and responsibility of the tenants for the administration of their housing areas, combined with a rather dense concentration, are favourable for a richer local cultural life and also a foundation for "aesthetic environmental conservation", the desire to maintain the area beautiful, well cared for and varied.

Activities and interests are shared by many people. It is a question of social co-existence, responsibility for the environment, mutual well-being and the improvement of the quality of life. The municipal housing company and tenant owners' associations are working with this. In areas of one-family housing there are pressure groups. The social administration has tried to stimulate activity in service houses and other collective living units.

In the schools there is a similar co-operation between teachers and parents.

THE AESTHETIC ENVIRONMENT

Örnsköldsvik has special natural qualities, with a pleasant combination of water, rock and green areas.

We have not properly utilized these qualities. The Örnsköldsvik bay has almost entirely been lined with industries, irrespective of whether they have special requirements for being placed there or not.

The park areas maintained by the municipality are the town park in the centre and the graveyards. In addition to these there are many green strips which have been left over in what is bad town planning. Instead of using available land in the centre to build houses on, it was decided that equally large sites should be exploited in the periphery, and so land was left over which nobody took care of - which is bad economy and poor environmental conservation. The democracy group proposed, and the legal representative authorized, that the site owners should be allowed to use and manage the adjacent green areas.

As a whole these green areas are included in the town plan, but there is no programme for how to use them. However, if both citizens and politicians are indifferent to the problem, then there is no problem. There is another aspect to this complex question, though. During the last 30 years land for building one-family houses has been exploited much more carelessly than before. Parks have been built on, or become parking areas.

There seems to have been a fear of trees in the town. However, in recent years the park administration has carried out active development of parks and street environments. The street aspect is otherwise uncoordinated and unaesthetic with respect to advertising and lighting etc, something Örnsköldsvik has in common with many other towns.

PUBLIC CULTURAL CONTACT

Initially the Main Street with the Town Square and the extension to the harbour were considered as a communal thoroughfare in the town. However, this is only during the daytime.

Örnsköldsvik is one of the towns in Sweden with the greatest concentration of social and cultural associations. What one does in one's free time, is often done in one or several associations. There are more than 1.000 associations in the municipality, how many more we do not know.

Many associations leave no trace. They do not report to the Archive for Popular Movements in Härnösand - the records just disappear. If we in Örnsköldsvik wish to keep our history and identity alive we must keep public records of our associations. There is space in the Nola School in rooms which the municipal archive has recently left.

The associations should also have a Club House - a meeting place where the various special interests could meet and discuss common issues.

There is a lot of activity going on in strongly diverse areas. This activity takes place at various locations behind closed, but not locked, doors. If the associations conducted their activities more collectively and more publicly the town would become much livelier and more interesting. The activities would inspire each other, and others also. Everything cannot be accommodated along Storgatan, and neither is this desirable. A certain concentration along the thoroughfare (between the Peoples Park and the Main Square, for example) or at other places (at the inner harbour, for example) would be a good solution.

In this respect the radius of activity we have within walking distance is of importance.

Such a cultural concentration also creates a better basis for temporary, both planned and unplanned, events on the streets and squares. One important future event will take place when the town celebrates its first 100 years in 1994.

CREATING A LOCAL IDENTITY

If we wish to build up or retain a local identity a number of different components are required, e.g.:

Preserved knowledge:

Archives, Museums, Local arts-and-crafts museums, etc;

Processed knowledge:

Books, Studios, Libraries;

Distribution of knowledge:

Courses, Seminars, Debates;

Activities for children:

Playleader activities, Handicrafts, The Children's 4H Association, Pre-school work;

Schools:

Local education in the schools - in Grundsunda the school together with the "local association" has initiated a praiseworthy programme for activities. The computerized work with material from "Björnaminne" is another way of working.

Theatre:

The theatre is a good way of spreading knowledge. It is important that schoolchildren's organizations receive help to do some of this - to depict life in dramatical form.

A local culture centre:

Per Råberg has sketched the structure of a local culture centre as an attraction point for local cultural life. Some of this exists, will exist and would be possible to maintain at the inner harbour, the service centre Arken and Torggatan - Town Square.

A local culture centre according to Per Råberg:

Museion:

Exhibition hall, Orangery, Meditation rooms, Music studios, Creative workshops;

Colloqium:

Meeting and discussion hall with information and newspaper shelves, Café environment in close contact with the other activities, Turntable floor for dancing etc, Arena or small theatre, Group rooms;

Refectorium:

Market for local foodstuffs (market hall), Restaurants, Handicraft shops, Unpretentious entertainment.

ÖRNSKÖLDSVIK
Town Centre

Area of 10 minutes walking distance to the Centre.

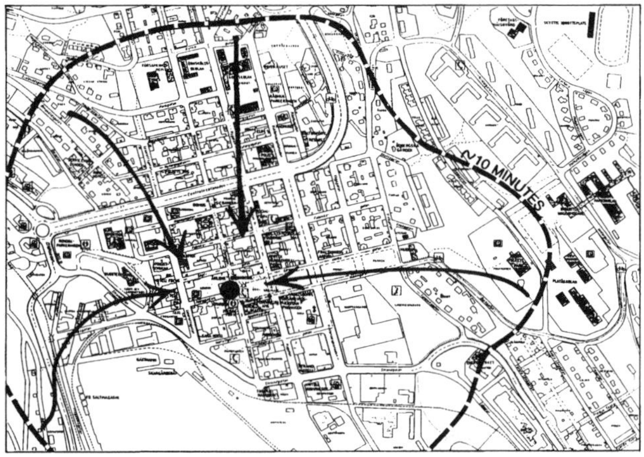

TOWN CENTRE
Existing and extended functions

1. Schools, primary and secondary
2. Storgatan, shoppingstreet
3. Extension of shoppingstreet?
4. Paradisbadet, swimming hall
5. Inner harbour
6. Arken, Conference centre
   Cultural development area?
7. Park esplanade
8. Playground
9. Orangery?
10. Sculpture park?
11. People's Park

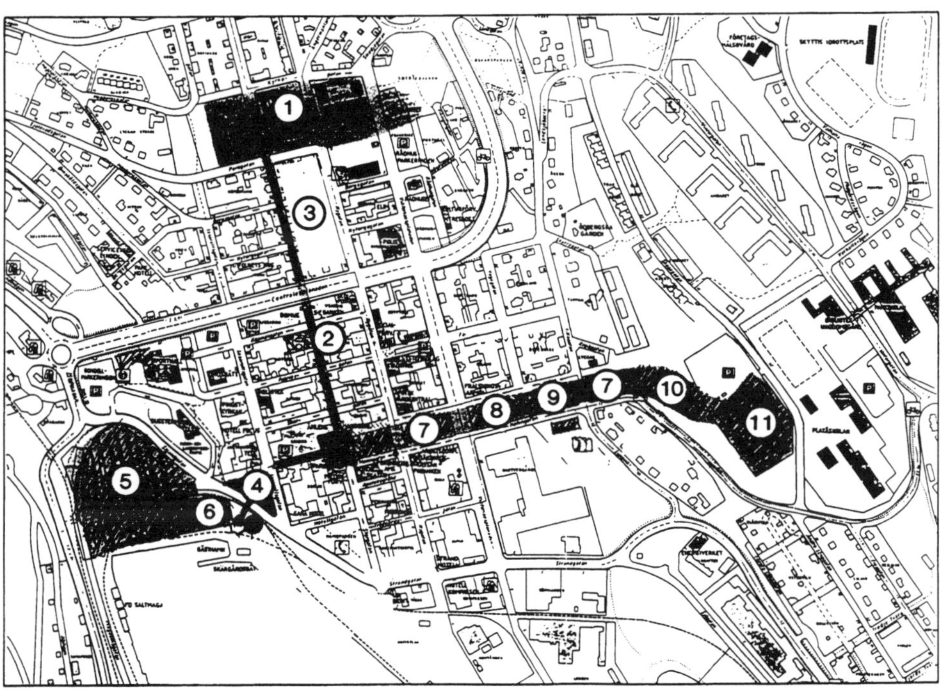

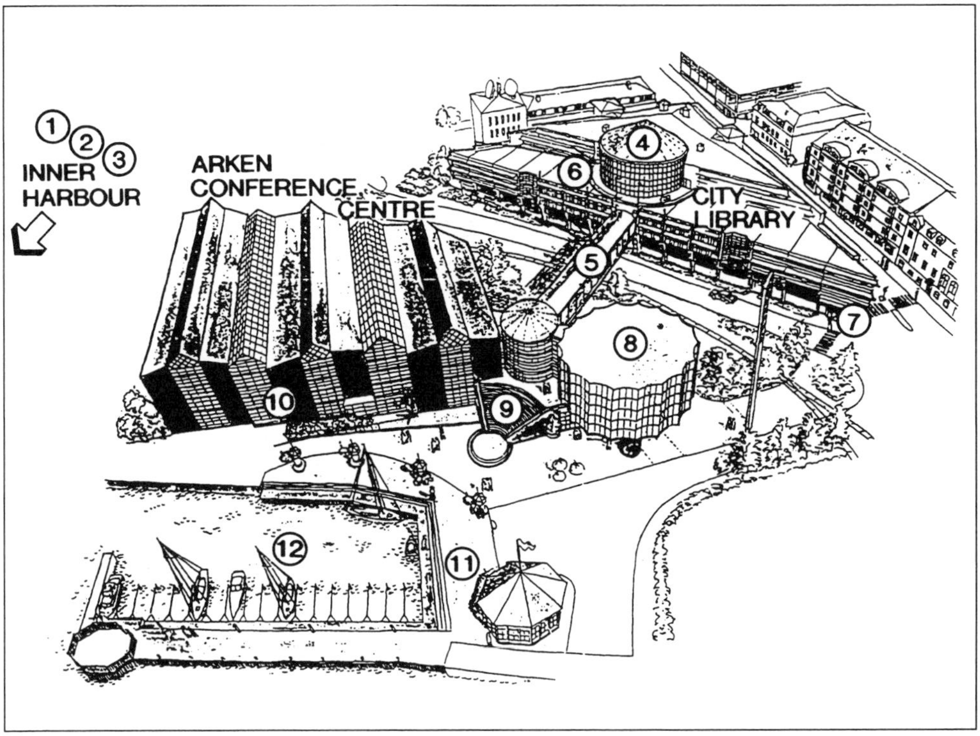

TOWN CENTRE

Inner harbour area in perspective.

1. **Music studios**
   Workshops for creative work
2. **Arena/theatre**
3. **Handicraft stalls, Restaurant**
4. **Exhibition hall**
5. **Meeting/conversation hall**
6. **Meditation rooms**
7. **Music studios**
8. **Group rooms**
9. **Open-air theatre**
10. **Pub**
11. **Café. Guest Harbour Service**
12. **Guest harbour**

Mosaic

An Ethnical Culture Forum in Örnsköldsvik

IDEA AND PROPOSAL

We live in a multicultural society. This is a popular slogan which often hides a reality where regions are opposing one another and ethnical minorities are forced into isolation in narrow quarters. It is now high time to complement benevolent rhetoric of coexistence with practical social alternatives.

We ought to start developing methods of ethnical collaboration which will make a multicultural dialogue possible in municipal practice, and which at the same time support the individuality of different groups. This task is especially important in many minor communities, where the levelling pressure at times is strongly experienced. We should also take into consideration that the daily confrontation between local regional culture and the new contributors provides a strong incitement towards a pedagogy for global coexistence.

The confrontation with foreign ethnical culture may not only help to improve our global consciousness, it may also contribute to a revival of the domestic culture and of local creativity. Old regional traditions may hereby be vitalized and given a renewed content. Crossing impulses from different areas and joint initiatives may result in completely new aesthetic events taking place in the popular culture of the region.

The proposal for an *Ethnical Culture Forum* in Örnsköldsvik here presented is an effort to form a free zone for creative meetings between local native culture and new ethnic groups. The Culture Forum of course should also be an open arena for possible original cultures located in the region.

The proposal takes especially well care of the aesthetic culture, and this is no coincidence. In the actual field a creative deficiency sometimes exists in the communities of the highly industrialized countries. On the other hand, many of the new minority groups possess a remarkable talent within the aesthetical sector. When this is activated, it may bring about a powerful increase in the overall quality of life in the recipient region.

PURPOSE OF AN ETHNICAL CULTURE FORUM

1. To create a free zone for ethnical culture, where native and new ethnic groups can discuss and practice genuine cultural activities and renew them.

2. To enforce aesthetic creativity within all areas of ethnic cultural life, such as music, poetry, dancing, art and crafts, but also lifestyle and culinary tradition.

3. To build a studio environment for ethnic creativity and collaboration. The Forum should include workshops for ethnic groups, and also venues for the groups' joint issues. Spaces for exhibitions and public functions should be provided. Forum is a meeting place for all local residents of the municipality.

4. To act for a central urban position of the Forum, by which means the importance of local culture in maintaining a dynamic life structure in the regional community is emphasized.

5. To carry out external activities, such as:
- Presentation of ethnical art, music, theatre and similar arrangements in Forum's premises.
- Catering and entertainment at private parties and public events in the area.
- Cultural activities in different places in the community and surrounding region.
- Official ornamentation assignments in the region.
- Co-arranging of regional and local festivals.
- Lectures and study circles on ethnical culture and local culture creation.
- Information and contact exchange with ethnical centres on a regional and global plane.

MOSAIC
Ethnical culture forum in Örnsköldsvik.
Design by architect Arne Wistedt

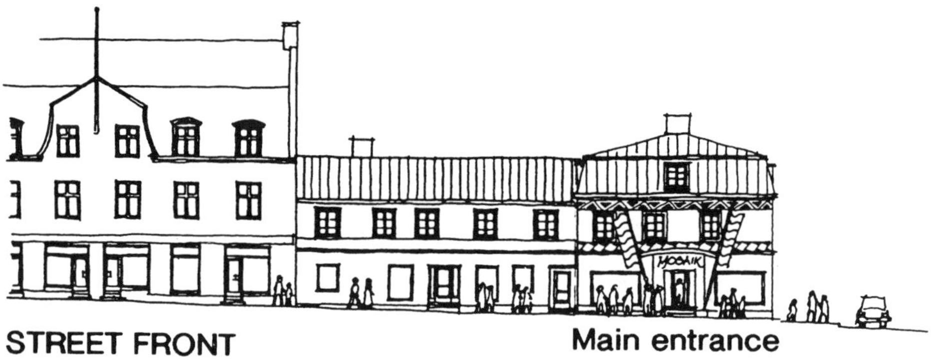

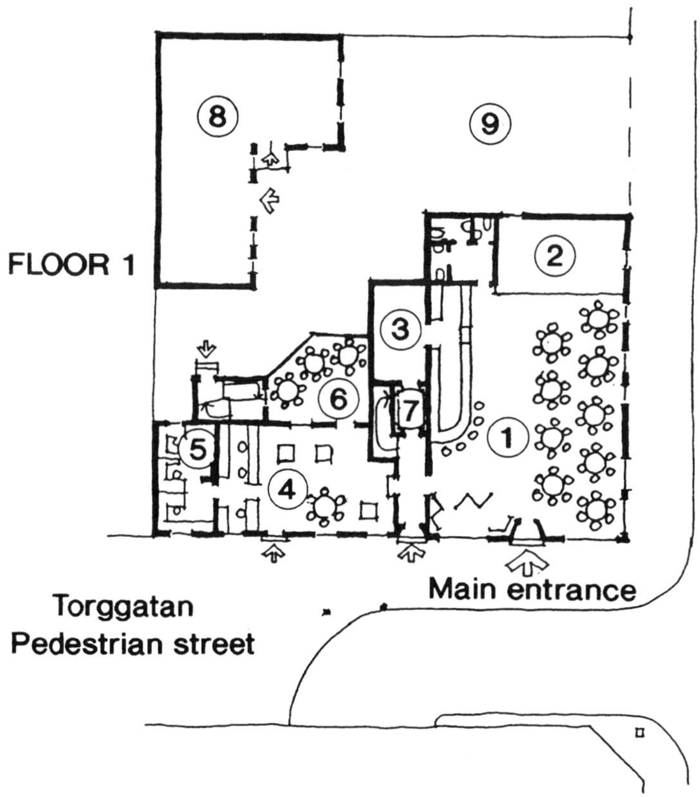

1. **Gallery and café**
2. **Stage**
3. **Kitchen**
4. **Inform-centre**
5. **Administration**
6. **Terrace**
7. **Elevator**
8. **Group workshops Store room**
9. **Yard**

City library

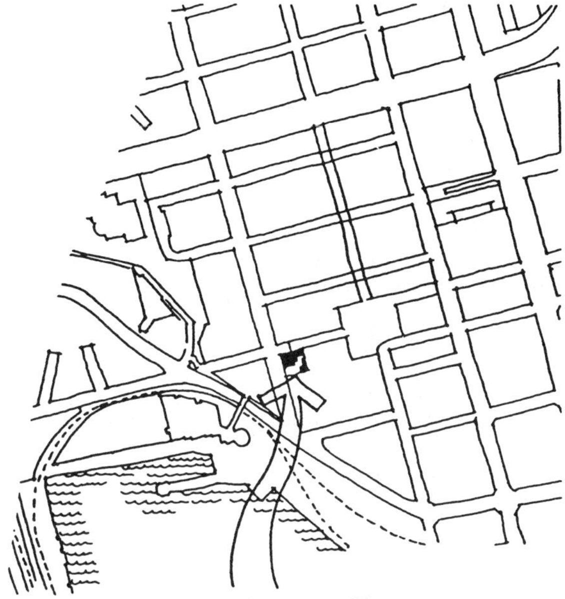

Ethnical Culture Forum
Site plan

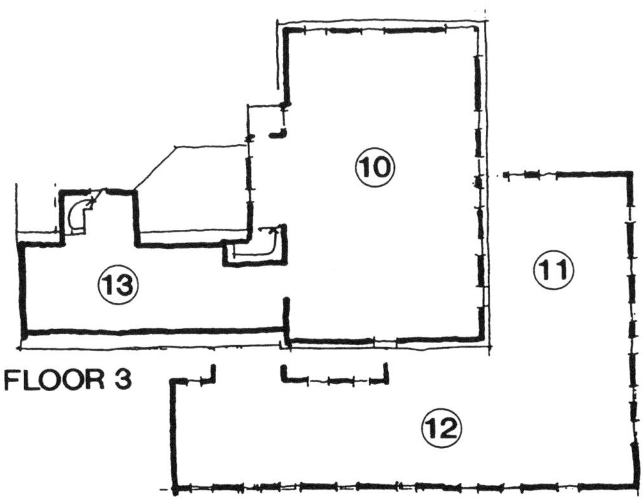

FLOOR 2

10. Group workshops Hygiene room
11. Meeting room
12. Administration Office service
13. Attic

The Role of Global Culture in Västernorrland

Global Culture Study Group:
Barbro Björk

DEFINING THE GLOBAL CULTURE

An expression like *global culture* can have a variety of meanings. On the one hand there are global traditional cultures, for example, as regards textile patterns, wood-carving traditions, folk dance and folk music. Many of these cultural manifestations can be more or less understood and appreciated the world over, as evidenced for example by the wave of so-called world music which has been engulfing the western world for a few years now.

On the other hand, there are global, advanced and sophisticated poetry and philosophy, avant-garde works of art in painting, dance, music, film, drama and so on. These latter cultural manifestations are characterized by their requirement of specially developed skills, not only on the part of their producers and performers but also among cultural consumers. The individual is a recipient of advanced high culture, whereas traditional cultural manifestations have a strong capacity for forging links between groups, peoples and nations. The manifestations of high culture can also have effects of this kind, but then the cultural processes have to begin on the individual plane.

In order for the cultural life of a region to be capable of developing so that new human and artistic experiences occur, so that frames of reference are expanded and art forms developed, the global culture must also find room in the local community. In order for the global culture to obtain that room, there must be practical and theoretical preparedness to receive it. This preconditions include:

Productive preconditions

Artists, musicians, actors, dancers, film-makers, authors and other cultural workers must be supported in their activities with assistance concerning artistic education programmes, studios, rehearsal facilities and simple premises for cultural experiments, and by the state and local and regional authorities engaging them and purchasing their good works. International cultural exchanges must be encouraged through international agreements, twinning arrangements, tax legislation and so on.

Cultural preconditions

The local habitat must be of a kind to encourage meetings between "ordinary people" and various people in the cultural fields; for example, a certain measure of density and liberty is necessary. A cultural base must be established to provide everyday insight and proximity to various forms of artistic expression in schools, within amateur culture and through continuous encounters with professional culture. The well-stocked, open library makes a big difference to people's opportunities for cultural renewal. During recent decades, events not least in the Eastern European countries have shown that the importance of the socio-political situation for the role of culture in everyday local life is not to be underrated.

Facility preconditions

There must be good facilities available for cultural events such as theatrical performances, concerts, films, lectures, exhibitions, festivals and so on. It is not usually the case that culture can be presented, for brief periods and in exceptional conditions, on makeshift premises.

Administrative preconditions

The regional administration of different art forms must have professional managements capable of helping, openly and in a spirit of humility, to establish contact between practitioners of the front culture and residents and temporary visitors in the region. The professional practitioner must be met by a professional administrator.

Private and commercial cultural establishments

Well-stocked bookshops and music shops are important channels for global culture on its way into the local community. Previously, commission systems were applied so as to maintain a high level of services to customers, even in areas of local population density. In a northern, sparsely populated region like Västernorrland, private museums and art galleries are hardly in a position to undertake large-scale, global cultural projects. As a matter of economic necessity, these will have to exist on a minor scale or be of brief duration, e.g. in the form of festivals, summer concourses and so on.

GLOBAL CULTURE IN THE SUNDSVALL DISTRICT

The Sundsvall region is the demographic centre of the Västernorrland region and is blessed with good communications. From many points of view, Sundsvall is in a good position for presenting front cultures.

The productive preconditions are supremely available here, through the Västernorrland Theatre, the Chamber Music Ensemble, an extensive municipal programme of school music, preparatory art education programmes and the County Audiovisual Centre, which has video editing facilities. The production facilities still have certain shortcomings, but improvements are in the pipeline for music and drama rehearsal facilities and for theatre workshop facilities. Permanent premises for cultural experiments are lacking but on several occasions it has been possible to make special arrangements, e.g. for exhibitions of works by young artists. Studio grants for artists, though, are virtually non-existent.

Thanks to multiple screening arrangements (with five or six films being shown simultaneously in one cinema), Sundsvall's cinemas have been able to maintain a relatively high programme standard. The films, film study material included, which are screened in Sundsvall in the course of a season definitely represent a global culture. A high standard is also maintained by the town's bookshop and one of its music shops.

To appreciate the cultural density of Sundsvall, one need only look at the monthly "arts bulletins" listing public cultural fixtures. For example, there is a big selection of music performances, theatricals for children and films, as well as exhibitions put on by the Sundsvall Museum at the Culture Store House (Kulturmagasinet). Amateur activities are flourishing in the theatre, music, dance and art, and perhaps they are beginning to develop as regards literature and film too. Sundsvall has a tradition of supporting the cultural activities of clubs and societies.

Through the establishment of the Culture Store House, "Tonhallen" and "Nordichallen", and thanks to the renovation of the Town Hall and the renovation and enlargement of the Sundsvall Theatre/Concert Hall, there are ample facilities available for cultural events. Nordichallen is so new that its capacity for cultural events has not yet been put to the test, but it should prove an appropriate venue for large rock concerts. We still lack a good, large building for art exhibitions, which makes it difficult for the Sundsvall Museum to do justice to its collection of 20th century art and to put on temporary exhibitions requiring more than 250 m^2 floor space.

If, in the future, Sundsvall can move up to a "higher division" in the matter of art exhibitions, this should also have a positive impact on the prospects of the best local artists making names for themselves outside the region. There is still not a single large exhibition hall to be found in Norrland, but certain plans for building one are also afoot in Umeå. However, many of the small or medium-sized art exhibitions shown nowadays at the Sundsvall Museum or, for example, at the Härnösand and Örnsköldsvik art galleries, can definitely be said to usher in a "breeze" of global front culture. The exhibiting artists, however, seldom come from further afield than the Nordic area.

At the same time, one has to realize that a large percentage of the people visiting, say, art exhibitions in Sundsvall, Härnösand or Örnsköldsvik also go to art exhibitions in Stockholm, added to which a certain percentage travel to Berlin, Paris, New York and so on. Unfortunately, for economic reasons, cultural producers at the local level are often less mobile. Seminars, workshops, travelling fellowships and suchlike therefore mean a very great deal to local cultural workers.

In recent years, the administration of cultural amenities in Sundsvall has moved towards a growth of professionalism, e.g. as regards music and theatricals for children, which as a result have required professional parity status with the museum and library. Sundsvall's reinforced media profile, with higher education programmes for information officers and journalists, as well as the activities based in Sundsvall by Swedish Radio and Swedish Television, should be an asset to cultural life. In Västernorrland, though, institutions like the theatre, museum or art gallery have to be something of a general store and cannot profile themselves by specializing exclusively in local or global culture.

A REGIONAL OUTLOOK

There is of course cultural competition between the different parts of the county. Even though, in our opinion, global culture belongs mainly in Sundsvall, we also find it elsewhere, for example, in Bror Marklund's studio in Örnsköldsvik and at the Junsele Festival. As regards musical events, sculpture exhibitions, theatrical performances and so on which take place outdoors in summertime, more care should be taken to find attractive venues which are distinctive and characteristic of the county, and to utilize encounters between nature and culture.

Traces of this kind can be instanced by the Villa Merlo Park, the Vivstavarv sawmilling community and the Svartvik area. As examples of places well used, I would like to mention the amateur pageant about Styresholm in Ångermanland, the Gnistor summer theatre at Gålsjöbruk, the Ballad Festival at Skuleberget and, at Ragunda, the summer play about Wild Huss to be performed on the banks of the Indal River. As far as these cultural events are concerned, there is no fixed boundary between local och global culture and the centre of gravity can fluctuate between the two directions, both of them equally justifiable.

Festivals and summertime theatricals are needed as points of focus and stimulus for regional culture, and as a means of engineering meetings between front cultures and local cultures. Festivals should be founded on a corresponding activity base in the region, for example a rock festival in Sundsvall with its many successful rock bands, a documentary film festival based on the film resources and media specialities of higher education and Swedish Television in Sundsvall, and so on. The abundance of musical activity has already engendered several festivals. This autumn (1992) Sundsvall's expanding children's theatre will be generating a festival of children's drama. The new organ of the Sundsvall Church is an outstanding instrument which should generate important organ recitals. And several of the historic churches in the county have an unquestionable place in the global cultural sphere, as do some of our archaeological remains. In recent decades our churches have also come to be used for an increasing number of concerts.

The local press and television have an outstandingly important bearing on regional culture. The worst thing that can happen - and, unfortunately, often does happen - is for cultural events to be passed over in silence. Public discourse concerning the various manifestations of culture must be kept alive, for the sake of the general public and cultural workers alike.

Mass media must not foment "the Stockholm complex" any further. Professional cultural workers should be entitled, also at the local level, to the attention of professional critics. We know that, for practical and personal reasons, this can often be hard to accomplish, but this is the aim which newspapers and other media should strive for.

Global culture and local culture are one another's prerequisites. The culture of events must rest on foundations of more modest cultural life flourishing, for example in schools, where pupils and culture are given a chance, in well-stocked stimulating libraries, in carefully and artistically designed housing areas and urban centres, and in communities where liberty and artistic creativity are valued highly.

Para-Museum Y

The Para-Museum is a utopian museum sketch originating in radical tendencies in civilization development. The name has been inspired by the inventive turbulance surrounding the ongoing shift of paradigm and the expectations which this awakens on the part of cultural renewal. But it also has a bearing on the vigorous expansion in the field of human consciousness towards parahuman dimensions connected with the new sensitivity of an emerging cosmic era. It is a museum model for transgressive exploration and experience between the genres and into new dimensions of reality now on the horizon. It is an observatory on the frontline of the changing world picture.

All thematic facets of the Para-Museum have a bearing on the cosmopolitan dimension. They draw our attention towards an experiential sphere which extends far beyond the home region. Still, despite the richness of content there remains one unifying theme: the search for universal experiential wholeness.

This striving for universality also directs our imagination towards an extraterrestrial cosmic sphere. In our imagination we may finally reach a distant point where experience and intellect amalgamate and where the cosmos seems to turn back into the eye of the observer. In our mind we then recover the home region as a distinct place in the cosmic totality.

The Para-Museum by means of its epistemological scope heralds a consciousness of reality, which we can name *cosmo-ecological.* A cosmo-ecological awareness is now under way in many places and triggers off changes in our experience of reality and scientifical viewpoints. The Para-Museum is congenial with this new, widely expanding, awareness. By means of its radical programme, the museum may be a strong source of power for building up a mundane regional identity beyond the stigma of provincialism.

Para-Museum Y is an application of this radical idea. It is an archetype originating in the efforts to design points of attraction which can emphasize the cosmopolitan identity of the life region. The idea has been translated into the architectural vision presented here in a congenial performance by the architect Abelardo Gonzalez. Gonzalez' drawings show a large-scale construction, articulated in an advanced international idiom and presenting itself as a conglomerate of cultural attractions. The various functions are both accentuated and united by the visual clarity of the design. There is an architectonic concordance throughout with the framing pedagogy of the museum institution.

The wide range of the Para-Museum's commitments also forms a broad contact surface towards the inhabitants of the region. This is necessary if the museum is to serve as a gathering place and as an attractive excursion spot for people in the surrounding communities. We have chosen to locate the museum on the island of Svanön at the mouth of the Ångerman river. Svanön is dramatically situated in a beautiful coastal area and is also at the geographical centre of the Västernorrland region. The nearest population centre is Kramfors municipality. The spot is easily accessible from the other towns and districts in the county.

Perhaps, on closer reflection, the idea of a Para-Museum will turn out to be just a passing fancy. Or maybe it will not.

PARA-MUSEUM Y.
Site plan

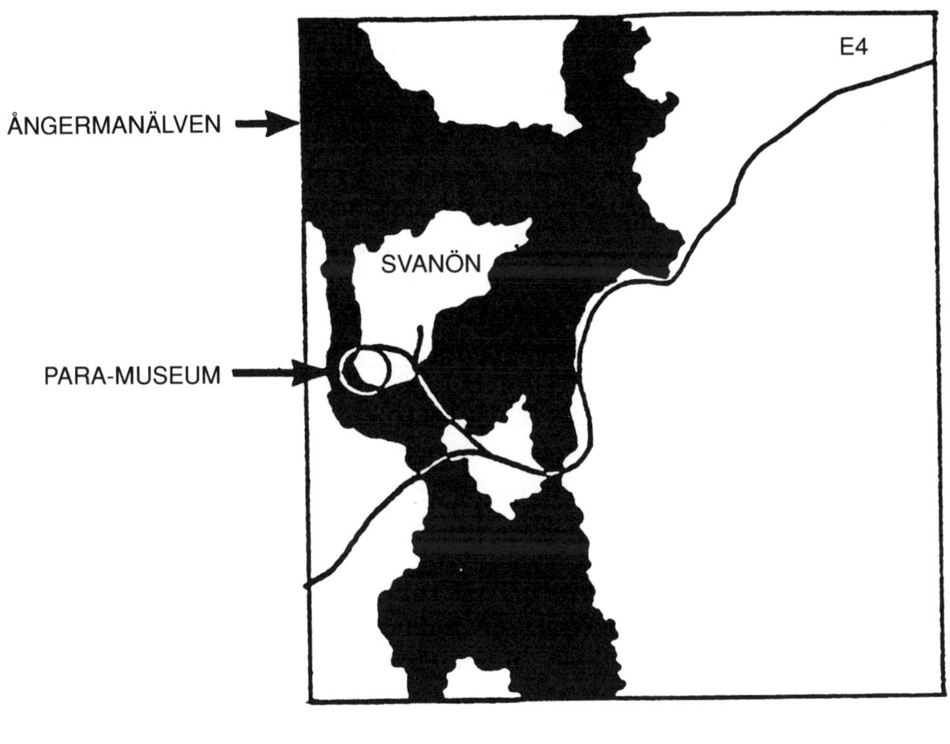

PARA-MUSEUM Y. Structural plan.
Architect Abelardo Gonzalez

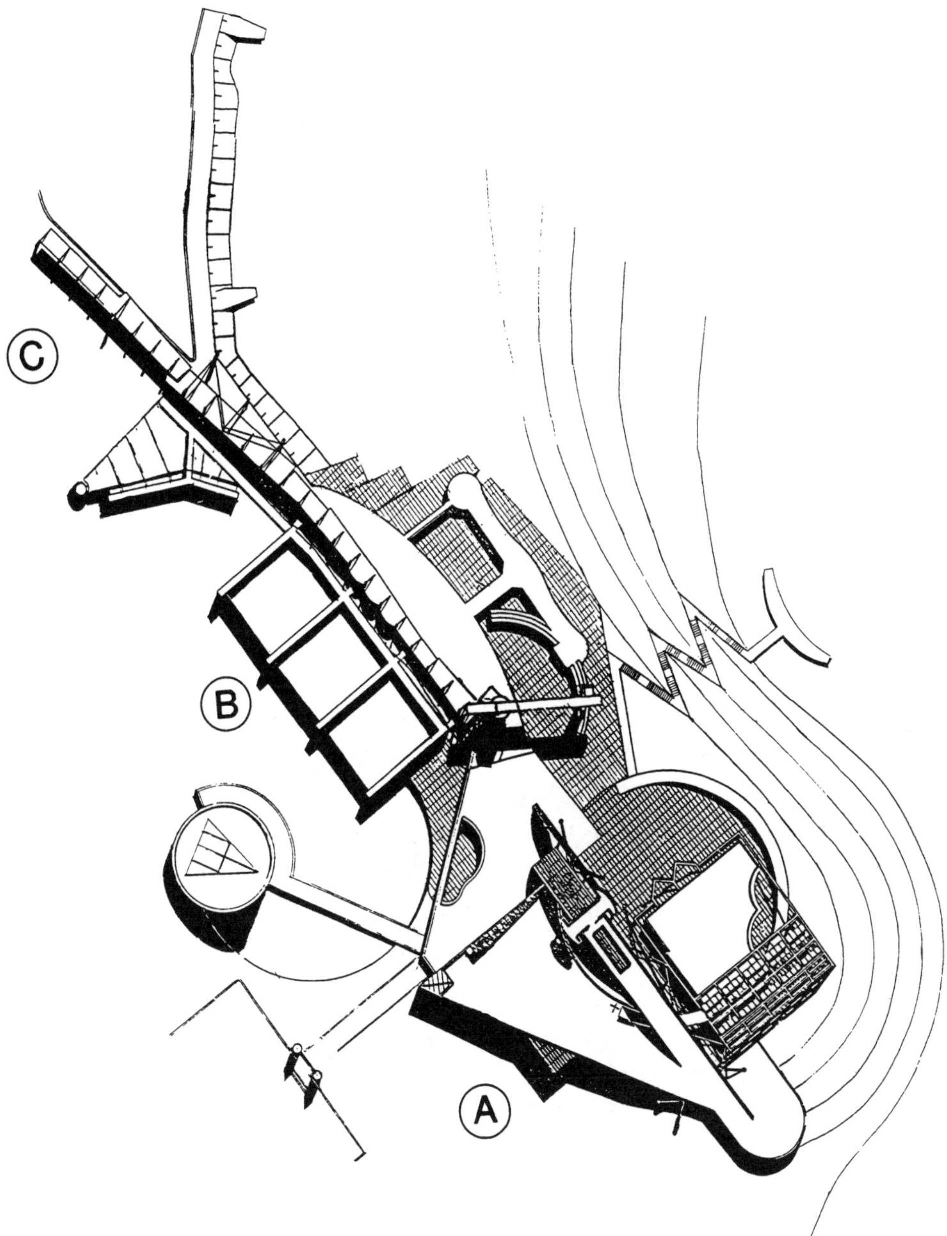

A. Museum section
B. Conference section
C. Hotel section

PARA-MUSEUM Y. Functional plan.
Architect Abelardo Gonzalez

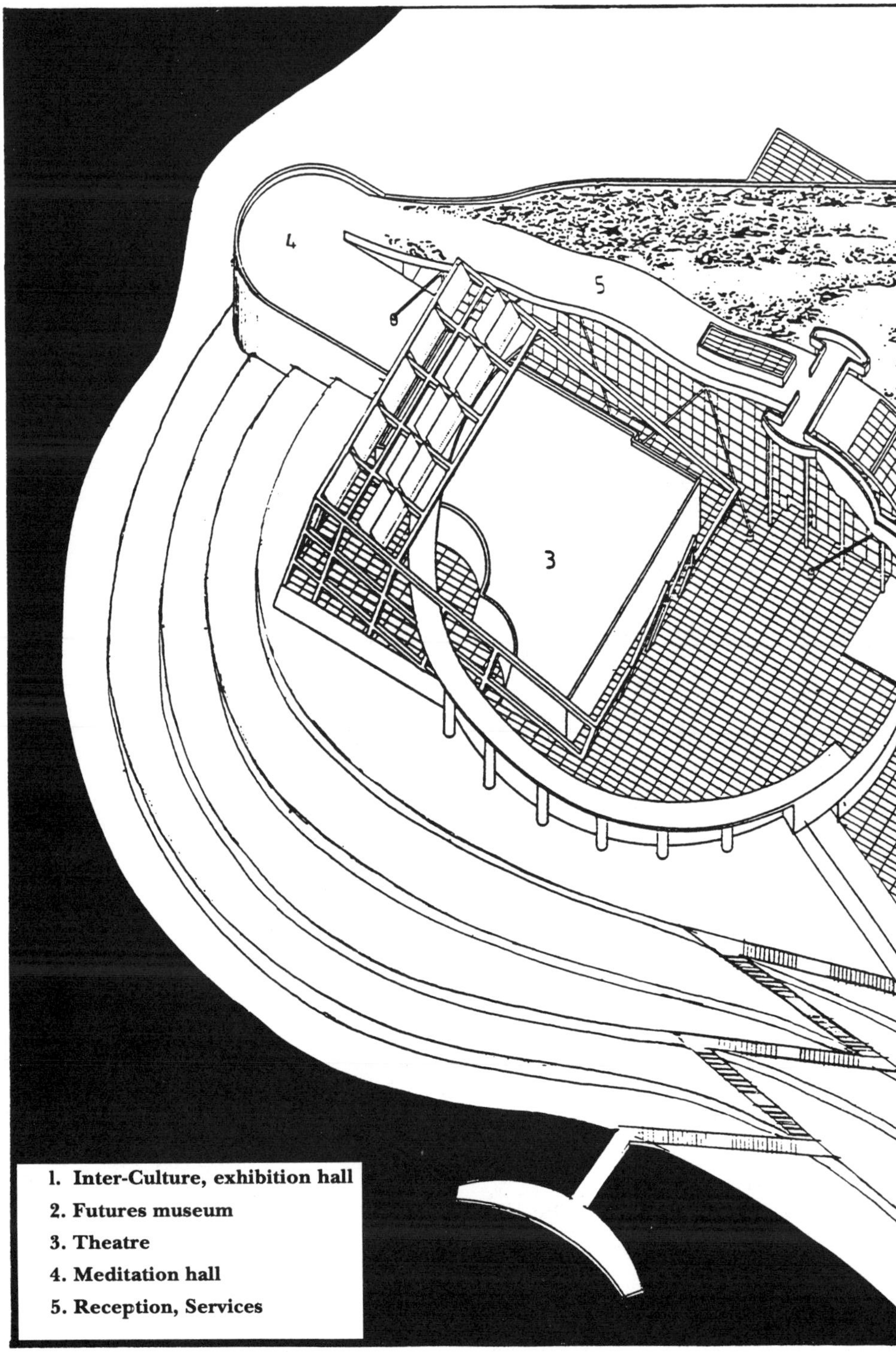

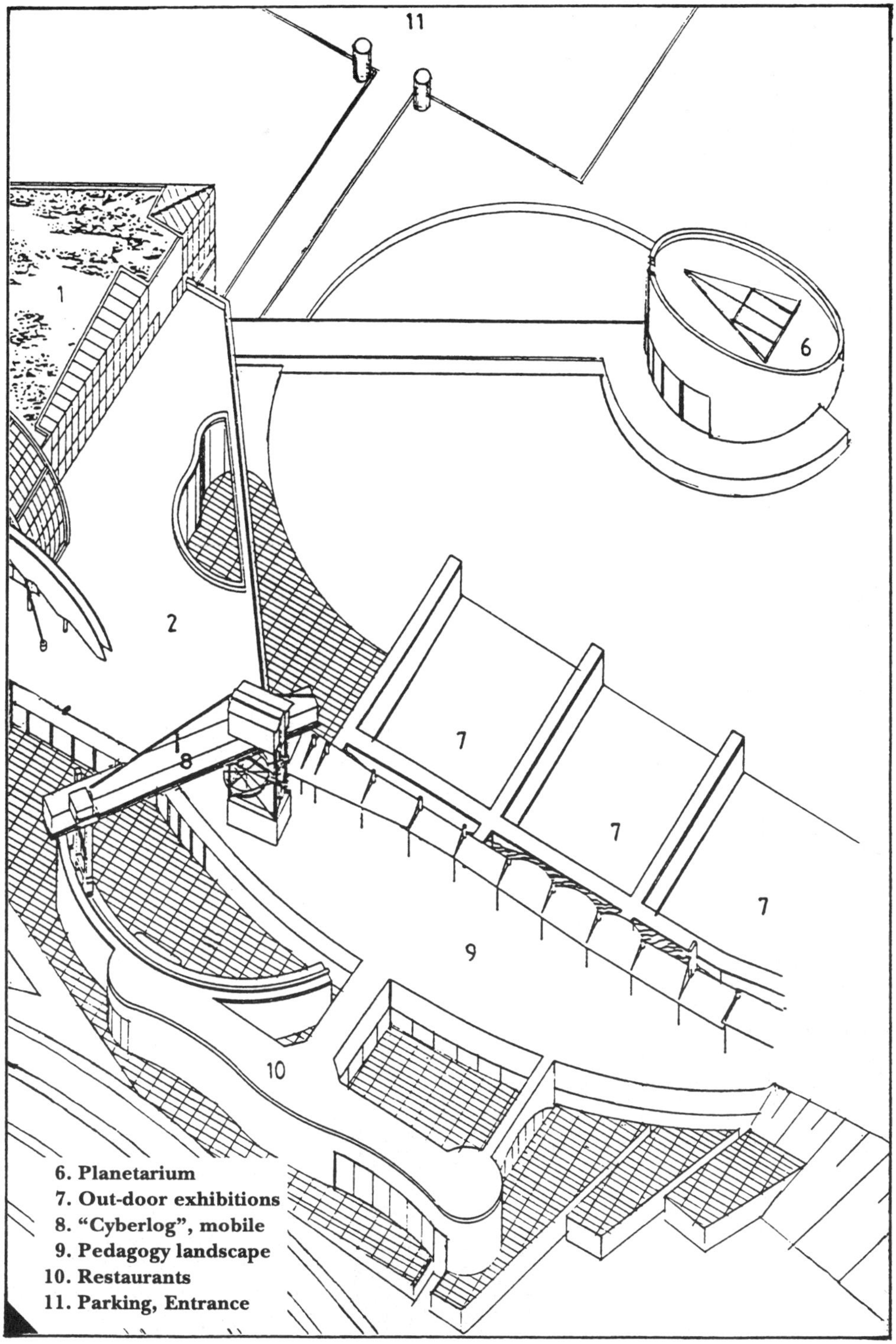

6. Planetarium
7. Out-door exhibitions
8. "Cyberlog", mobile
9. Pedagogy landscape
10. Restaurants
11. Parking, Entrance

Cultural Renewal and the Regional Landscape

Regional Culture Study Group:
Inger Lilliequist

THE CULTIVATED LANDSCAPE

The study group's task has been to undertake a review of *the regional natural and cultivated landscape* of Västernorrland and point to the aesthetic cultural values which it includes and which deserve to be highlighted.

In this connection the group has also discussed the value and advisability of incorporating *aesthetic viewpoints* in regional community planning and heritage conservation. By heightening aesthetic awareness of the cultivated and natural landscape, a new dimension can be added to environmental policies, and at the same time additional weight will be lent to arguments for heritage conservation.

Briefly, the ideas launched by the project "The Ecological Life-Region" come close to activities in Swedish heritage conservation. One highly topical issue is the closure of the historic agricultural landscape; a discussion is in progress concerning suitable strategies for safeguarding the threatened landscape values. In several counties, Västernorrland among them, cultural landscape maps have been compiled in which environmental qualities of different kinds are classified. It seems as though heritage conservation and the cultural-aesthetic approach could cross-pollinate one another in terms of ideas but also support each other with arguments. The establishment of a dialogue in the future between these two approaches is clearly an urgently needed initiative.

The following is a summary of the discussion group's improvised and preliminary arguments.

HISTORICAL DEVELOPMENT

The two provinces of Ångermanland and Medelpad did not become a county, Västernorrland, in their own right until 1762, but intermittently they already belonged together when county boundaries were established in the 17th century. Both provinces derive their main character from the deep river valleys and the indented coast line. This is a far-flung county, including both a coastal region and a forested interior.

Most human settlement has followed the main routes, the valleys and the coastal region. Here there is cultivable soil, but the lake networks further inland were already of interest for hunting and fishing during the Stone Age. Agriculture and fishing were continuously developing settlement down to historical times. The parochial structure and the building of churches gave the settlements an organized form, and interest came to focus on strong parochial centres. The cultivated environments of the countryside are rooted in this social structure, with village formations consisting of individual homesteads and with churches as significant buildings.

The landscape was extensively changed in the 19th century with an expansion of cultivated land and the opening up of the countryside into complete, continuous settlements. Villages were spread out as a result of the land distribution reform, and settlement acquired different formats and greater variety. The long strips of settlement following the edges of the fields and lakes extended further and further inland and bifurcated.

The coast and the river estuaries occupy a central position in the landscape, both ethnogeographically and economically speaking. This interest has been manifested by the strong establishment of settlements in these areas.

Ångermanland is famous for its magnificent, dramatic scenery, its many fine views and the nip river landscape of Ådalen. The interior has sweeping, grandiose contours with a backcloth of heavily forested mountains.

The Ångermanälven River flows into a long, fiord-like arm of the sea. These narrow inlets have given the coastal landscape a distinctive character, reflected by the ancient place names including "anger". This part of the region has a special place in historical tradition. The transition from coast line to heights is very rapid here. Höga Kusten (The High Coast) and the Nordingrå area squeeze mountains and valleys together into a glorious pageant.

The two great rivers of Medelpad enter the sea close together in the Sundsvall Fiord, which forms a large natural harbour with the island of Alnön as its breakwater. This creates an important centre of communications inland and further towards the western sea. The rivers drain a large part of the Norrland forests. The Ljungan valley is more open and generous in its grandeur, while the Indalsälven River is cramped, offering bolder perspectives from dizzying heights. The river landscapes have been greatly affected by man's harnessing of the water through the construction of hydropower stations. Thundering falls and rapids have been replaced by more tranquil sequences of lakes.

The first industrial activities in the county - iron manufacturing and water-powered sawmills - were also based on the assets of the forest, but they did not entail any extensive modifications of the landscape. Instead the manufactories could provide important centres for the dissemination of new cultural patterns. The first towns in the county Härnösand (1585) and Sundsvall (1621), were founded as a direct result of centrally controlled initiatives and in traditional, strategic positions for trade and administration. They were linked, respectively, to the estuary of the Ådalen River and the sheltered position of the Sundsvall Fjord in the Selånger valley. Örnsköldsvik, Sollefteå and Kramfors were founded in the late 19th and early 20th centuries and were built up round different societal patterns for communications and trade.

The urban environments in the county were established relatively late, but the coming of towns meant a new community structure which eventually developed a cultural pattern of urban life.

ERA OF INDUSTRIALISM

The dramatic transformation of the landscape came with industrialization during the second half of the 19th century. New technology - steam power - made it possible for industries to be established in the coastal region, within easy reach of seaports.

Surrounding the industrial centres, communities sprang up and entire coastal stretches became urbanized, as for example in the case of the Sundsvall district between the two river estuaries. Industrial development created entirely new types of community which facilitated new cultural patterns.

At the same time, a new system of communications was constructed across the country. The railway, cutting straight through the cultivated landscape, could unite settlements and broaden perspectives. Community expansion followed this new line as well.

Modern social development is based on the structure of the first great phase of industrialization, with expanded nuclear areas. Industrial development has changed and has created a more differentiated picture.

PROTECTING THE CULTIVATED LANDSCAPE

Social developments have created new tasks for heritage conservation. Instead of preserving individual buildings and archaeological sites, heritage conservation is today concerned with the preservation of complete environments. This calls for an integration of heritage conservation with other planning activities in our environment.

This work must take in the entire history of society, with emphasis on bringing the heritage to life. Heritage values are today an important factor in the design of the physical environment, e.g.in connection with land use changes and the formation of infrastructures.

The actual outward changes correspond to new conditions of life and residence in the region, generating a need for stronger roots in knowledge of the past and present. Before urbanization there was a natural and powerful interaction between the natural and cultivated environments. The older settlement environments, as a rule, have a strong local character which has evolved from their preconditions and traditions. Large-scale developments today often mean cultural impoverishment and the substitution of anonymity for local identity. What is more, we are moving rapidly through time/space with the aid of high-technology communications. It is important that the infrastructure should harmonize with the landscape, the scale of things and environmental values.

The composition and variety of the public space can create experiences of security, belonging and continuity, but equally well excitement and drama. The cultivated environment is constantly recounting historic events and phases of development. Great changes to the landscape can seriously damage these values and result in the loss of important relationships. Heritage conservation must work actively to bring about a number of measures for the protection of the cultivated landscape. Local and regional authorities have drawn up preservation programmes so that community planning will be able to show consideration for valuable environments.

HISTORIC AND AESTHETIC VALUES

Heritage values are important to a highly developed society. They are connected with the everyday environment in which we work and live, and they can be expressed as qualities of habitation and life. Much depends on by whom these qualities are evaluated. The results can vary, depending on whether terms are defined by politicians, officials, media or users. It is important that common objectives should be formulated for the guidance of heritage conservation and for the work which will have to be undertaken on many fronts. The growing attention which is being paid to the cultural heritage is connected with the fact of increasing numbers of people taking part in and influencing social development.

Experience can be defined as aesthetic, but it is hard to define limits for an aesthetics in multiple environments which include expressions of completely different values. In this context, heritage conservation lacks a number of inspirational concepts. The project entitled *The Ecological Life-Region,* focusing on the qualitative, aesthetic core of the regional concept of culture, may help in providing new perspectives for heritage conservation.

Although today this sector is guided by a holistic view, its work proceeds within he frames of traditional community planning, the focus of which is on the rational structuring of the physical environment. Rational community planning is not making sufficient allowance for cultural values. Design and selection of materials are governed primarily by building standards, credit regulations and the construction industry.

Insights concerning the historical and aesthetical environmental values transform our attitudes, causing us to assess them in relation to purely short-term economic profit considerations. Cultural conservation also needs an instrument of systematic valuation for its planning tasks. In the holistic perspective - a synthesis of all interests in a community or region - heritage conservation should play an important part beside the prevailing system of rational and economic planning.

In the socio-cultural perspective, cultivated environments are breeding places for the existential needs, in which respect they form the natural basis of social development. Heritage values also convey a sense of identity and coherence. With the dual living horizon of our time, activities should focus both on the local and the global horizon. Such an approach would enabale heritage conservation to perceive its relational connections more clearly and to create a better understanding of different kinds of culture.

The Umeå project stresses the aesthetic and experiential aspects as a planing factor and a creator of cultural identity. Educational work plays an important part herein. Experiences and dramatization of historic events or environments also can contribute to a better understanding of regional cultural values. Participation in the work of preservation can be made more interesting and more people may be induced to make contributions of their own. Above all, the Umeå group´s project can help to supply heritage conservation with new good arguments. Increased knowledge of the life-region concept, therefore, ought to be actively sought by the experts of heritage conservation.

REFERENCES:

Blomkvist, N: Kulturmiljö. Historien i landskapet (The Cultivated Environment. History in the Landscape), 1990.

Forskning för kulturminnesvård (Research for Heritage Conservation). Riksantikvarieämbetet 1986.

Kulturmiljövård (Cultural Environment Protection). Government Bill 1987/88:104.

Lindberg, G: Odlingslandskap och livsform (Cultivated Landscape and Life Style). Bygd och natur 1979.

Odlingslandskapet - en lång markanvändningshistoria (The Cultivated Landscape - a Long Story of Territorial Exploitation). Riksantikvarieämbetet 1991.

STRUCTURES OF THE CULTIVATED LANDSCAPE

The forest landscape

e.g. forestry
timber driving
hunting and fishing

The agricultural landscape/fisheries

e.g. villages, homesteads
forest holdings
crofts
shielings
mills
fishing hamlets
churches

Urban environments/urban settlements

e.g. the stone buildings of Sundsvall
Härnösand, a county town
Örnsköldsvik, a trading town
Kramfors - industry and communications
Sollefteå, a garrison town

Industrial environments

e.g. ironworks
sawmilling communities/sawmills
pulp mills/papermills
chemical industry

Coastal landscapes

e.g. fishing hamlets
harbours/shipping/navigation
Höga Kusten
lighthouses/pilot stations
summer homes

The river valleys

e.g. viewing points
communications/small craft

Communications

e.g. the railway
shipping/small craft
main roads
aviation

Recreation

e.g. places of amusement
sports facilities/bathing
popular movements
museums/museum environments

Eco-Museum Västernorrland

THE CULTURAL LANDSCAPE

To consider the regional landscape as a cultural and existential resource is a foreign thought for metropolitan man, who is the ideal of our time. The countryside areas in his eyes belong to a tourist culture, which he visits sporadically for recreation purposes or nostalgic reasons.

This, however, is an outlook which has only affected the inhabitants of these provincial regions to a limited extent; a healthy outdoor life is still a natural part of their lifestyle. On the other hand, this is primarily a question of traditions and habits, and the deeper existential meaning is seldom clearly explained in words. Therefore, when the regional environmental values are today threatened with pollution, forest damage and the close-down of cultivating areas, values of the urban lifestyle play a decisive part. No really convincing arguments can be mobilized to defend the cultivated landscape. As a consequence, a dismantling of indispensible culture and life values takes place in many countries these days.

An important aspect when discussing an alternative cultural policy for the regions, is the articulation of strategies capable of more effectively defending their environmental values. But we also require incentives which will entice the inhabitants to make better use of the experiential qualities of the home districts than hitherto. When taking stock of various ideas which can market the regional cultural landscape, one concept in particular stands out for its congeniality with the purpose. That is *the eco-museum idea.*

The idea of an eco-museum was launched for the first time in the 1960s in museum circles. Its birth can be described as the fruitful union of two separate trends. The proposal emanates partly from the crisis of the regional and local museums and from their search for structural reconstruction. Partly also, there was an inflow from the contemporary alternative youth movement, with its demand for popular participation and ecological consciousness. From this constellation the idea was born that the museum's historical and regional memories should be studied in situ, in their natural local context, and using

"Window Vision". Photo by Hans Marklund

the local population as caring and researching participants.

Instead of an expert-ruled museum institution the complete local community was recognized as a museum. Visitors were supposed to move between marked stations in the community, and the regional cultural landscape was seen as a pedagogical frame of reference. The description given to this radical concept was the name *eco-museum*. The prefix eco- associated both to the new global environmental consciousness and to the local history as an interplay between a population and its unique living environment.

ECO-MUSEUM PIONEERS

The idea of an eco-museum was met with approval in museum circles. The matter was discussed at several conferences held by the museum society's leading international body, ICOM, in the early 70s (1971, 1972). The first task to accomplish was to draw up the basic principles for the eco-museum and to develop a concrete work programme. The campaign soon led to the founding of several eco-museum institutions. As a pioneer example Le Creusot/ Montesceau-les-Mines in France is usually mentioned. Le Creusot was followed by other examples in France, South America and Canada. Similar projects also began to appear in Scandinavia during this period, mostly in Norway. "Eco-museum Bergslagen" is a Swedish example.

This is a development which has carried on during the 80s, and which has resulted in the eco-museum idea reaching worldwide recognition. The eco-museum has also become a central theme in the new museology movement which was started in the early eighties, with the international organization MINOM as its principal agent. During the time that has since passed, the scheme has developed widely and has specialized in several directions. Today at least four different types of eco-museums can be defined; the boundaries between them are somewhat indefinite.

1. The eco-museum as an extended central institution under expert responsibility.

2. The eco-museum as a decentralized and participating institution under direct civil rule.

3. The thematic eco-museum (e.g. a community's specific branch of industry).

4. The eco-museum as a mobilization centre for the people.

The debate on eco-museums carries on. The idea continues to fascinate and give rise to new proposals. The division in several directions should not be seen as a sign of weakness, but is instead evidence of the richness and prosperity of the scheme. It looks, in actual fact, as if a new vein for museology and the care of cultural memorials has been uncovered. The one thing still lacking, is an effort to widen the theoretical placing of the eco-museum. Argumentation up till now has often been conducted with the aid of catchwords in a limited museological debate. There is a risk that an initiative, which could be profitable for the whole of society, is hereby imprisoned in new museological routines on a level below the overall social and cultural potential of the idea.

THE AESTHETIC HERITAGE

When we now launch the idea of an *Eco-Museum Västernorrland*, the most important motive is the powerful support for a regional aesthetic culture which the eco-museum conveys. In this new light, the visitor's attention is drawn to the abundance of experiential values inherent in the localities of the regional districts. He is encouraged to pay respect to these values and regard them as aspects of a history and an evolution, where he is himself a natural link between past and future times. But the central message is found in the overtones of this regional pedagogy: the visitor learns to consider the natural and cultivated landscape as an aestethic source of life in the regional community. Maybe man's striving for a regional quality of life has its deepest sounding board in the aesthetic experiences which he meets with in the rich and varying culture of the regional landscape.

The interpretation of the eco-museum idea advanced here, partly differs from the explicit motivations given in the professional museology debate. But that does not mean that such an interpretation has never crossed the minds of museum spokesmen. Probably the silence on the issue is due to the fact that aesthetic themes are difficult to put into words and give credibility in a rationalistic era like ours. An audacious presumption is that the aesthetic motivation constitutes a major driving force behind the eco-museum idea, even with museologists.

If this is so, it may not entail any major modification of the regional cultural pedagogy. The themes of regional life are intertwined with one another and interact in mysterious ways. It is only important to keep the deeper undercurrents of meaning in mind when reorientation of the project comes up and major decisions are to be made, so that cultural policy failure can be avoided.

THE "TIMBER LAND" PROJECT

With the purpose of discussing an establishment of the eco-museum idea in the region of Västernorrland, the Umeå Futurology Group contacted the Project Office in Härnösand, well known as a development centre in the county. The Project Office is a foundation with several trustees, of which the County Administrative Board is the largest.

As a result of this contact we received information on the project *Timber Land* (Träriket). Timber Land is a grand propaganda project for the county, based on the role that the forest and wood processing industries have played for a long time. A regional tourist venture is part of the project and includes monuments from the history of the forest industry during two centuries. The aim has been to launch complete production environments where the production plant, technical systems and the workers' living quarters have been preserved. The office has produced a tourist map for international and regional tourism, where the Timber Land´s most important regional monuments have been inserted on a number of suitable routes.

As far as we can understand the Timber Land concept in its present form, and together with the efforts that are planned, is most suitably termed "eco-museum". From a radical museological point of view, one should just underline the potential of the project as a support for the regional identification of the population. An increased participation by local citizens in inventory tasks, administration and the production of ideas may increase its significance within the region.

The Timber Land concept is also useful in order to animate the aesthetic environment view launched by the Umeå project, as a part of a radical culture policy for the region. When we say this, we do not think only of the selected localities and their art historical value. The communication links that connect the different monuments are no less important from an aesthetic point of view. They take us through the regional natural and cultivated landscape, and during our transportation we conceive a complete picture of the region as an aesthetic space and a qualitative framework for our lives. Our awareness of existence as a regional process is strengthened, and identification with the region as a place worth living in is strongly supported.

We publish here - unaltered - the map of the Timber Land project, to which we have only added the title *Eco-Museum Västernorrland.* This came about after inspiring discussions with representatives for the Project Office in Härnösand (Lars Guvå and Anders Hult), and after the mutual value of an exchange of ideas had been confirmed. The Project Office has supplied us with a map and a register for this presentation.

REFERENCES:

Davallon, J., Grandmont, G., Schiele, B. (1992): The Rise of Environmentalism in Museums. Musée de la civilisation, Quebec.

Gjestrum, J.A., Maure, M. (eds.), 1988: Ökomuseumsboka (The Book of Eco-museums). Norsk ICOM, Tromsø.

TIMBER LAND.
Eco-Museum Västernorrland

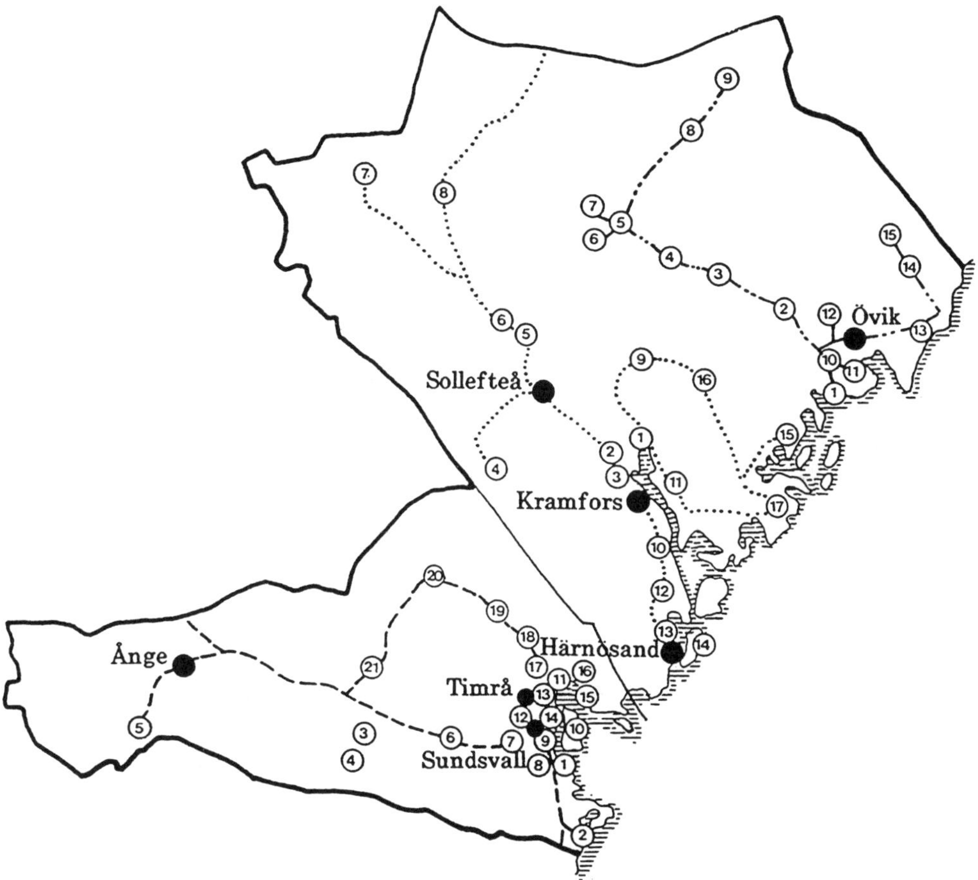

- - - - Medelpad: Green route
······ Ådalen: Red route
- · - · Nolaskogs: Yellow route

ECO-MUSEUM VÄSTERNORRLAND

MEDELPAD

Green tour

1. **Svartvik**
   is one of the best preserved old industrial areas in the Sundsvall district.
2. **Galtström Bruk**
   is the oldest known foundry in Medelpad, licensed since 1673.
3. **Stöde**
   is a centre for transportation in Västernorrland.
4. **Flatklocken**
   the central point of Sweden. Flataklocken near Munkbysjön is the geographical centre of Sweden.
5. **Östavall and Haverö forests**
   Östavall has one of the most modern sawmills in Sweden.
6. **Töva terminal**
   is the centrepiece of the new transport system that was developed when timber-floating came to an end.
7. **Kulturmagasinet**
   housing Sundsvall Museum, Sundsvall Library and the Cultural Office, consists of four grocery warehouses dating from the 1890s.
8. **Ortviken**
   one of Europe's largest fine-paper mills, produces newsprint and LWC (lightweight coated paper).
9. **Tunadal**
   is one of the first steam sawmills in Sweden, founded in 1849. Today the sawmill is the last out of a total of 43 in the Sundsvall district and one of the most modern plants in Sweden.
10. **Alnö**
    outside Sundsvall was at the end of the last century an area which had the highest concentration of sawmills in Sweden, with some twenty steam sawmills in operation. It contains several relics of considerable importance for the working-class movement at Tor's Room, the first Folkets Hus (People's House) in the countryside.
11. **Sunds Defribrator**
    one of the most international companies in Västernorrland, is a modern engineering industry. Products include machines for the paper industry.
12. **Merlo Castle**
    is the summer residence in Timrå of Fredrik Bünsow, the timber baron, and one of the most striking buildings in Norrland.
13. **Östrand**
    (now SCA Wifsta-Östrand AB) supplies pulp to various paper mills, including Wifstavarf and Ortviken.
14. **Wifstavarf**
    on an islet in Klingerfjärden was named after the first large shipyard in the district in the 1790s. The industrial community is untouched, with workers' cottages lining the street and the manor house in the background.
15. **Bogrundet**
    is the largest nursery in Europe. SCA grows pine and spruce seedlings here for planting out in its forests.

16. The Indalsälven delta
is the largest living delta in Sweden. A river landscape with rare flora and excellent angling waters.

17. Älvens Hus
or River House at Bergeforsen is dedicated to the Indalsälven. Here you can visit a distinctive exhibition and obtain information about the river, the power station and local fishing.

18. Lögdö Bruk
was licensed in 1685. The buildings, spread over a large area, include an iron-ore store, a roasting furnace and a blast furnace at nearby rivers. A manor house, chapel, workers' cottages, a warehouse and school are all situated by river Ljustorpsån.

19. Gudmundstjärn, Lögdö wilderness
At Gudmundstjärn in Indal lies a colony which was transformed in the space of 165 years into a viable forest homestead and is one of the last remaining self-supporting households in Sweden. Gudmundstjärn is at the southern edge of Lögdö wilderness, a large area of lakes and forest covering more than 40,000 hectares which has been laid out for angling and recreation.

20. Österström
is a reloading site for timber and sawn wood from the inner water systems to the river Indalsälven. A rail link carried the timber up the mountain, where it was released in a flume to travel downhill to the Indalsälven, to be floated downstream to the industries at the coast. Here you can see the "cognac saver", a rotating pavilion driven by water power.

ÅDALEN

Red tour

1. Sandslån
was for many years a centre for timber-sorting in the river Ångermanälven and is now on its way to becoming a centre for tourism in Västernorrland with a museum, a smithy, a hostel and a restaurant.

2. Bollstabruk
is a foundry and sawmill community with traditions going back to 1737. It is still a living industrial environment.

3. Väja/Dynäs
The pulp mill and paper mill at Väja owned by NCB Dynäs has over 70 years' experience of manufacturing cellulose pulp and unbleached sack paper.

4. Graningebruk
was an important ironworks in the 18th century. Workers' cottages, a chapel, a tilt hammer, a smithy and a mill have all been preserved giving an excellent picture of life as it was in the industrial community.

5. The primeval forest of Oringsjö above Näsåker
is one of the most interesting primeval forest areas in Västernorrland and perhaps the most easily accessible.

6. Mobodarna
In the vicinity of Oringsjö lies Mobodarna, a mountain pasture which has recently been restored with help from volunteers.

7. Nässjö mill and pover station
In Nässjöån. The existing mill, together with its saw and planing machine, was built following a fire in the 1880s.

8. **Ruskån and Kvarnån**
   Ruskån, north of Lake Betarsjön, is a nature reserve with old brooks used for timber-floating. South of Junsele runs Kvarnån, a secondary river with repaired timber channels.
9. **Gålsjö Bruk**
   is a well-preserved industrial community from the foundry era in Ådalen with a manor house, workers' dwellings, a smithy and a beautiful and captivating chapel. Today, Gålsjö Bruk also functions as a lecture and conference centre.
10. **Lunde**
   at Sandöbron is considered by many to be the heart of Ådalen. It is the site of Lenny Clarhäll's memorial to Ådalen 31, at the spot where five workers were shot by soldiers in May 1931 during a demonstration against strike-breakers housed in Lunde.
11. **Lungvik sawmill**
   in lower Ådalen, one of the most up-to-date in Sweden, belongs to SCA Timber.
12. **Utansjö Bruk,**
   about 20 kilometres north of Härnösand, has produced sulphite pulp since 1900 and mechanical wood pulp since 1972.
13. **Murberget County Museum**
   Murberget's open-air museum on the top of a mountain with a view of Härnösand is part of the cultural history of Norrland. Murberget is a living museum with occupied farmhouses and animals in enclosures. Activities take place throughout the summer.
14. **Östanbäcken**
   A few steps from Härnösand's modern shopping centre lies Östanbäcken, one of the oldest parts of the town and a favourite place for a walk. Östanbäcken is as old as the town, which was founded in 1585. Most of the well-preserved wooden buildings were built in the 18th and 19th centuries by fishermen, merchants and artisans.
15. **Svartnäsudden**
   one of the many primeval forests in Västernorrland, is situated by the large Degersjön lake.
16. **Skuleskogen National Park**
   is the county's well-known reserve with magnificent scenery formed over millions of years, and many miles of paths for hikers including breathtaking climbing paths. At the foot of Skuleberget is Skule Naturum natural history centre with exhibitions, computerized information and expert guides.
17. **Mannaminne**
   At Häggvik on the High Coast lies the distinctive Café Mannaminne. The artist Anders Åberg is busy creating a whole league of museums around the café. Open to visitors are the Coast Museum, the Art Museum, the Emigrant Museum and the Agricultural Museum.

NOLASKOGS
Yellow tour

1. **Köpmanholmen**
   was an industrial community formed at the beginning of the steam saw era. It includes the dwellings of workers and clerks, a chapel and much more.
2. **Brynge sawmill and linen mangle**
   at Nätraån between Bjästa and Sidensjö have carefully restored 18th century buildings.
3. **Anundsjö Trä**
   is a modern sawmill with a small-scale planing mill.

4. Sörflärke saw and mill
is a water-driven frame saw which is still in use. The mill, built in 1914, is in an extremely good state of repair.

5. Gammelgården in Myckelgensjö
is a fine example of a north-Sweden farmstead from the age of self-supporting households with twenty-four log cabins from the 18th and 19th centuries.

6. Torrvattsån
In Torrvattsån south of Myckelgensjö there are relics from the timber-floating era. The bottom of the brook is paved with boulders closely packed together to make floating easier.

7. Östalbodarna
on the other side of Myckelgensjö belong to Östalbyn. One of the largest and best-preserved mountain pastures in Ångermanland.

8. Rödvattnet
is one of several old forestry-worker villages with relics from traditional forestry. Also found here are irrigated meadows that are mown every year.

9. The Stockholm Streets
a canyon in Ångermanland, created by a prehistoric glacier-river, with caves, passages, rocks, shelves and giant's kettles.

10. Domsjö Mills
The community of Domsjö developed in the second half of the 19th century from an agricultural village into an industrial community belonging to Mo and Domsjö AB. It contains industrial installations with a sawmill from 1912, a pulp factory and a chemical industry, Svensk Etanolkemi, together with MoDo's county headquarters. A manor house and workers' dwellings from the old industrial era can still be seen here.

11. The MoDo Museum
in an annexe to the manor house at Domsjö portrays in pictures, sound and recreated settings the development of forestry and industry from the 19th century to the present day.

12. Örnsköldsvik Museum
reflects the early history of northern Ångermanland, peasant handicraft and historical art of the region. Next to the museum is the studio of the sculptor Bror Marklund, which has been moved here from Funäsdalen.

13. Husum Mills
one of the largest examples in Europe of an integrated pulp mill and paper mill, grew from a steam sawmill in the 19th century. The rise of the industry has left its mark on the tiny community north of Örnsköldsvik.

14. Gideå Bruk
is a well-constructed industrial community setting from the early 19th century, with a beautifully situated manor house beside the dam gate. The industrial community contains pavilions and summerhouses, a mill and dwellings.

15. Gideå nursery
is the largest of MoDo's three nurseries. Each year 21 million seedlings are grown here ready for transplantation in the forests.

IV REGION AND COMMUNITY

"Angelica Silvestris". Photo by Margareta Klingberg

Social Life and the Living Region

Per Råberg

THE SOCIAL CASE FOR TRIAL

One of the most frequent passwords of our epoch, for a long time even the most widely used of all, is the term *social* with its many derivatives. In political debate, in the tutorial rooms of the universities and in the headlines of the media, the word has recured regularly for several decades as an obvious positive touchstone, an incantation with the aid of which the modern age demonstrates its all-embracing humanitarian devotion.

It is true that the passwords of a culture often give a good guide to an understanding of its deeper intentions. But quite often in our epoch, appearance is misleading. Little reflection is required in order to confirm that the designation "social" is also one of the most debased and hackneyed words in our language. Take the references in EU rhetoric to the "social dimension" of the European vision. The slogan is open-ended and calculated to reassure a public alarmed by the one-sided economic and bureaucratic presentation of the European integration. But the actual meaning behind the term has long been the creation of a common, mobile labour market that will strengthen the pan-European economy. It is not difficult to figure that such a policy may have negative social implications in a more elementary, humane interpretation of the term. The word "social" has been given a meaning opposite to that which it conjures up in the minds of the broad public, yet it retains its positive connotation. This is a typical example of the manipulative "newspeak" described by Orwell in his futuristic classic "1984".

The same observation applies to many other well-known terms containing the prefix "social". Socialism, social democracy and social liberalism, for example, are modern political currents whose rhetoric has appealed strongly to our social instincts, but which have in practice taken the form of mass movements surrounding abstract technocratic power structures.

The notorious manipulation of political language ought of course to be corrected by the critical authority of the relevant branches of academic learning. A number of closely related disciplines exist, e.g. sociology, social psychology, social medicine and social work. But critical scrutiny is very seldom effected, as the content of the term "social" has been diluted almost to the point of invisibility even in academic circles. Beneath the positive connotation - which is anxiously preserved - an anaemic descriptive study of various societal circumstances prevails which lacks most of the creative energy permeating the fields of modern science and technology.

We can go on to examine area after area of contemporary society, from the supposed "social" community planning carried out by architects and town planners to the "social" reportage of the media and literature, with the same depressing result. Undeniably we will encounter both social indignation and genuine sympathy in many places, but it is a sympathy which offers no substantial alternative and shows no serious will for communitarian action. Thus spontaneous criticism is effectively deflected instead of being mobilized. A convincing illusion is created that we live in a society in which - on the whole - social happiness and harmony prevail.

DEFINITION OF THE TERM

Before we continue this intricate discussion, we must pause a moment to confront the question of definition itself. What does the term social mean historically and linguistically? A consultation of several dictionaries, amongst others "The Oxford English Dictionary" (1989), allows us to formulate the following comprehensive answer.Roughly interpreted, the word denotes society as a community of citizens, or in reverse; citizenship welded into community form. In its core meaning, however, the word marks a clear boundary, both towards society as a state-controlled, authoritarian system and towards citizenship as an isolated private community. The word is perhaps best described by using a synonym which has come into more frequent use during the last decade. It is the term *civil society*. In using this term, the aim has been to denote the genuine life sphere of citizenship, which is wedged between the various public and economic structures in social life.

The word social, used independently or in various contexts, has in the general consciousness been almost notoriously surrounded by a positive aura. The positive charge of the word is closely linked to notions concerning the altruistic spirit of the protective neighbourhood and of collective co-operation. Social community is perceived as a network of warm, compassionate relationships, and as a primary existential form of indispensable value to the individual, and to public society.

Side by side with this elementary interpretation, appear others which can be traced to specific political movements (for example state socialism) or scientific areas (for example biology) into which, however, we do not have cause to delve here.

SOCIAL DECONSTRUCTION

If we examine the social field of contemporary society in accordance with the civic definition presented above, a picture very different from the official view will appear. Not to our surprise we will find that events have moved in the opposite direction to that asserted by official rhetoric. We can clearly see that what has long been taking place in the modern community is not a build-up but a systematic dismantling of social networks and security. We recognize that a cold undertow is flowing silently and invisibly through our society, irresistibly undermining its basic social ideals and collective initiatives. This anti-social current of the time has a friendly face when accidently showing up, but is mostly hiding at great distance from the social life which is affected. It has led to the disintegration of primary communities, to growing generation gaps everywhere and to unbridgeable geographical or mental distances between individuals. Its logical end result is the almost total dissolution of the neighbourhood community with its tight net of human contacts and informal services. A community instead has been erected whose crude hallmark is the alienation of the individual from the living collective and of the living collective from influence over its basic living conditions. Often it shows some technical perfection which is pointed at in order to legimate the efforts.

The process of social dismantling has not only brought about individual and collective alienation. On several levels of community organization it has provoked pathological reaction. This new *sociopathology of our age* presents a broad spectrum of symptoms. At the individual level the symptoms often take the form of a flight into the dream-world of artificial stimuli with the aid of perception-enhancing drugs. But they are also manifested in a wide range of mental disturbances, with suicide as an increasingly common final exit. At group level, social alienation is often expressed in aggressive behaviour, finding outlets in bullying, environmental destruction or organized crime. Violence among young people, which today is erupting throughout the world, is the most tragic aspect of this threatening situation. On more sophisticated levels, to continue our scrutiny, we encounter the disciplined and silent violence of society's administrative powers. Widespread as it is, it represents one of the strongest repressive forces in contemporary society. The ultimate and most atrocious revenge of the alienated social consciousness is found in systematic military violence, which has become a notorious feature of our age. Here social frustration culminates in pure excesses which, ironically enough, often bear the seal of official legitimacy.

Most of these pathological symptoms have long been classified as individual-related. Drug addiction was stated to involve only the maladjusted and wars were supposed to be caused by crazy dictators. The social erosion however has now reached such proportions that the stigmatization of the individual appears ridiculous to almost everybody. With this the climate is created for a new understanding which is gradually permeating the public opinion: that the serious symptoms are the result of a malign sociohistorical course of events, the ultimate cause of which is a system error, a basic anomaly in the established social system itself. We are now confronting the fact that we live in a society which is badly suffering from ill-health. Its fever and shivering affects all of us, and some even severely. But how should we characterize those people who actively and apparently in very good health continue to promote the sick system?

SOCIAL CRISIS ANALYSIS

With increasing humanitarian awareness in widening circles new questions are raised. How shall we proceed in order to cure a disease of such immense proportions? How do we explain the socio-pathology of the contemporary community logically and in scientific terms? Is it possible to identify the pathogenic core of the social crisis? We may find useful guidance towards a deeper understanding in different intellectual camps such as the social criticism of the Frankfurt School (e.g. Herbert Marcuse and Jürgen Habermas) and the theoreticians of the alternative movement (e.g. the Americans Murray Bookchin and Theodore Roszak). Several of these critics regard *the dismantling of the civil core of society* (the life-world, as Habermas puts it) as a general destructive tendency of the modern movement which is the great project of our era. In their writing they describe this alarming process as a logical consequence of the contemporary modernization project and its overstressed rationalistic world view.

The dogma of modern rationality has been realized in its purest form in the scientific and technological disciplines, we are told, but the apologists of modernity also wished for society as a whole to be be reshaped by the rational purism of those disciplines. The social consequence hereof has been a concentration of the population in metropolises where a cosmopolitan life style, very different in kind from that of the provinces, has come into existence. By its intrinsic logic a hierarchical organization is established, with deep class divisions. The humane qualities of the civilian life-world, so typical of the historical society, are eroded step by step. Instead, the libertinistic ideals of mass society invade the human stage. Acclaiming the achievements of the cool intellect - fostered not least by the school system - leads to an increasing disinterest for the aesthetic, humanist and social manifestations of life. So far the judgements of the classic critical schools.

The critical analysis of these philosophical schools seems, by and large, accurate. Still, certain amplification is required if the picture of the crisis is to obtain full clarity. Some results of the Umeå University Group's socio-ecological studies (refered to in several places in the book) may deserve mentioning here. The success of the rationalist paradigm, according to the Umeå Group, is indissolubly linked to the process of globalization, the culmination of which we are experiencing in this century. By the cultivation of rational thinking and modern technology a radical expansion of the human living space has been made possible. As a consequence, a cosmopolitan civilization is now taking shape and new organizational principles for the international society are being devised. The new global space opening up is mastered by advanced networks of rapid IT-communications; they are maintained by impersonal contacts between easily replaceable human parties.

All these happenings stand in striking contrast to the historical and even evolutionary preconditions of the human society, say the Umeå researchers. The characteristic form of the historical society was *the stationary habitat,* conceptualized through a development similar to that of the other species' on the planet. The local community with its sensory and spatial values, its loyal contacts and natural responsibility, in our days has become the very antithesis of the advanced cosmopolitan life: a place to disdain and to escape from. To cosmopolitan man, the local community is a blind spot on his consciousness. It represents no more than a pre-modern stage of development.

This, however, is a very fatal event, claims the group. What is forgotten is that the local sphere represents not only an earlier stage of evolution, but also a more fundamental organic level in man's consciousness. Sensory and aesthetic life, intimate social communication, but also the daily interaction with natural surroundings - all these are values which we can only reject at the cost of stripping our society of its human identiy, its basic moral content and also, perhaps, of a vital social communication structure. Modernity has long tried to build a society in the open air, throwing off all its historic foundations. No wonder that these castles in the air have lately gone adrift.

Still, this is not the final word of the tale. At the same time it remains an undeniable fact that we possess *a cosmopolitan awareness,* which extends far beyond the boundaries of the local community. This awareness, too, refuses to be subdued and pushed aside, and very rightly. The logical conclusion to these conflicting observations is obvious. It is the very duality of the human needs structure that causes the great trauma of the present stage of man's evolution. It is the disinterest - or inability - of ageing modernity to devise a social system which resolves *the evolutionary dilemma of polary conditioning,* that has led to our socio-cultural schizophrenia with its many strange symptoms.

THE CAUSE FOR SOCIAL RECONSTRUCTION

Diagnosing the sociopathology of the times is one thing, but curing the social sickness is another, rather more difficult, matter. This calls for innovative work, possibly leading to more credible solutions to the new social challenges raised. Closer examination shows that radical work in line with the new considerations has already been taking place for some time. Symptomatically the contributions do not come from the political or social scientific domain. They mostly derive from forces at grass-roots level and consist of a long series of community experiments, which have sprung up in unanticipated patterns during the last few decades. Many of these alternatives have distinct precursors in the intellectual history of man (these are mentioned in other chapters of the book).

The explicit object of the grass-roots experiments varies, but the social intention has quite often been to recreate the generous humanity of the primary local neighbourhood. With intuitive accuracy the alternatives, by way of their spontaneous initiatives, have succeeded in laying bare the primeval germ of human fellowship. It is an astonishing but relieving thing that the instinctive social forces are still capable of manifesting themselves in a technocratic and system-dominated epoch like ours.

A more articulated variant of the grass-roots perspective exists in the communitarian movement, which stems from the same ideological roots in 19th century popular history. In this case the alternative models often are realized within the frames of the established political system, instead of from a rebellion angel. Another initiative, now with a clear academic profile, is the concept of civil society and the discussion of its revival. Partly this tendency is caused by the transformatory process in European politics following the break-down of the Soviet system. But there has been a parallell debate in the West since the 80s with liberal democrats as the mainspring.

However, most of these radical tendencies suffer from a limitation which makes them inadequate as a contribution to a winning humanitarian vision for the future. This is their one-eyed concentration on the local level and lack of interest in trying to organize the global space in accordance with civil principles. In this half-finished version the social alternatives are easily absorbed by mainstream techno-society - but without correcting its pathology at all. If the civic idealism is to retain its radical force, it must eventually take up also the challenges posed by the cosmopolitan situation and pit its ideals against contemporary global logistics. It is a fair guess that only from such a confrontation can socially viable political solutions arise in the post-modern age.

THE NEW PUBLIC HEALTH

It is not possible, in this short paper, to do justice to all the humanitarian alternatives just touched upon. Instead, I would like to mention one remarkable initiative which has not, up till now, figured frequently in the social development debate. This is the World Health Organization and its radical "Health for All" programme.

This is a policy scheme which has been launched by the WHO's European office, but which has already attracted worldwide attention among professionals. The WHO radicals maintain that major deficiencies in global public health today are of a psycho-social character and have their roots in structural shortcomings in the established living environment of our society. These health deficiencies are a consequence of the careless social planning of the advanced industrial society and cannot be remedied either by individual health care or sectoral intervention of the type normally provided in public health work. Instead they require structural initiatives and, above all, a holistic view, which places public health in the overall societal context. The conclusion drawn is that public health should also be regarded as a matter of social policy, not only at the local level but in a long-term global range. What the world society needs in order to avert the new environmental health risks is, in other words, a policy which has placed *sustainable social development* on its agenda.

The core of the radical WHO ideology described here consists of an "ecological health draft". In this, public health is regarded as a human right and at the same time a social resource which society has an unconditional duty to provide. The quality of the physical but also of the social and cultural living environment is the key to success or failure. In a series of international conferences from the mid-80s onwards the WHO has perfected the health-ecology draft. An important milestone was the Sundsvall Conference in 1991 (in Västernorrland County), which was devoted to the theme of "Environment and Health". The conference is one of the main manifestations so far of a holistic approach to human health. A holistic and environment-oriented view is also a characteristic of several ambitious projects recently launched by the organization (i.a. the Healthy Cities project).

WHO's agenda for an ecological people's health is a source of inspiration to all who are engaged in matters of social development. To the Umeå Group it has brought both new insight and renewed courage. The initiative has spelled out clearly the need for *a global platform* for the discussion of a social and humanitarian future of our society. Simultaneously the health perspective brings an awareness of medical risk to social crisis analysis that has previously been lacking. We now begin to realize that humanitarian negligence may lead to ethnic threats to whole populations and even evoke genetic dangers to the species.

NEW CHALLENGES TO SOCIAL SCIENCE

The conclusion of radical debate is that the discource on society's social and civil development must be given the highest priority in long-term politics at the national and planetary level. But after the long period of social ignorance, society is lacking both professional competence and relevant knowledge to tackle the demanding tasks. The sector is far behind the strategical position of the leading powers of today. This state of things is deplorable, but at the same time represents a challenge to new generations of researchers in the human and social sciences. The construction of *a new social knowledge base* and of new society models are now strongly called for in order to handle the provoking issues.

The first task in such a project must be to formulate the humanist scientific arguments which may legitimize a more prominent position for the social dimension in the political strategies of society. So far in this paper, I have been satisfied to ride on the traditional consensus which surrounds the term. This has been the behaviour of social debaters for a long time, whether they have sprung from politics or the social sciences. However, the factual contents have gradually come to stand in reverse proportion to the rhetorical emphasis. In fact this is the normal method used by technocratic ideology (or sub-ideology) to realize its ideals. The illusion of a society with human and cultural norms is upheld by defending its value concepts, while at the same time these concepts are filled with a foreign content which is strange to our species.

This constitutes, of course, a misuse of language, which entails an undermining of our political understanding and in the end a misrepresentation of factual reality. We find ourselves in a situation where all realities seem to be afloat and where rhetoric alone seems to grant some kind of luminosity to the political discourse. This situation is an expression of the state of impotence to which the political debate has come, and which makes it into a powerless victim of the development forces which dominate the scene. The concept "social" is a notion that has become a linguistic relic in the clearance sale of value concepts which is bound to the contemporary phase of modernity.

When we therefore try scientifically to recapture the living substance of sociality, we must start from the beginning, from the very bottom of our knowledge and insight, in the hope of finding there the human truths which so mockingly elude us. Aided by the organic fragments that we discover, we can start a reconstruction of man´s constitutive needs, which may eventually form the basis for a supportive course of development. Giving a nuanced description of the social dimension is one of the most important tasks here.

After presenting this long harangue, I am almost obliged to make one or two specifications of concept. I would start by claiming (the truism?) that the social dimension is a key factor in an alternative human scientific theory and makes up a highly original field in the spectrum of existential needs. The social dimension can be described as the mental element through which the interaction between the individual and the surrounding spatial asset is mediated. The social element is a necessary medium to all species which have passed the stage of reflex and are carriers of a heritage of species behaviour acquired through genetic heredity or environmental influence. The social dimension does, however, also have a communicational task in the service of the species. To man, in his advanced stage, well-developed social dynamics is more important than ever for the successful interaction of the vital functions.

The single individual rests in the social element almost in the same way as in the natural element from which he came forth as a biological manifestation of life. The individual is both an integrated part of the medium and an independent actor. He influences the system at the same time as he lets himself become influenced. These interactive dynamics within the social element are of decisive importance to its creative functions and are upheld through the adaptation of the subject and the element to each other. Above all, the social system must adapt to the scale, capacity and other conditions of man in order to obtain the stimulating vital injections from his active commitment. If this co-operation fails, a system breakdown soon takes place, entailing mutual distress for both parties and widespread effects on other parts of the vital system.

IMPORT OF SOCIAL DYNAMICS

All these are important observations in the new social scientific approach, but they are followed by still more intriguing questions. Why is maintenance of a dynamic social system so essential to a vigorous human society? There seem to be two main explanations for this; one is structural and the other is functional. The *structural* explanation is the most fundamental and is closely linked with the role of social communication in society's reproductional processes. The reproduction of society is one of the basic aspects of evolution and implies, to the side of biological functions, a number of institutions that maintain and improve the social system, at the level of man for example schools, social administration, physical planning, the care system, higher education and research. If this indispensable structures are not adapted to the social abilities which each living subject possesses, it will quickly be reduced to dead matter, a gigantic rubbish dump. This will soon suffocate man and society through its sheer weight.

The *functional* explanation for the indispensability of the social system is linked to the communicative processes in living systems, and, above all, in the human community. I have in mind the kind of communication which takes place in the myriad of interactive encounters between people and groups in the various situations of daily life. These comprise actual physical encounters, verbal communication, written messages and direct contact via networks over the whole planet. In particular, they include language, well-articulated and communicating meaningful messages. Social communication has, in our time, evolved an intensity unequalled throughout history.

This communication structure has, however, already begun to be weighed down by its own tremendous mass, and seems more and more impenetrable. Chaos and anarchy loom when people finally abandon their efforts to bring order to the system. Chaos also looms because the communication structure of society has ended up well out of the reach of the single individual and the living pattern of language built up within the historial collective. The language and its content become impoverished, which in a fatal way also affects the capacity to articulate and the creative thought of the individual. Contemporary educationalists have a lot to tell us about this phenomenon, which constitutes one of most shocking elements in the modern school system. There seems to be a logical connection between social communication and the advanced intellectual processes, a connection which is threatened if society alienates itself from the organic foundation for language development. In the long term, then, the ingenium of human civilization is threatened by the development forces that it once conjured up. All the proud flagships of the Cartesian paradigm will sooner or later be threatened by this impoverishment. The natural sciences, technology and the industrial economy will not be spared.

There is yet another argument for the maintenance of a vital social dimension in society. I have saved it for last, since in the present paradigm, many people see it as dubious. It concerns the role of the social dimension as *fluidium humanum*. Man is, at the basic level, a social being, and life in the social context satisfies fundamental existential needs within him. The social needs, however, are of a qualitative scope, and the social environment must show a harmonious scale and content in order to satisfy man. The close but open relations of the civic neighbourhood with its creative communication and sensory, emotional and reflective contacts make up an optimal fluid for man's wish to reach social identification. When this identification is achieved, he feels deep joy and security - in other words, he experiences satisfaction of his existential needs.

A THEORY OF ECOLOGICAL MAN

One might claim, as I have done above, that the social dynamics are in the service of species reproduction and functional communication. From an individual action perspective the situation, however, appears in a different light. The subjective inclination to communicate and reproduce, and the personal joy engendered by social processes, are the immediate driving forces behind social dynamic in the community. It is through the living human being that the species' social functions are acted out. All this has to be kept in mind by the architects of a radical social theory.

But let us now take the final step in the argument of this paper. The various perspectives on a living social dimension which I am refering to here have their origin in reflections which have been exchanged over a period of time within the University Group for Humanist Futurology in Umeå. They end up in a *theory of social ecology* which describes the interaction patterns of human existence in the living environment. An evolutionary approach is behind the multi-layered model of identity which has been constructed. The aim of the University Group is to contribute to the creation of a solid fund of ecological knowledge through the intensive study of human needs.

Arguments for this strategy have been put foreward earlier in the book and will not be repeated here. In order to increase understanding for the role of the social dimension, and to prepare its acceptance in long-term politics, I will, however, try to expound, in the following pages, the social part of the Umeå model. As a start, though, the position of the Umeå model in relation to the contemporary social development debate must be stated. I therefore repeat the message inherent in the previous pages: the Umeå project wholeheartedly endorses the radical efforts to reconstruct the civil society and its humanitarian ideals. We believe in a social reform which again gives civil citizenship the initiative and which liberates it from the bonds of a disproportionate official and political repression. As to the facts and arguments which we put foreward in order to achieve this goal, they have mainly been culled from the modern life sciences. Still they aim at constructing a community at the highest level of human experience.

I start my presentation of the Umeå group's social model by recapitulating the table for the socio-ecological hierarchy of identity. By doing this, I will be able to describe the special characteristics of social identity, and at the same time specify its place in the total eco-identity.

Systemic (or Natural) Level of Eco-Identity

The basic level of man's eco-identity is occupied by the natural identity. It comprises man's consciousness of his personal ego as a nucleus of organic life within the natural environment. In the guise of natural identity the ego acts as a dynamic life-force unfolding itself in natural space in both self-confirming and self-protecting patterns.

The existential task of the natural identity is to uphold homeostasis in the interplay between individual and natural environment. Herein, the life-identity (as we may also call it) plays the part of an indispensable control and steering centre in the comprehensive system of the eco-identity. At the advanced level of man the nature-identity is a platform for what can be called an *eco-political reflection and responsibility*. In a viable community this is the point from which the dynamic processes of the human and natural setting are best surveyed and can be directed towards qualitative goals.

The politicization of the natural identity is the key to introducing ecological reason and judgement in social and environmental planning and to guarantee the maintenance of a sustainable society in the broad sense.

Structural (or Species) Level of Eco-Identity

The second systemic level of eco-identity is based on the species identity. At this level the individual sees himself as a representative of the group of species-neighbours from whom he is descended and with whom he is deeply involved both physically and mentally.

An important factor in the promotion of species behaviour and in securing species survival, is the territorial habitat which encapsulates the interactive nets. This can be observed with many species in bio-existence, but is of special importance at the advanced level of mankind, because of the great complexity of life functions here. Within the structure of his habitat man erects stable institutions of many kinds which serve the mediation of traditions and other long-term needs of the species population.

Human civilization, to all appearances, obtains its primary inspiration for social organization from the level of species identity and from the structural caring about human survival which is at the center of concern here.

Functional (or Ego)
Level of Eco-Identity

The third level of ecoidentity is occupied by the functional consciousness. While the more fundamental identity levels are concerned with the comprehensive and long-term issues of ecological life, the task at the functional level is to deal with the short-term dynamics of daily life and the nutritive and material needs of the individual. This is the main level of activity for the processes of "the civil society".

We can observe three different stages within the functional eco-identity: **1.** *the eco-cultural (aesthetic)*, **2.** *the eco-social (communicative)*, and **3.** t*he eco-instrumental (productive) stages*. Probably the three stages reflect a logical evolutionary course of events.

Functional identity level 1:

The first of the three functional sub-levels in man contains *a cultural programme* whose goal is the qualitative satisfaction of civic social needs in terms of aesthetic and sensory experience. By way of its sensory calibration this level remains the motivation center of the functional ecosystem. The sensory stimuli of the surroundings activate the individual in creative projects.

Functional identity level 2:

Secondly, the functional ecolevel includes *a socio-communicative programme*. The purpose of this is to uphold current collective care and exchange of ad hoc information within the civic habitat. Communal group identification is the goal and at the same time a motivating factor.

Functional identity level 3:

Thirdly, functional existence in the civil society comprises *an instrumental action programme*. The objective here is the systematic exploitation of natural resources and their preparation in order to secure material survival in the short term. Egoistic satisfaction of needs is the main driving force, but the productive processes may take the form of cooperative action.

Dimensional (or Logo) Identity Perspective

It should be remembered that eco-existence is acted out in two partly contradictory dimensions: the spatial (sensory) and the temporal (cognitive).

Unlike most animal ecosystems the human ecoidentity is capable of consciously exploiting both dimensions. As a consequence human eco-existence takes the form of a dual process. It is realized through the individual's alternation between a local sensory and a global cognitive action level. This fact has important consequences for the organization of man's ecological habitat. Social action plays an important part at the global level, but is decisively coloured by cognitive reflection.

THE ARENAS OF SOCIAL LIFE

How do we translate our knowledge concerning social needs, encoded in the ecological model of identity, into plans for collective and political action? This is the logical second step in the construction of a model for supportive social development. As we have observed, man's eco-identity also contains institutional and political parameters. A central question is how these parameters are to be satisfied without violating the hegemony of civil society in the existential whole. Another question is how the civil structure is to be organized in order to best meet the qualitative criteria of the individual's ecological judgement.

The answer of the Umeå Group is simple and unambiguos. An adequate civil structure can only arise if respect is shown for another postulate for social formation, namely the demand for territorial organization of human processes. The territorial element is the basic medium for the interaction of ecological man with the outside world, and the territorial limit marks the extent of the habitat which a collective is capable of administrating, without losing control of the system. Such an order is part of the rules governing a biological structure according to principles of vital dynamics.

A social organization which is surveyable in extent logically entails a population which is limited in number and possible to overview. In such a society, comprehensive human contacts between individuals can be upheld and a warm, humanitarian climate can be created. Hereby the foundation is laid also for well-functioning communicative structures in other parts of society. A local society, limited in this way, will also maintain a transparent structure, making it possible for intersectorial co-operation between the civil core and the institutional and political complements to take place without conflicts of system. It is the same people, but in different roles, who are active in all sectors.

Thus in a civilian and ecological vision of society, local society, today so threatened from all directions, stands out and assumes a wider significance. The cosmopolitan needs which once led to the dissolution of traditional local society are also satisfied by the model, but in a new way. Partly, local society is expanded both territorially and functionally, becoming a regional living space with a more comprehensive structure. Amongst other things, the region is empowered to give its citizens an active role as cosmopolitan agents. This can, however, only take place under the condition that the ecological region also has counterparts in the cosmopolitan field. The total human living space appears in a global perspective as a network of life-regions, which encircle the planet in geopolitical patterns and form functional and administrative coalitions of a kind that is today unpredictable.

This is the scheme of thought which makes up the foundation for the second tentative aim of the Umeå University Group. This is to draw up a model of society that provides a framework for the ecological identity to achieve comprehensive self-realization. Our designation for the model of society which has crystallized is the *ecological life region*. In the final part of the article we will show how the different facets of social eco-identity are realized in the ecological living space and thereby form characteristic patterns. This is done by demonstrating the various arenas of the social dynamics of life.

The Local Arena

The primary arena of social life in eco-society is the local network, which we describe as the grass-roots level of communal dynamics.

At grass-roots level community life is a behavioural fabric of interpersonal relations. The relationships may be emotive and centre on the provision of care, but may also be the carriers of services and information. Together they create a communal unity with which the individual tries to identify, as it corresponds to his basic social instincts. One main determining factor is his drive for social anchorage within a particular territory. It is only by the ties to a specific place that the network acquires the existential significance that man requires of basic social life. The primary local network becomes a fertile ground for feelings of love and friendship, and for the assumption of neighbourhood and family solidarity.

The mobilization of local solidarity and dynamism is mainly the responsibility of its citizens. Citizens should be given the opportunity uncompromisingly to carry through communitarian projects, as these represent quality-enhancing initiatives of the first order. Several areas for communitarian initiatives can be distinguished. Maybe the focus of interest should be, initially, at the neighbourhood unity, with the primary aim to re-establish its lost social vitality. Inventing routines for a more creative neighbourhood life is one of the most urgent tasks ahead.

The dismantling of the local community networks is perhaps the biggest mistake of all in the structural transformation of modern society. It is a fact which has long been denied (if it has even been noticed) by the political and intellectual elite. The mistake has been all the more clearly revealed by the alternative movement of the last few decades. The most important lesson that we can learn from the grass-roots is that the local network is not the exception but the fundamental rule of the social system. Setting out the social living principles of the local commune will therefore be the first task of a socio-ecological policy.

The Global Arena

But protecting the grass roots level is not enough to realize a modern social vision. Local life has to be complemented by a global arena.

No human being of our days finds his need for civic communication satisfied via the primary networks alone. We are all members of a global civilization and take part in a global communication process. At this level different rules apply and other social needs, mainly of a cognitive nature, are satisfied. As receivers and contributors of useful information we take part in a planetary exchange of ideas. The school and modern pedagogy have shaped our cognitive identity and provided us with the instruments for participation in a global interchange of knowledge and news.

Despite the level of technology achieved, society's patterns of global communication have long remained very traditional. The prevailing routines have been those of physical movement via the transport systems or one-way media offerings (incorrectly described as mass communication). These are patterns which are rapidly changing today. Modern information technology contains all the prerequisites for a broad two-way mass-communication structure over the entire planet, particularly if it is connected with the audiovisual media. An intensive development of qualified multimedia applications is taking place, which will shortly transform the regional stage into a splendid vantage point overlooking the whole world, regardless of the particular locality of the region.

This new situation creates a hitherto unseen opportunity for the provincial regions - even those with remote localization - to take part in global affairs on equal footing with the metropolitan areas. But, above all, for the first time in history, it opens the door wide for *a civil invasion* of the global communications scene. The potential fields of civil electronic communication are innumerable; they extend from personal to interest-based exchange between citizens of all ages, and cover matters ranging from hobbies to research and philosophy. They also include collective ad hoc participation in a global entertainment culture. If the municipalities collaborate in regional networks they can finance large multi-media installations giving them full planetary presence. By erecting special public environments the social character of the new communication structure will be well underlined.

Global and local communications are, today, regarded as incompatible components both by the civil movement and the political elite, albeit on mutually contradictory arguments. A sustainable social vision is based on the replacement of this conflict by combinatory solutions. In the network of the ecoregion each individual can comfortably alternate between his two social platforms. He is active in the *local civil culture* which defends the sensory root system forming one part of his identity. But at the same time he may participate in *the global civil culture* which is associated with his modern cognitive personality.

The Public Arena

Ecological social behaviour covers still further arenas. Beneath the local and global scene of daily life more fundamental spheres are to be found. I am thinking under this heading primarily of the action field related to the structural (species) level of mans eco-identity.

This arena handles long-term social processes of an institutional kind, such as the intersectoral circulation and the maintaining of the local and regional infrastructue. We describe this as the public arena of social life. This is the level at which active sociality is transformed into the concept of society, i.e. a permanent frame for collective living. This, to a considerable degree, consists of the social setting in a physical sense, i.e. by cities and regions, but also includes the socio-cultural traditions of the community. Language preservation and the cultural heritage are among these traditions.

Described in familiar political terms this is the domain of the public sector. It consists, on the local level, of a handful of sectoral organs which deal with social services and care, recreational activities, schools, environment protection, health care and social planning. The public sector in the 90s is regarded - not least because of its size - as a wasteful factor, and its right to exist is questioned from several quarters. But to draw the conclusion that the sector should be scrapped is based on ignorance of its ecological significance. As the administrator of the institutional level of civil living it represents an indispensable aspect of social life. It possesses a great unexploited - or rather neglected - development potential.

The renewal of the public sector in accordance with civil and ecological ideals may be accomplished in several ways. One first step should be to hand over (not just delegate) responsibility for many of its routines to the drained civil society.

This would set free substantial institutional forces which can be directed towards more creative tasks in line with the new humane lojality of the sector. One of these tasks is to formulate guidelines for an ecological social policy. This means a policy which identifies with the civic level and is prepared to take a comprehensive responsibility for the well-being of the citizens. But the commitment also includes defending legitimate civil interests against other sectoral interests in the overall political spectrum. Good contact with avant-garde human and social research is a prerequisite for performing this demanding task.

In practical community development two areas seem to be of particular significance. The first is the promotion of a viable life style in the region. Measures should be taken which strengthen the citizens' identification with the regional community. The most important area for radical reform in this context is probably the school system. Contemporary schoolteaching is excessively marked by the pursuit of the ideals of cognitive knowledge. This tendency should be modified by a pedagogy which draws attention to local and social values and to the aesthetic and qualitative needs.

The second main task is to introduce deeper civil perspectives in community planning. Many elementary inadequacies exist in today's social environments and additional efforts are needed, especially in neighbourhood planning. Another object is the district and municipal plans of the region. Both the municipal and regional territory should be designed with the objective to promote the growth of social communication nets. Also, it is important for the social networks to be made visible in the physical environment, and for the whole community to be identifiable at individual level.

The Political Arena

However, social life in the Umeå model embraces one more level which lies even deeper than the already mentioned. This is the arena of man's systemic eco-identity, which includes the control and steering center of the human ecosystem.

This is also the level of overall systems responsibility and the point where the eco-society's efficency in the service of civil and humanitarian needs is supervised. An important task here is to survey the maintenance of balance between the social community and its natural resource setting. We may call this the eco-political arena, and thereby we raise the claim that the latter remains a necessary part of the social life-world and should be profiled in accordance with this.

Which are the special tasks of the eco-political level as compared with the established politics of our time? Beside the major tasks to defend the natural ecosystem against ruthless exploitation and protecting the human ecosystem against natural disasters, it is to launch *a deepened civil rights perspective* in local and global politics. This explicitly means to claim that *fulfilment of human existential needs* be placed as the central and overriding goal of political commitment, to which other sectoral goals have to submit. But it also is a claim for introducing, at last, *qualitative principles* in politics, both in a humanitarian and an aesthetic sense. Everybody who can distinguish philistine luxury from true quality of life will understand the importance of this distinction for human welfare and a sustainable development.

The existing political system, in its rhetoric, draws heavily on conceptions like those maintained in a socio-ecological analysis. They emanate from pre-political notions of grass-roots type, which are officially never uncovered. By the seducing appeal of these humanitarian slogans the public interest in politics has been upheld for roughly a century in the different camps.

But the practical politics of the day - in all parties - has distanced itself miles away from truly human basics. Its ceaseless attempts to ascribe social, democratic, liberal and humane credibility to technocratic solutions discharged of all social responsibility and far from the domains of human welfare bear witness of the deep alienation with political powers from the primary tasks of their mandate. The basic civil duties and the human meaning of the term democracy seem almost to be forgotten. In the prevailing ideological vacuum the Napoleon syndrome flourishes in every little nest of the political structure. A renewal of the political system, i.a. in accordance with socio-ecological principles, is highly called for.

SUSTAINABLE SOCIAL DEVELOPMENT

There are two final questions that we must pose when faced with the model of sustainable social life outlined here: Is the model credible and is it realistic? The first question can only be answered by empirical testing in practical community experiments. In this way the qualities of the alternative model may be a subject for comparison with present routines.

But even if credibility is accomplished, there remains the question of realism. Is it possible to accomplish the alternative model, given the visions entertained today by those in power and, even more, given the public infrastructure which has long been cemented in society? My personal belief is that the prospects are quite good. Acceptance and breakthrough of civil society models may very well occur over a couple of generations, provided that the models keep their promises with regard to a) the claim for high quality of life and b) the pretentions to ecological reliablity.

The most important reason for optimism is that the prevailing social paradigm is showing itself increasingly clearly to be an obsolete model with fairly outworn ideals. Its social ideal is that of the mass society, since long embodied in the cosmopolitan metropolis. This is a social system which has in our epoch assumed more or less monstrous dimensions. As a living environment the metropolitan community represents a growing sociomedical health risk, and more and more of its countless sensory stimuli turn out to be dubious substitutes for real civic quality of life. Regarded as an ideological environment the situation is still worse. In the modern metropolis the aged visions of autocratic absolutism and class society are cemented and continue to flower in never-ceasing new forms.

But, strange to say, this monopolistic system has at the same time created ideal conditions for the realization of alternative community models in the civil society style for which I am arguing here. The conditions are provided by the many peripheral and intermediary regions of the national territories. These are today expanding into enormous problem provinces all over the world, as a consequence of the draining of population and resources caused by metropolitan magnetism. The provincial regions have, up till now, almost totally lacked the support of theoretical visions and autonomous objectives. In actual fact they constitute ideal areas for the application of alternative social models. They represent a fertile ground for basic ecological interaction and possess discernible socioterritorial frames, both of which encourage the will to achieve local identification. This is the best conceivable prerequsite for building loyal and transparent social structures centred around civic ideals.

Developed with care the undervalued qualities of the provinces may bring about an alternative regional life style, which might set the tone for the new global society. With its qualitative advantages the ecological life-region will attract increasing numbers of people. A self-sustaining social development may get started, relieving the burden on today's disoriented regional policy, and certainly to many of the sorely tried world metropolises. From this a reconstruction of political life and the political structures of society may also start, based on a more truthworthy human and civil interpretation of the concepts of democracy.

Whether the vision of a viable regional development is to be realized in the end depends on the regions themselves and on the life force that they can mobilize. But for them to succeed it is necessary for today's blind faith in the blessings of higher authorities to give way to an independent comprehensive view, which takes responsibility for the regional house and cares for the whole regional family. Only from the shared social and humane commitment can the spirit for true renewal be obtained.

REFERENCES:

Ambjörnsson, Ronny, 1984: Socialismens idéhistoria: en inledning (An Introductory History of the Ideas of Socialism). Gidlund, Stockholm.

Another Development. The 1975 Dag Hammarskjöld Report on Development and International Cooperation. Development Dialogue 1975 1/2. Dag Hammarskjöld Foundation, Uppsala.

Archibugi, D., Held, D. (eds.), 1995: Cosmopolitan Democracy. An Agenda for a New World Order. Polity Press, Cambridge.

Arvidsson, Håkan et al., 1990: Det civila samhället (The Civil Society). Timbro, Stockholm.

Arvidsson, H., Berntson, L., Dencik, L., 1994: Modernisering och välfärd - om stat, individ och civilt samhälle i Sverige (Modernisation and Welfare - State, Individual and Civil Society in Sweden). City University Press, Stockholm.

Avineri, S. and De-Shabit, A. (eds.), 1992: Communitarianism and Individualism. Oxford University Press, Oxford.

Cohen, J.L. and Arato, A., 1992: Civil Society and Political Theory. The MIT Press, Cambridge, Mass.

Elmér, Åke, 1994: Svensk socialpolitik (Social Policy in Sweden) (18. uppl.). Liber, Stockholm.

Friedmann, R., Gilbert, N., Sherer, M. (eds.), 1987: Modern Welfare States. Wheatsheaf Books, Brighton.

George, Vic, 1985: Ideology and Social Welfare. Routledge & Keagan, London.

Giddens, Anthony, 1989: Sociology. Polity Press, Cambridge.

Giddens, Anthony, 1991: Modernity and Self-Identity: Self and Society in the Late Modern Age. Polity Press, Cambridge.

Giddens, Anthony, 1992: The Transformation of Intimacy: Love, Sexuality and Eroticism in Modern Societies. Polity Press, Cambridge.

Giddens, Anthony, 1994: Beyond Left and Right. Polity Press, Cambridge.

Habermas, Jürgen, 1981: Theorie des kommunikativen Handelns. Suhrkamp Verlag, Frankfurt am Main.

Heywood, Andrew, 1992: Political Ideologies. MacMillan, London.

Keane, John (ed.), 1988: Civil Society and the State. Verso, London.

Langby, Elisabeth, 1993: Vinter i välfärdslandet (Winter in the Land of Welfare) (2. uppl.). Brombergs, Stockholm.

Lichtheim, George, 1972: Socialismens historia (A Short History of Socialism). Wahlström & Widstrand, Stockholm.

Marklund, Staffan, 1982: Klass, stat och socialpolitik (Class, State and Social Policies). Arkiv, Lund.

Max-Neef, Manfred et al., 1989: Human Scale Development. An Option for the Future. Development Dialogue 1989:1. Dag Hammarskjöld Foundation, Uppsala.

Mårtensson, B., Alfredsson, B., Dahlgren, L., Grahm, L., 1988: Det hotade lokalsamhället (The Local Community under Threat). Statens Råd för Byggnadsforskning, Stockholm.

Nygren, Lennart, 1994: Trygghet under omprövning (Security under Review) (2. uppl.). Fritzes, Stockholm.

Olsson, Sven E., 1990: Social Policy And Welfare State in Sweden. Arkiv, Lund.

Parsons, Talcott, 1966: Societies: Evolutionary and Comparative Perspectives. Prentice-Hall, Englewood Cliffs, N. J.

Strzelewics, Willy, 1983: De mänskliga rättigheternas historia (A History of Human Rights). Ordfront, Stockholm.

States of Disarray: The Social Effects of Globalization. 1995, UNRISD, Geneva.
Turner, Stephen P., 1990: The Impossible Science: An Institutional Analysis of American Sociology. Sage, Newsbury Park, Calif.
Wolvén, Lars-Erik, 1990: Jakten på det goda livet (In Pursuit of the Good Life). Rabén och Sjögren, Stockholm.

Grass-Roots in the New Age

Experiences from Västernorrland

SOCIAL ASSUMPTIONS

An elucidative discussion of social mobilization in the regions has to start by uncovering with care the vital core of social life and its germinal conditions.

The starting point of most forms of social life in our species is the basic network of personal eye-to-eye contacts in its many varieties. Approaching the subject uncritically, it is easy to imagine that the nuclear family is at the heart of a functional definition of collective life. But in our era the family community has been stripped of so many vital functions, and is so loaded with prejudices, as to have become virtually unusable for a progressive discussion. Irrespective of our opinion concerning its intrinsic social value it is recommendable to look for the subject of a constructive discussion in other quarters.

The most striking manifestations of primary community life seem to be found today in the popular trend that has long been called *the grass-roots movement*. The trend is one amongst several which express citizens' protest against the ongoing technocratization of society, such as the civil society and communitarian movements. While the civil society debate opposes the hierarchical stratification of society and communitarians primarily its socioterritorial organization, the grass-roots lay stress on functional matters. Their object is the mobilization of the individual or group in distinct projects, preferably in the local community.

As such the movement is more pragmatic than the other two, and therefore more efficient in the short run as an expression of citizens' will. Whether the aim of social research is to investigate the way in which primary communities function or to discuss strategies for social mobilization in a global context, the grass-roots remain a most rewarding phenomenon to study. This is why the committee of the Örnsköldsvik Conference on Social Development chose the theme "Grass Roots in the New Age" for one of the study groups which took part in the conference.

The group which was formed for the purpose included the following members: Gunnar Holmberg - recreational services consultant in the Municipality of Örnsköldsvik and for several years coordinator of the Gullänget District project; Elinor Bergkvist - youth activities officer and member of the grass-roots association "Husrådet" (Housing Council) in Köpmanholmen; Per Lennart Ledin - of the Municipality of Sollefteå, chairman of the "Sambygd" (Village Co-Operative) network in the County of Västernorrland; and the vice-chairman of that network, Klary Vågström, of Örnsköldsvik. The group also included Per Råberg, from the Group for Humanist Futurology in Umeå.

The result of a couple of brainstorming sessions of the group during the spring of 1993 is reported on below.

GRASS-ROOTS HISTORY

The prefix *grass-roots* in the phrase *grass-roots movement* is a metaphor which indicates two of the fundamental characteristics of the movement. While the first element hints at the broad popular ideal, the second denotes its location in a specific place. Genuine grass-roots action can thus be described as a civic initiative which has a more or less permanent local base. We shall henceforth adhere to this definition, to which we add only the functional perspective mentioned earlier. The apparent modesty of the term, as a matter of fact, conceals a conceptual core with a revolutionary potential. More will be said about this later.

The modern grass-roots movement arose in the 1960s, and represented an important aspect of the new radical mentality which culminated in the youth revolt of that decade. The movement was actively engaged in the showdown with the technocratic welfare model, the authoritarian style of which reached its climax at this time in ruthlessly industrial forms of social administration. As an alternative to the mechanical visions of the systems society, quality of life, smallness of scale and local democracy were held up as new ideals, and alongside the demonstrative internationalism of the day the local community was asserted to be the place for a responsible social life.

In this guise the alternative quickly became a vital - though rather irrational - factor in the expansive socialist mass movement. Notwithstanding this, it was only in the 70s that the alternative movement really began to flourish. This was partly due to its connection with contemporary ecological currents of thought, which soon led to its radical message spreading far and wide. Local action groups were established all over the industrialized world, and still further away. They became the popular vanguard in the struggle against widespread environmental destruction, the arms race and unbearable social discrimination. "Think globally, act locally" became the well-known slogan of the many radical youth groups.

Grass-roots idealism also became a model for many social civic initiatives, whose aim was to create a more viable community around the family and the

“Angelica Silvestris”. Photo by Margareta Klingberg

primary collective. Village organizations, housing collectives and neighbourhood associations of many kinds arose. In the burgeoning environmental parties the local initiatives were a constant feature, not least because of their ability to arouse public emotional involvement.

The success of the grass-roots during the 60s and 70s was to no small extent a consequence of the almost symbiotic relationship with first the socialist and then the environmental movement. At the same time it must seriously be asked whether this close relationship led to a loss of the genuine communal identity in the process. Neither the ecological nor the socialist movement, in their classical form, is ideologically bound to the local perspective; their approach is, on the contrary, inclined to be global, in the latter case also with a programmatic environmental alignment. In the policy discussions of that time the intrinsic values of the local and communal perspective always take second place to the "great" issues of the day. The attractive examples of the grass-roots initiative were certainly indispensable as a part of the propaganda. But even radical people were not ready to take on board and work through in their minds the inherent potential for a radical systemic change in society which the initiative implied.

A similar ambivalence between intellectual and emotional reaction was characteristic of the establishment's attitude to the movement in the 70s. Grass-roots thinking and planning relatively quickly gained a foothold in practical community work and earned much appreciation. The indifference of the political and intellectual elite was in striking contrast. To their hidebound way of thinking the local idealism of the grass-roots people was little more than a nostalgic futility on the margins of society. In no way did it shake the visions of progress through growth of scale and power hierarchies and of a future technological world culture. No new systemic suggestions for social change were to be found in initiatives at the grass roots-level.

THE MOVEMENT OF THE 90S

For a long time many people regarded the alternatives as a transient phenomenon - and some still do. This particularly includes politicians, whatever their hue. Yet as we approach the midpoint of the 90s, the grass-roots syndrome is still alive and well and acting as a vitalizing force in society. Not only do its ideals live on in people's dreams of the future, they have also been channelled into development work in wider fields of activity. In many quarters alternative work has become an accepted, indeed an absolutely essential, feature of practical community work.

This is a trend which is connected with the current crisis of the state but also with the fact that radical initiatives have been launched in new areas. Whereas the movement of the 60s and 70s - despite the attention attracted by the so called "green wave" - was highly an urban phenomenon, the progressive drive of the 90s is chiefly linked to the provinces. It forms a dynamic component of mobilization strategies which population and community activists are testing in many crisis-threatened parts of the countries in order to keep their local community alive and their region viable.

In the green wave of the 90s, ideals such as local democracy, self-reliance and quality of life count for as much as before. But a more heterogeneous body of activists is now expressing solidarity with the idea: pensioners' and youth groups, entrepreneurs and co-operatives, representatives of municipal and regional authorities. As now practised, grass-roots are slowly moving from the position of a protest movement to a progressive current in the mainstream of development.

Under the label of experiment, radical initiatives occasionally gained legitimacy back in the 80s. Now there is less reservation in delegating not only functional but also administrative responsibility to alternative groups. The municipalities themselves even take the initiative in what might be called communitarian administration and civilian planning. Specialist authorities are shouldering the task of initiating research and launching projects to develop the ideas - in Sweden the National Rural Area Development Authority and the National Board of Housing, Building and Planning are two examples.

GRASS-ROOTS' WORK

The future looks - at the time of writing - fairly promising for the grass-roots tradition. This is a forecast which will become more authentic according as the actions acquire the character of popular movement and the confidence to assume a social responsibility.

To accelerate this process and avoid pitfalls, however, it is important at last to pluck the radical topic resolutly from the category of marginal exception and to place it centre stage in the debate on our future. Even if only to be spared the misplaced arrogance of the political and scientific establishment, it is a matter of urgency to demonstrate that the approach is rooted in an *epistemological systems perspective,* though one which is based on a humanist and ecological stance rather than on increasingly anachronistic mechanistic and technocratic ideals.

Above all, however, the problems of the 90s confront the movement with a series of new tasks of a practical nature; it is now important to take an active part in responsible regional work for the future. This is a job which involves both short-term and long-term perspectives and which demands the capacity for new radical thinking. Both perspectives raise the question of a review of the strategies for social action and an open discussion of new themes in grass-roots work.

Such a discussion is in fact already in progress in several regions. As a modest contribution, we here bring forth some ideas which have arisen during the study group discussions prior to the Örnsköldsvik Conference. The ideas embrace themes which in our opinion ought to be central to progressive alternative work in the 90s. Some of them have found their way onto the agenda, whereas others are still waiting for an initiative.

Grass-roots co-operation

The establishing of regional networks is a current trend, which may serve to stimulate civilian initiatives in a county like Västernorrland, particularly in its rural areas. Such networks may be used for supportive individual contacts, exchange of practical experience, planning of concerted action etc.

Communication and exchange are strengthened if the network builds up an IT net with local terminals. The establishing of IT nets is a major common interest of the municipalities, by virtue of the social inspiration which such projects entail.

Grass-roots awareness

Deepening of civilian awareness and increased knowledge of theoretical fundamentals is important if the radical effort is to provide the impulse for regional development. Regional centres can be set up for conferences, courses and the dissemination of information, possibly in collaboration with a sympathetic folk high-school.

A "grass-roots academy" of this kind may also run projects of the "dig where you're standing" type, or pursue critical social journalism. Maybe a regional members' newsletter can be published.

Specialization in rural community matters

Specialization in rural community matters and social development is a characteristic of the grass-roots activity of the 90s. There is reason to encourage this tendency in the County of Västernorrland also. It both represents a productive contribution to the development of the county and helps to reinforce regional self-reliance. This kind of initiative will serve long-term development best if the radical profile can be retained in the projects.

Grass-roots administration and policies

Delegation of administrative responsibility to the local level is an expedient which is now being tried in rural area politics. Self-management committees, composed of local consumers and associations, manage the budget and the planning of certain fields (e.g. in the social sector). But there are also more informal committees with responsibility for development. One example is the "Husrådet" (Housing Council) at Köpmanholmen, in the municipality of Örnsköldsvik.

Both versions represent interesting examples of the return of influence and responsibility from the administrative level to the users. There is an important social learning function in this, of value both to individuals and to the future of the region. Perhaps, given a greater local awareness, a political dimension can be added to the administrative one. This presupposes that the time is ripe for the idea of a political party whose main task is to represent the local community level and its values.

A RURAL COMMUNITY NET IN VÄSTERNORRLAND

Developments in grass-roots work in Västernorrland are sufficiently interesting to deserve more detailed discussion. The main focus of interest, according to the Umeå group, is the new network "Sambygd" (District Coop) and its activities.

"Sambygd" was formed in February 1993 at a meeting at Hallstaberget Hotel in Sollefteå. The initiative came from various local interest groups, which formed a working party and forged links with the county administration. Representatives of all the village organizations and local interest associations in the county had been invited to the meeting. These were the local activists and live wires of the rural community, and here they were brought together for a joint discussion of the regional future. "Sambygd - Västernorrland's Rural Community Council" was the full name adopted by the network at the meeting.

The primary task assigned to the Rural Community Council was to support the village organizations in their radical efforts to ensure the survival of the communities and to strengthen local identity. Among concrete village projects which were considered may be mentioned the establishment of shops and local companies, the provision of assembly rooms and recreational premises, the retention of the postal services and schools, good telecommunications and road transport, and the preservation of public health services in the local area. The coordinating of all these wishes in a common rural area policy for the county was seen as a fundamental task, where the rural community council ought to take a leading part.

The attached map gives a view of the distribution of the local associations in the county. However, the map and the accompanying register contain only a selection (53) of the regional groups. The total membership is at present just over 100 village organizations and action groups.

Activation of a dynamic village policy in the municipalities is another part of the "Sambygd" programme. In addition, acquiring the status of a consultative body on rural community issues was set up as an objective. Collaboration between village organizations and their parent municipalities has also started in a number of cases. As an example of municipal integration we present a diagram of municipal neighbourhood organization in Sollefteå (Fig. 1). As this shows, representatives of several authorities belong to the local co-

Fig.1. **MUNICIPAL VILLAGE COOPS ORGANIZATION**

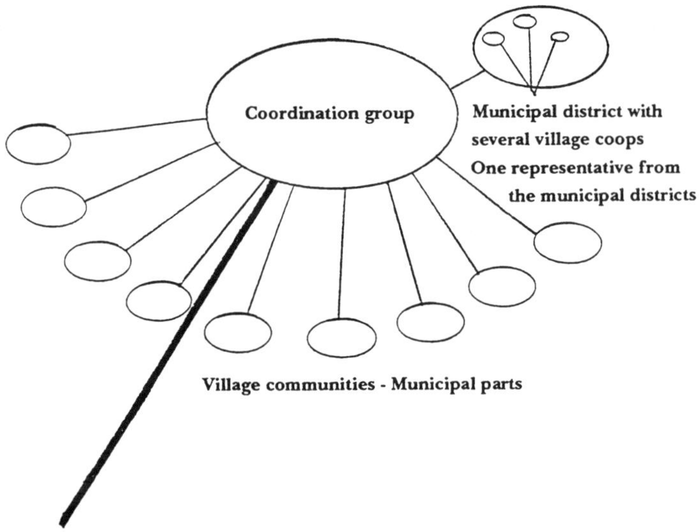

COORDINATON GROUP

One representative from each village coop
One municipal councillor
Trade and industry company M.D.
Chief municipal officer or head of information
Coordinator, municipality's representative in county district council

ordination group (including the director of the local department of trade and industry, the chief information officer and a municipal commissioner).

Another task of the council is to keep the regional rural community network alive and to promote purposeful communication between groups. Collaborating in the taking of liaison initiatives is high on "Sambygd's" list of priorities. Meetings with local groups take place and ideas for joint activities are ventilated. Current proposals include a training programme in project management and a catalogue of successfully implemented neighbourhood projects. A special seminar debate took place at "Sambygd's" annual meeting in 1994 and resulted in proposals for a joint rural community policy.

The county rural community council "Sambygd" is a genuine and inspiring example of local grass-roots mobilization in the mid-90s. Considering the short time during which the council has been active, it is an impressive start. Perhaps the most important result so far is that the local village organizations have become visible to the outside world in all their impressive diversity. This has strengthened the motivation of many small groups and increased their willingness to take action. At the same time this greater visibility has already caused the media and the public sector to show a new respect for popular initiatives. A certain reassessment of the potential for rural development (and therefore also of future rural policy) is in progress.

NATIONAL PERSPECTIVE

"Sambygd" is a gratifying phenomenon from the point of view of social and democratic community forces. What is equally remarkable is that it is not an isolated occurrence. Similar networks have sprung up in many parts of Sweden over the past few years.

Whether these are autonomous initiatives or a part of a chain reaction is hard to say, but their appearance in widely separated parts of the country has been almost simultaneous. Among the very earliest examples is the Västerbotten rural community council, which began discussing co-ordination in the autumn of 1992. By March 1994 there were eight county rural community councils in the country (Norrbotten, Västerbotten, Jämtland, Västernorrland, Dalarna, Örebro, Skaraborg and Kalmar), and a council for the Stockholm archipelago. Activities and profiles vary between the counties, but local development issues are the focus of interest in all of them.

The efforts of the regional rural community councils to coordinate their work deserves comment. As early as in November 1993, the movement called its first national meeting in Östersund. The primary purpose was to ascertain what common interests the associations shared - or perhaps what differences there were between them - and whether there was a basis for joint action. The feelers suggested a fruitful basis for collaboration, and the meeting led to the statement of a joint position and the drafting of an open letter to the Swedish government. In its statement the meeting demands that municipal general plans be drawn up, which take account of the needs of rural area policy in consultation with the village organizations. The letter to the government was formulated as an appeal for a co-ordinated rural area policy, one of the main issues being the retention of basic services to the communities. The need of new patterns of social thinking, starting from local conditions instead of from the central metropolitan view, was underlined.

Another advance at national level took place with the 1994 "Rural Area Development Conference" (Landsbygdsriksdagen) in Växjö, which was arranged by the Popular Movements Council for Rural Area Development in Sweden (Svenska Folkrörelserådet). In recent years this body has been dominated by the classical popular movements and a number of heavyweight rural area organizations. There has been a certain tension between the thematic and the new, locally based, groups. At the Convention the united rural area movement presented a demand for access to half the places in the committee and thus to an influence proportionate to the real importance of the movement today. The meeting decided to draft a motion incorporating this demand for the next Convention.

By their participation in the Rural Area Development Conference, the rural community groups contributed to the formulation of the twelve points of an active rural area policy which were adopted by the meeting for presentation to the Riksdag and the government and, not least important, for consideration in the regional policy bill being prepared at the time. The twelve-point programme is as follows:

- Integrated instead of traditional sectoral thinking in all social planning;
- Co-ordinated services;
- A bottom-up planning perspective (with the emphasis on greater influence for young people);
- Development of the local labour market;
- Local advice adapted to local conditions;
- Allocation of funds to women's interests;
- Recycling;
- Agriculture, forestry and fishing - the basic rural industries;
- A fully developed communications network throughout the country;
- Stimulation of housebuilding in rural areas;
- Culture and tourism as development factors;
- Research and development.

The rural community movement has attracted attention for its solid and vigorous efforts, which have in a short time earned its admission to the committee room of the national organization of the popular movements and even to the government's Regional Policy Council. But what is happening at the same time on the home front of the village organizations? Is there also creativity at social grass-roots level, and to what concrete initiatives is it leading?

It is difficult, if not almost impossible, to venture an opinion concerning the situation in the country as a whole. Random snippets of information show that there is an enormous capacity for initiative in many quarters and that ambitious projects with considerable local viability are taking shape. It is fascinating to see how in many places the rural area initiative has inspired hope and a will to survive in the rural population, and how a sudden glow of enthusiasm can suddenly suffuse a whole community, with its farms and fields, its wooded hills and winding streams.

In Västernorrland, as elsewhere, the vigour of the local initiative naturally varies between village organizations. Some are showing the way, while others are following in their slipstream. In some village organizations there is a lot of discussion, in others action is what counts. Which of them will find themselves on a sustainable course, only time will show. To obtain a glimpse of grass-roots life, however, we shall conclude this survey with a concrete example. We choose the Recreational Activities Association at Lungsjön, in the Municipality of Sollefteå; Lungsjön is located 17 km from the nearest sizeable centre of population at Ramsele and 92 km from Sollefteå.

SOCIAL MOBILIZATION IN LUNGSJÖN

The village of Lungsjön lies near the boundary between the counties of Västernorrland and Jämtland, in a typical rural area of the inland part of northern Sweden. The village dates back to the seventeenth century and is surrounded by extensive forests whose undulating topography, diversity of lakes and numerous watercourses give the countryside both a magnificent and a picturesque character. Forestry, small farming, hunting and fishing have until recent times been the main sources of livelihood.

The rural community, with its patches of cultivation, its road network and buildings, consists of a number of scattered farms, with traditional Swedish timber buildings and Västernorrland type facades. In the 30s and 40s the population was approx. 200 inhabitants, but it has now shrunk to about half that. The figures conceal surges of emigration, particularly during the 60s and 70s. Recently the trend has turned, and the current tendency is towards a return to the community, particularly on the part of young people.

The emigration of the 70s was accompanied by the disappearance of several important amenities from the area. The school was closed, followed by the local postal office and the shop. The loss of basic services meant a sharp deterioration in the quality of life in the village and was a direct threat to the survival of the community; here we have the reason for the first efforts at mobilization in Lungsjön. A local interest group was formed in 1977 on the initiative of a few vigorous spirits, and soon the municipality and the county council's rural developers had been prodded into action. One early step was to transform the closed village school into a camp school.

When we tested the local commitment in the early spring of 1994 the tone was optimistic and the enthusiasm for new initiatives seemed very high (interview with Staffan Magnusson, Lungsjön). One important reason is the return of young families, which has led to a rise in the number of children in the village from three in 1981 to thirty in 1991. But another reason is the formation of the regional rural community council. In many matters affecting the survival and viability of the village the county rural community council is credited with an impact which the particular group could never dream of achieving by itself. The capacity of the rural community for action has thus increased dramatically.

Lungsjön's local grass-roots have adopted the name "Lungsjöns fritidsförening" (Lungsjön Recreational Activities Association). The name reveals frankly the limitations of the ambitions. The inhabitants of Lungsjön have no plans to attract to the area companies which might offer its 100 inhabitants a permanent and secure livelihood. Most of the population travel daily to work in the local centre of Ramsele. Forestry and farming on a limited scale constitute the local subsidiary occupations.

Mobilization of the rural community has had results in three fields. The first is the maintenance of local basic amenities, the most important matter here being the reconstruction of a village shop with a basic range of goods. The second field is tourism for nature lovers, and over the years a lot of voluntary work has been done (long-distance footpaths, illuminated ski trails, chalet holidays, fishing and canoeing etc.). The third field is perhaps the socially most interesting, involving the conversion of the camp school into a community centre for the whole area (meetings, parties, lectures etc).

The implementation of these projects in the face of higher powers and through periods of hope and despair is in each case a long story. Together they attest, in particular, to the vitality and resilience of the population of a rural community which has consciously chosen its own way of life. The core of this is the close contact with nature for either recreational or utilitarian purposes; the inhabitants of the rural area are prepared to pay a high price in order to retain their identity.

EXPECTATIONS

How are we to summarize the impressions of our contact with the contemporary rural community movement; what hopes may be placed in the phenomenon and what misgivings may there be reason to harbour?

The hopes from a social research perspective are surely first and foremost that the initiatives will be able to infuse a social viability into the local community which will also light a flame in other groups and rural areas, and perhaps finally illuminate whole municipalities and regions. When we ask Anneli Bohman from Gårdskläppen in Västerbotten (a leading light of the movement) about her vision of the future, her answer shows that she sees the grass-roots effort as a national concern, not merely a rural matter. In her bold vision the next step is the invasion of urban areas and major conurbations by the rural community networks, in order to recreate a social vitality which has been lost in contemporary culture.

In this formulation the project undeniably looks slightly utopian. But the carrying message of this utopianism is not the fantasy but the vigorous optimism which emerges from an unbroken life force. This is an optimism which has the power also to inspire other groups and agents, residing in other quarters, to strengthen their objects. In a joint initiative grass-roots activism may merge with the movement for the civilian state and realize maybe the communitarian vision of a civic society. Such a scenario strengthens our hope to be able to reconstruct a living social culture on humanist foundings, which is a prerequisite of a well-functioning global community.

It is the closeness to the vital forces which decides whether such a project will be successful or not.

REFERENCES:

Berg, Micke och Lundström, Stefan, 1983: Skogsnäs. Ett arbetskollektiv (Skogsnäs. A Work Collective). Nordan, Stockholm.

Case, John, and Taylor, Rosemary (eds.), 1979: Co-ops, Communes and Collectives: Experiments in Social Change in the 1960s and 1970s. New York.

Christensen, Erik, 1990: Nye værdier i politik og samfund [New Values in Politics and Society]. Hovedland.

Ekhagen, Sven et al, 1984: Närkooperation - en strategi för samhällsförändring [Neighbourhood Co-operation - a Strategy for Social Change]. JordEco Förlag AB, Stockholm.

Friberg, Mats and Galtung, Johan (eds.), 1986: Alternativen [The Alternatives]. Akademilitteratur, Stockholm.

Friedmann, J., Rangan, H. (eds.), 1993: In defense of Livelihood: Comparative Studies in Environmental Action. Kumarian Press, West Hartford.

Ghai, D., Vivian, J. M. (eds.), 1992: Grassroots Environmental Action: People´s Participation in Sustainable Development. Routledge, London.

Herlitz, Ulla and Hahne, Gunnar, 1986: Glesbygd med livskraft [Rural Community with Vitality]. LTs förlag, Stockholm.

Jonsson, Britta, 1983: Alternativa livsformer i sjuttiotalets Sverige [Alternative Life Forms in the Sweden of the 1970s]. Sociologiska inst., Univ. Uppsala.

Kvaerner, Sverre, and Nakkerud, Mads, 1986: Idébok for en bedre framtid [Ideas for a Better Future]. Framtiden i våre hender, Oslo.

Pedersen, Knud P., 1984, 1986: Den nordiske udfordring (1-2) [The Nordic Challenge 1-2]. Grevas, Århus.

Rexroth, Kenneth, 1974: Communalism: From its Origins to the Twentieth Century. New York.

Roszak, Theodore, 1969: The Making of a Counterculture. Anchor, Garden City, N.Y.

Satin, Mark, 1978: New Age Politics - Healing Self and Society. Delta Publishing Co., New York.

Sundh, Kenneth, and Turunen, Päivi (eds.), 1992: Social mobilisering. Om samhällsarbete i Sverige [Social Mobilization. On Community Work in Sweden]. Publica, Stockholm.

Svensson, Ronny, 1984: Social planering i praktiken [Social Planning in Practice]. Socialstyrelsen, Stockholm.

DISTRICT COUNCIL "SAMBYGD".
Village Coops in Västernorrland County

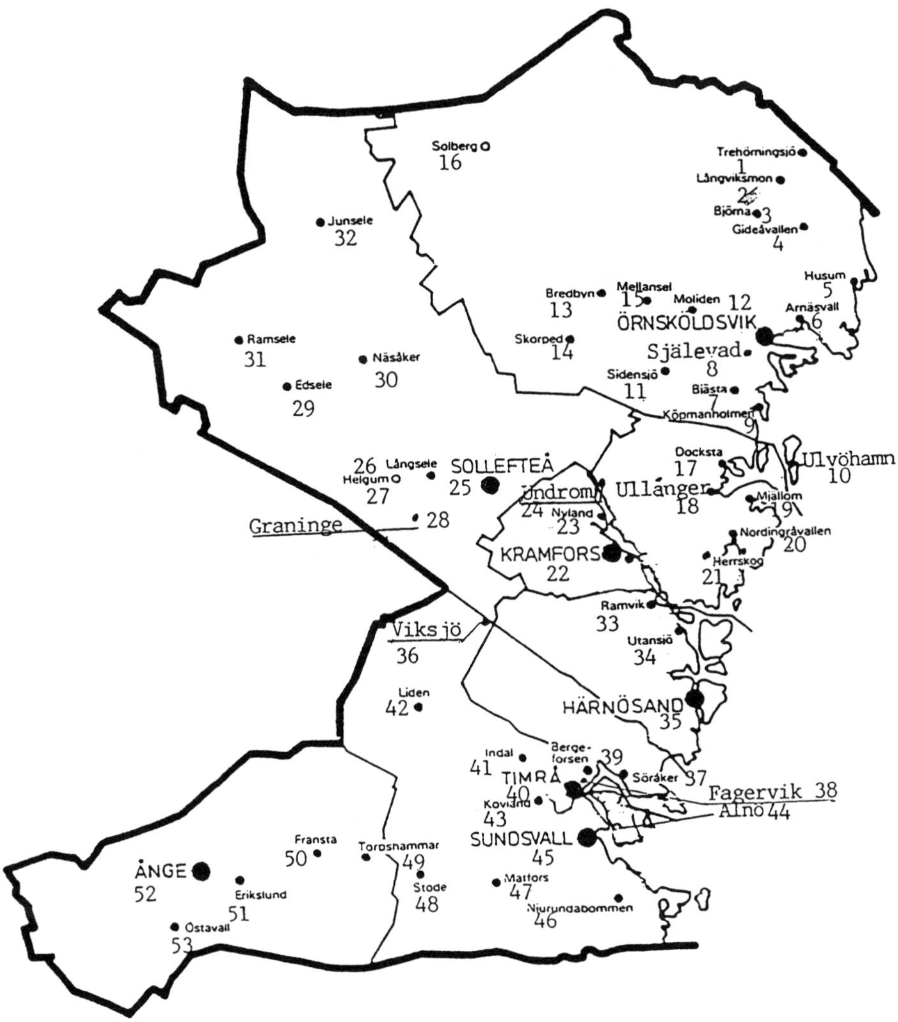

- ⬤ **Municipal centres**
- • **Other population centres**
- ○ **Rural areas**

MEMBERS OF "SAMBYGD" DISTRICT COUNCIL

1. Trehörningsjö
   - Trehörningsjö Interest Association
   - Kärrsjö Interest Association
   - Forsgården Cooperative
2. Långviksmon
   - Non-profit Enterprise Development
   - Långviksmon Interest Association
3. Björna
   - Campaign Committee in Björna
   - Björna Interest Association
   - Uttersjö Activity Association
   - Nyliden Interest and Sports Association
   - Hemling Interest Association
   - Remmarn Interest Association
4. Gideå
   - Gideå Community Committee
5. Husum
   - Campaign Committee in Grundsunda
   - Vallen School and Community Centre Association
6. Arnäsvall
   - Torsböle Interest Association
   - Campaign Committee in Arnäs
   - Ravesta Village Coop
   - Banafjäls Interest Association
7. Bjästa
   - Bjällsta Community Association
   - Utbygden Interest Association
8. Själevad
   - Västerhusorten Interest Association
9. Köpmanholmen
   - Trysunda Fishing Port Association
   - Köpmanholmen Society Association
10. Ulvöhamn
    - Ulvö Chapel Association
    - Ulvö Forum
11. Sidensjö
    - Campaign Committee in Sidensjö
12. Moliden
    - Alnö Village Coop
13. Bredbyn
    - Campaign Committee Bredbyn
    - Kubbe-Norrflärke Interest Association
    - Rödvattnet Tourist and Recreation
    - Långsele Interest Association
    - Anundsjö District Association
    - Campaign Committee Myckelgensjö
14. Skorped
    - Interest Association Skorped's Future
    - Djupsjö Interest Association
15. Mellansel
    - Gottne Area Society Association
    - Campaign Committee Mellansel
16. Solberg
    - Solberg Interest Association
    - Sjö Interest and Recreation Association
17. Docksta
    - Docksta Table Tennis Club
    - Käxed Interest Association
18. Ullånger
    - Ullånger Interest Association
19. Mjällom
    - Project Mjällom Activity Hall
20. Nordingrå
    - Part-time Activity Coop
    - Nordingrå Interest Association
21. Skog
    - Skog Campaign Committee
22. Kramfors
    - Gudmundrå District Association
23. Nyland
    - Ångermanland Home District Union Kramfors-Inland
24. Undrom
    - Boteå Village Coop
    - District Development Project Boteå/Styrnäs
    - Styrnäs' Future
25. Sollefteå
    - Nipstan Village Coop
    - Övergård Village Coop
    - Project Hunters
26. Långsele
    - F.L.U.
    - Campaign-90
27. Helgum
    - Helgum Association Council
28. Graninge
    - Östergraninge Interest Association
    - Graninge Estate Interest Association

Graninge Future Group

29. Edsele
    Edsele Village Coop

30. Näsåker
    "We Want" Committee
    Resele Village Coop

31. Ramsele
    Lungsjön Recreation Association
    Ramsele Future
    Skognäs' Incorp. Association

32. Junsele
    Junsele Parish
    Vallen Village Coop.

33. Ramvik
    Högsjö Trade Association

34. Utansjö
    The Utansjö Group

35. Härnösand
    Västernorrland Sports Union
    Häggdånger New Community Centre Association
    Gussjö Community Centre Association

36. Viksjö
    Viksjö Home District and Tourist Association

37. Söråker
    Söråker Forewards
    Tynderö Community Centre Association

38. Fagervik
    Medelpads Home District Union
    Indalsleden Association

39. Bergeforsen
    Lögdö Estate Interest Association

40. Timrå
    Bergeforsen Community Guild
    Laggarberg Comrades Association
    Timrå Handicraft and Industry Association

41. Indal
    Campaign Committee in Indal
    Indalsleden Project
    Indal Home District Association
    Gudmuntjärn Home District Association

42. Liden
    Västanå/Korsåmon Community Centre Association
    Liden Districts Future
    Liden Interest Association
    Campaign Committee in Liden
    Campaign Committee in Holm
    Holm Data Pool Association
    Holm Home District Association
    Liden Home District Association

43. Kovland
    Kovland Sports Union
    Sättna Home District Association

44. Alnö
    Alnö Home District Association

45. Sundsvall
    Selånger Home District Association

46. Njurunda
    Njurunda Home District Association

47. Matfors
    Tuna Home District Association
    Attmar Home District Association

48. Stöde
    Stöde Home District Association
    Ede Village Coop
    Campaign Committee in Stöde

49. Torpshammar
    Östra Finnbygden Interest Association
    Torp's New Tourist Association
    Mittpunkten Allköp Incorp. Association
    Naggen Incorp. Association
    Munkbysjön Interest Association
    Mittpunktsvägen Association

50. Fränsta
    Treklövern Roggafors-Finstaviken Interest Association
    Torpsbygden Interest Association
    Torp Home District Association
    Naggen Interest Association

51. Erikslund
    The New Hub
    Östby Community Centre

52. Ånge
    Borgsjö Home District Association
    Youth -90

53. Östvall
    Haverö Home District Association
    Haverö Data Pool

The Eco-School in Härnösand

Education for the Sustainable Region

Elementary schooling is a major instrument for implementation of the prevailing social norms and values in the young individual.

This is a basic observation which ought to be of interest in many provincial regions that are today being drained of human capital and skills because of the state of the national economy, and to serve as a reminder in their long-term political visions. Specially designed educational initiatives might awaken the local interest of the new generation and strengthen their feeling of regional identity. By such means young people may be induced to put down roots in their own home region and to make a contribution to living conditions there. Instead of, as now, packing their bags as soon as things get tough.

But what is actually happening? Does today's municipal school policy help the rising generation to identify with their home region? Does the work of educationalists in any serious respect produce a true feeling of local responsibility? If the answer to these questions is negative, which there is good reason to suspect, the next question is: Is there any point in trying to achieve a change in the direction indicated, and, if so, how should a project with such an ambitious objective be realized?

These were questions which the group carrying out the Västernorrland Life-Region R&D project considered interesting enough to merit special investigation. A suitable opportunity was presented in connection with the University Group's social conference in the spring of 1993. After testing professional reaction in the region, a joint project was started with a group of educationalists and school administrators in Härnösand municipality.

This article summarizes the result of that limited study. A provisional account was given at the conference in Örnsköldsvik in June 93.

THE ORIGIN OF THE MODERN SCHOOL

Let us begin the report with a quick look at the history of the modern school system. In order not to complicate matters more than necessary, we will confine ourselves to the European scene. The more generally applicable intentions of the project ought to remain quite apparent despite this restriction. The modern European school system, as still exemplified in the countries of Scandinavia and in Västernorrland County, has its origins in the early nineteenth century. Its growth and initial development show a close connection with the efforts to consolidate the new nation states which were accomplished all over Europe at that time.

The introduction of a state-controlled public school system was an important point in the political programme of the nation states from the very outset. The new concept, as a matter of fact, draws inspiration from two competing sources in the field of political ideas of the time. The first was the desire of governments to underpin the control systems of central bureaucracy in all corners of social life. The other was the aspiration to disseminate the ideals of the rising democracy and to make all citizens into members of the emerging national and industrial culture. At the crossroads of these conflicting intentions the modern school gradually took shape, and in the clash between the ideals the educational ideas which are still guiding the school curriculum of the 1990s began to develop.

Two main educational traditions emerge from these beginnings and they continue to struggle for power today. One is the rationalist orientation, which sees the school's main task as that of equipping its pupils with the knowledge and skills for operational duties in the labour market. The other is the cultural tradition, which regards the school's main purpose as that of imparting knowledge of the classical cultural heritage, but which in addition considers personal self-fulfilment a goal. It is easy to trace these two educational ideals back to the disparate objectives which in their day provided the impetus for the emergence of the modern school.

In the tug-of-war between educational ideals that has been taking place for far more than a century, it is the rationalist orientation that currently has the upper hand. Achieving the productive goals of our modern technological society calls for ever rising levels of specialist competence, and the school system is under constant pressure to adapt to developmental forces. In curricula and expert reports educational theorists and administrators fight a rearguard action to defend the classical cultural tradition, although actually offering no arguments other than a historical-philosophical rhetoric, which we begin to perceive as highly repetitive.

The ideological quandary described here is, for the most part, obscured in day-to-day debate by considerations of educational methodology. Discussion of methodology has been fuelled over the years by theoretical writings culled from the different stages of educational history: Pestalozzi, Kerschensteiner, Herbart, Dewey, Köhler, Piaget, Bourdieu are some of the best-known names. In the methodological debate we can follow the fashioning of increasingly sophisticated methods of imparting knowledge and training skills.

From crude beginnings in the authoritarian rote learning of the nineteenth century, school methodology gradually undergoes a modernization, reaching all the way to modern interactive teaching, group work, project orientation and pupil responsibility. This process has led to a humanization of instruction, which has often been praised. However, it must be pointed out that the humanization is confined to the methodological level; it hardly touches the thematic content of the syllabuses.

HIDDEN AND SUBLIMINAL CURRICULA

From time to time there has been discussion of the school's "hidden curriculum". By this is normally meant the role played by the teaching situation itself in relaying the educational message. The general idea is that the division of roles between teacher and pupil, the rigid regularity of the school timetable and the sterile school environment - regardless of the course content - imprint a conforming pattern of behaviour on the pupils, and thus discourage a liberating personality development.

This discussion has always been very relevant, but there is good reason to go into the question even more deeply. Alongside the hidden curriculum the alert listener can detect a *subliminal* (or *subconscious*) curriculum. This subliminal curriculum is based on socio-cultural norms that are so deeply embedded in the educationalists' mind that they are perceived as beyond question - where they are even noticed at all. In any case, the subliminal curriculum exerts a powerful influence not only on the behaviour of the teachers but also on the syllabuses. Despite this, it is hardly ever subjected to scrutiny or debate.

What, then, is the message of he subliminal curriculum? Its intrinsic purpose in today's school is to inculcate a cognitive perception of reality in the young. By *cognitive reality* is meant the picture of reality which appears when we take a reflective and purely intellectual approach to the world around. This is a world of facts and empirical data which are tied to causal processes and multi-layer hierarchies. Faced with this objectified structure of reality, cognitive man adopts an instrumental attitude; he reveals and exploits it. To assist him he has a series of technical tools, which he has refined to a dazzling degree.

The subliminal curriculum mirrors, as we can easily see, favoured ideals in the sophisticated information society of the late twentieth century. This is a high-technology world, which prizes cool intellect and rational talent. The unstated purpose of the subliminal curriculum is to prepare us for life in this specialized technological society. Silently but adamantly, therefore, the official curriculum is aligned to provide the instrumental and cognitive mediation of knowledge. Instinctively and as a matter of routine, the human material is sorted into a technical elite, an intermediate layer of administrative technicians and a bottom layer of operatives for the inescapable toil of an industrial culture. At the same time the objectivity and universality of the school's agenda is proclaimed to the children and the public.

Moreover, the cultivation and subsequent grading of cognitive talent by the school does not take place only in subjects of a mathematical, scientific and technical character. It is also practised in languages and in social or cultural subjects. Admittedly the techniques are more sophisticated there. But in the invisible staff meetings of the subliminal curriculum, the cultural wing very often shakes hands with the rationalist wing. The inclusion in the timetable of subjects such as physical education, music and crafts does not modify this judgment very much. Their purpose in the curriculum is often of a recreational, not a personality-forming, nature.

SOCIO-ECOLOGICAL PERSPECTIVES

How is the subliminal curriculum of the school to be appraised from a socio-ecological perspective of the kind tried in the Umeå Futurology Group's project? Is it reasonable or unreasonable, is it progressive or reactionary?

Our first and decisive reaction must be that apart from its striking elusiveness, it represents a cultural step forward, even a revolutionary one. The cognitive school, hidden or visible, confirms the progress of human civilization from an instinctive and intuitive stage of development to a reflective one. This step is based on an evolution of consciousness that has been taking place over millennia, but which has always remained the privilege of a limited elite.

Only with the advent of organized schooling is cognitive consciousness incorporated in the general educational ideals of society and in its institutional structure. This is an innovation which initially involved an enormous quantitative leap. But the quantitative leap opens the way to a second stage - a qualitative leap. Presumably we have seen only the very first manifestations of this process.

The other side of the coin is - as we have pointed out in greater detail elsewhere in this book - that unfortunately the gains in cognitive consciousness have been accompanied by a loss of interest in man's more fundamental forms of consciousness and their natural expression. These include elementary social communication and also sensory perception of the world around. They must also be considered to include the primary instrumental consciousness that we call “practical aptitude”.

All these are mental expressions of life which are deeply rooted in our nature. During earlier stages of man's evolution they formed independent networks of interaction between the individual and the surrounding world.

They were intimately connected with existence in a local or regional territory and together with this constituted a functional system, an ecological habitat. The science of ecology has described how these processes unfold in the life of the species on our planet.

The modern teaching system behaves as if these sub-cognitive levels of consciousness no longer existed and as if their realization in a social habitat were a stage that had long since passed. The local community, with its primary values and genuine life forms, has been effectively repressed in the minds of the educationalists and squeezed out of the curricula. For more than ten of the individual's most formative years his consciousness is instead occupied with cognitive training, which directs his attention to abstract worlds far beyond the horizons of local experience. *The modern school is in several respects the tool of a fatal deschooling of our elementary common sense which is taking place in society in our time!*

This is a provocative statement which needs to be clarified in order to be properly understood. The rigid insistence on cognitive training was probably a sensible measure when the democratic school was first introduced. The local community, with its social networks, sensory stimuli and popular ethics, was absolutely the dominant way of life and its despotism needed to be broken. Today the situation is almost the opposite throughout the developed world. The cosmopolitan technological society has hegemony and the local community has been almost eradicated as a social living environment. The cognitive curriculum strengthens this tendency and has in our epoch led to the near-total skewing of preparations for life towards the rational and technical sphere.

This alarmingly accentuates the trend towards a one-sided technological society. Other types of talent and attitudes to life are forced out into the margins of society and into sheltered activities.

There is good reason to note the entirely new set of problems for education which has arisen with the altered historical situation. And there is cause to start discussing the introduction of changes of a paradigmatic nature in the school system. Many desperate and destructive characteristics of modern youth culture are probably due to the lack of an integrated education which promotes an all-round personality development. Mainstream teaching today unintentionally creates an alienation of the individual, both from society and from his fellow-beings. An authoritarian insistence on the one-sided cognitive ideal can easily have destructive implications for the whole of society.

In these circumstances it seems a vital task of society to design as soon as possible educational plans which are suited to the primary world, to life in the habitat which forms the foundations for man's ecological existence on our planet. But even more than improved existential balance is to be gained. Such an initiative could in fact make a frontline contribution both to restoring the lost humanist identity of our society and to regaining the disrupted balance in man's environmental ecosystem.

A PEDAGOGY FOR REGIONAL IDENTITY

It is against the background of this set of problems that the prospects of an education designed to strengthen regional identity and solidarity ought to be considered. To what extent is traditional schooling conducive or inimical to a sound development of the regions? The first thing we should acknowledge is that here, too, cognitive teaching is an indispensable tool for progress. It represents one of the preconditions of the regions' ability to take an active part in cosmopolitan life with its networks of knowledge, ideas, services and goods.

Still, one problem is that the fruits of the cognitive curriculum do not necessarily benefit the home region. Schooling to produce a cosmopolitan competence and to prepare for a global labour market arouses, in the most gifted pupils, expectations which the home region may not be able to meet and, among the many, a demand for remuneration which may more easily be satisfied somewhere else. What often happens is that the prized elite of regional education say "thank you very much" and move to the conurbations or to other industrial centres favoured by the current economy. In its lopsidedness cognitive pedagogy strongly encourages the disloyalty of the young to their native region.

As a result of the outflow of its vital force, the level of competence and the quality of life of the provincial region will inexorably decline. No true economic or cultural progress can be based on such premises.

Many experts are nowadays taking part in a deterministic debate on a development which is draining the provincial regions and overfeeding the metropolitan areas. In the project "The Ecological Life-Region", we take a more constructive attitude to these regions and their future. The reasons for this are both humanistic and ecological, and have been explained elsewhere in the book. Actually, we believe that many provincial regions possess the potential for a community life of a quality which would place them far ahead of today's socially and ecologically sorely tried metropolitan areas. But they seldom seem to be fully aware of their advantage.

One complicated question is how the attention of politicians and population is to be drawn to the dormant regional assets. This question is closely linked with another: how is the school to be reformed to make it an instrument for local identification and sustainable regional development? Is it possible to devise a schooling which reinforces the cultural identity of the region? Can this schooling be introduced into the school system in such a way that its values can really assert themselves? And, at the same time, can this be done without the content of the cognitive curriculum being discarded?

Is it possible, in other words, to construct a dual pedagogy which satisfies the complementary needs of both cosmopolitan life and the local microcosm?

THE UNIVERSITY GROUP'S SCHOOL PROJECT

The project group in Umeå saw these provocative questions as an interesting challenge. We felt that the University Group's socio-ecological theory contained a number of ideas that were worth trying in the actual context. Socio-ecological theory certainly recognizes cognitive reflection as the most advanced level of human consciousness. But at the same time it identifies more fundamental and inalienable levels. An education that ignores the elementary levels is regarded as being in conflict with man's indispensable ecological nature.

The Umeå group has devoted considerable time to the designing of a societal model which bridges the gap between the fundamental and the advanced levels of human existence. In the various studies resulting from the group's experimentation, there are included many suggestions regarding a complementary "ecological" schooling. Perhaps the new ideas can be used in a curriculum designed to strengthen regional identity.

Admittedly, to some extent the theses of the Umeå group are in opposition to established education theory. This is true, for example, of the group's evolutionary view, which must be anathema to the classical cultural education theorists. It is also true of our notion that the child goes through several stages of development, which has often been denied by the rationalist wing.

Nevertheless, there are wellknown currents in the schooling tradition which harmonize with the ideas of a socio-ecological pedagogy. These include classic examples such as Maria Montessori's humanism, Celestin Freinet's work school, Ivan Illich's libertarian programme, Paolo Freire's dialogue teaching and of course the aesthetic idealism of Rudolf Steiner and the Waldorf School. Mention should also be made of the developmental psychology of Sigmund Freud, Jean Piaget and Erik H. Erikson.

Some inspiration is obtained from the modern school curriculum, especially from the subjects Art, Home Economics and Civics in the compulsory school. Various aspects of life in the local living environment are touched upon, and some interest is shown in sensory and intuitive talents.

The school project of the Umeå Group aspired however to produce more than a historical and theoretical discourse. The group hoped to go on to some form of practical experiment. After an elucidation of the ideological position, therefore, the search began for suitable partners for collaboration within the school world of Västernorrland. This turned out to be no easy task. We quickly found that the pedagogic climate of the 90s is very different from that of the lively 70s, when school experiments flourished in many municipalities. In today's staff rooms and professional journals, an ideological silence seems largely to prevail. The educational rebels and their attacks on all established systems are forgotten. And this is despite the fact that the liberalization and decentralization of the school has opened the doors wide to educational alternatives. The initiative today seems to lie almost entirely with school administrators with a brief to make cutbacks, even at the expense of patiently constructed curricula.

Still, there are exceptions to the dismal contemporary picture. The Umeå group was lucky enough to encounter one in its research territory. In Härnösand municipality, the local education authority and the staffs of the schools have been engaged for some time in lively discussion of the school system. The emphasis of these discussions is on an individual-centred, but also more holistic, elementary school in the municipality. Many suggestions have been made, one of the most interesting ideas being that of bursting out of the school's traditional classroom environment and moving certain parts of the teaching to alternative settings in the community. These were ideas not far removed from the project group's conceptions of an interactive and environmental schooling.

Mutual curiosity led to the formation of a discussion group. Several joint seminars were held in the local Education Offices in Härnösand. The main ideas were then presented at the conference on social development by the municipal director of education (Dan Holmberg). The preparations for a reform of teaching at the Johannesberg School, which were carried out by a team under the school's director of studies (Gerd Björk), were of special interest. The contact with the school led to an invitation to help in the designing of an alternative teaching environment. Plans to move the school created a special opportunity for an attempt.

The dialogue between the Umeå research group and the Härnösand group took place on two levels: one theoretical and one practical. In the remainder of this paper we shall summarize the results of the dialogue. But first a comment is needed on the strategic philosophy underlying the initiative. The Umeå group wishes to underline that ecological schooling is not to be regarded as an emergency measure, devised only to rescue regions threatened with crisis. The idea is proposed with a considerably greater vision in mind, and finally aims at the establishing of a curriculum which will help to strengthen the prospects of a sustainable global development.

If the objectives of the 1992 Rio Conference are to be fulfilled, it is within the system of social norms, and primarily by the education of the young, that the solid new foundations must be laid.

GUIDELINES FOR AN ECOLOGICAL EDUCATION

The main objective of our theoretical discussions in Härnösand was to specify the basic principles of an ecological education. This was a discussion which led our thoughts in several directions, but one in which the University Group's model served as a catalyst. It resulted in a number of propositions, which we boldly designate "Guidelines for an Ecological Education".

An interactionist approach

The new education takes an interactionist stance. The individual is seen as an ecological being, whose nature has been formed by the species' interaction with the living environment in an evolutionary process.

The purpose of the ecological curriculum is to promote the human qualities which are required for a natural interaction with the elementary living environment.

A developmental psychology perspective

Ecological schooling is based on psychological ideas of development. Development during childhood and adolescence is taking place in distinct stages. New aspects of consciousness come to the fore one after another and show maximum receptivity or creativity.

In the actual model the years of schooling cover three periods. The first (6-10 years) is primarily a period of holistic-sensory perception. In the middle phase (10-14 years) a physical-operative interest is the focus. Finally a rational-cognitive consciousness becomes more decisive (14-18 years).

An environmental teaching theory

Eco-schooling unfolds within a local and regional habitat. The adolescent's eco-identity is formed through intimate involvement in the matters of the surrounding community. It is realized in patterns which possess a social, aesthetic, technical or economic significance.

In eco-schooling the living space thus is more than an alternative classroom. It also contains the main educational matter.

A broad introduction to life

The task of eco-schooling is to give the child an all-round introduction to regional existence. In this it differs from the cognitive school, whose emphasis is on intellectual knowledge and specialized skills.

Schooling for life does not mean that the individual should be treated as an irresponsible infant for more than a decade. Every period of the life-cycle possesses unique qualities, which should be respected in the social life of the community.

Social responsibility of the educators

One element of an ecological school is the educators' responsibility for maintaining a dynamic relationship between school and society. This means more than merely overseeing the adaptation of the school to the outside world. It is equally important for society to be in harmony with the educational models, as long as these are founded on knowledge of the basic existential needs in man. Educators should help to keep this knowledge alive in society.

A dual pedagogy

Eco-schooling is not a total revolution against the cognitive education - but on the other hand it is no mere minor revolt. For the cognitive culture to flourish, necessary respect has to be shown to the evolutionary roots of man's nature, where the basic forces of life and social living are anchored.

The special task of the ecological curriculum is to underpin awareness of the fundamental human and ecological consciousness in the individual and thus in society.

OUTLINE OF A HUMAN-ECOLOGY CURRICULUM

The practical part of the deliberations of the study group concerned the design of the syllabuses for a well equipped eco-school. In view of the special profile of the alternative education, this was seen as being a matter of spatial organization as well as of educational content.

The question therefore arose whether architectural expertise should be engaged for the design of the educational programme. Contact was made with Krister Wiberg, an architect from Lund with unique experience of human-ecology planning matters. The task which we wished to discuss with the architect was the creation of teaching environments for the different age groups designated in the ecological syllabus. The following three schemes were provisionally proposed:

- For the age group 6-10: *an experiential workshop;*
- For the age group 10-14: *a work farm;*
- For the age group 14-18: *a future lab;*

Krister Wiberg was sympathetic to the group and produced the desired drawings with enthusiasm. The result was a series of architectural drawings of an eco-school, with special teaching environments provided for the three age groups. As may be seen from the drawings attached to the paper, the architect proceeded from the assumption that the eco-school was a complement to the new Johannesberg School in Härnösand. A concession to the utopian is that the school has been extended with a unit for upper secondary level. In actual fact Johannesberg is an elementary school (ages 7-16).

We shall end this overview of radical schooling ideas by presenting a draft of the projected eco-school's curriculum. It should be regarded as a counterpart to the instigating drawings by Krister Wiberg.

EXPERIENTIAL WORKSHOP

Ecological teaching environment for age group 6 - 10

Primary educational aim

The task at this level is to promote aesthetic life experience and the natural growth of sensory creativity in the individual. The child regards the surrounding life-world as an experiential whole, which is filled with sensory stimuli, empathic relations and imaginary pictures. He roams the space like an explorer and maps it out with the aid of his imaginative genius. Experiences and artefacts are captured and reshaped into subjective visions. Creativity is expressed in mimicry, words, images and scenographic installations.

The local and natural surroundings are the primary source of inspiration, and experience is coloured by strong imaginative wishes and intentions. The implicit goal of activity is to achieve an experiential identification with the basic living space. Identification with the local environment is seen as an important source of the individuals existential security.

Educational themes

Primary knowledge of the local district

In this part the child obtains experiential knowledge of the local community and its streets and spaces, different social activities and unique events. Imaginative curiosity also extends to the local cultural landscape and its perceived activities. A historical perspective is automatically included.

Social life knowledge

Here uncomplicated practical knowledge of the elementary forms of community life is given. The processes of family life is a basic source of inspiration. There are meetings with personalities and social groups in the local community, but communication within the peer group itself is also a part of the curriculum. During the eco-schooling process a democratic consciousness is spontaneously established, and respect for human freedoms and rights is awakened. A humanist social ethic is thus implanted.

Basic ecological knowledge

This stage promotes the formation of a basic ecological awareness in the individual. Knowledge of the local biocyclical rhythms and insight into the individual's organic involvement in the ecosystem, are the educational goals. These are achieved by unpretentious field studies in the surrounding area. A good starting point is the study of local forms of life in their natural setting.

Community profiling

The pupils' human-ecology study is presented in workbooks and displays, which are also addressed to a wider public in the local community. Emphasizing the intrinsic value of the age group's personal means of expression is an important objective.

The educational programme consists of external activities in the community alternating with internal teaching in locations close to the cognitive school. The internal environment is shaped to meet the age group's specific need of accommodation, but also to reflect architecturally the current level of development.

WORK FARM

Ecological teaching environment for age group 10 - 14

Primary educational aim

The task at this eco-schooling level is to consolidate an active interest in the local and regional operational environment in the individual. The adolescent regards the surrounding territory as a resource space, whose useful features he can exploit for processing and possible consumption. The mobile and manual resources of his own body constitute the primary instruments for the different operations, but are supplemented by a set of technical equipment.

The educational programme is realized in the local living environment. From practical experience the individual learns the devices and techniques of daily subsistence. The examples are taken from agricultural livelihoods, the local craft tradition and the family economy, in other words all the irreplaceable sectors of a sound ecological way of living.

Educational themes

Elementary eco-technology

Knowledge of the technical instruments of everyday life, their manufacture, use and, possibly, historical development, is essential to a civilized livelihood. To this stage can be added an ethical discussion of energy sources in the local community, both of an ecologically renewable and of a high-technology type.

Social knowledge

Social knowledge is a main theme, which at this stage is elevated to a practical standard. The focus is on working life both in its co-operative and administrative aspects. The elucidation of democratic and socioethical aspects of life also continues at a higher level and is related to the main social conventions, such as the family and the neighbourhood. A debate on political forms of organization now begins and is illustrated with simulation experiments.

Regional culture

The object here is development of the individual's aesthetic creativity. The training programme is given in the decorative arts, the theatre, music and dancing. Local and foreign popular culture serves as sources of inspiration. Here, also, activities are put in a living community perspective. Public exhibitions are arranged. The aesthetic living environment is a subject of discussion.

Community profiling

Community profiling continues to be an important aspect of eco-schooling. The profiling may be achieved by addressing activities to the public. Assignments are carried out for different customers, and shops are set up for the products manufactured by the eco-school.

The many educational activities belonging to this stage have been placed in a unit that we call the "Work Farm". Activities are divided into central handicraft and peripheral agricultural functions.

FUTURE LAB

Ecological teaching environment for age group 14 - 18

Primary educational aim

In the final phase of the eco-school the aim is to put the universal cognitive competence to the service of the home region and the local community. There are two sides to this: first, what creative contribution can the home region make to the cosmopolitan society; second, how can the home region benefit from the offerings of a global culture?

This brings a new dimension to eco-schooling: the dimension of modern science, technology, industrial production, and also of cosmopolitan culture and the mass media. At the centre is the body of electronic communications, by means of which the regional society can maintain an unrestricted contact with the global environment. One basic theme of this final educational stage is the development of competence in multimedia technology and inter-regional network communications.

Educational themes

The region in the global ecosystem

A high level of ability in global environmental and socio-ecological matters is built up via inter-regional IT networks. Study and research projects are designed in partnership with young people in other regions and countries.

The region in the inter-regional civilization

Similarly, IT networks are established between youth groups in the economic, social and cultural spheres. Dialogue projects are set up to discuss important social issues. In these projects the role of the region as a transmitter and receiver of cultural impulses is illuminated. The projects may be given a utopian, futurological character.

The region in the scientific cosmos

The study groups keep reasonably well informed - via the IT networks - about scientific progress all over the world. Applications in the local and regional community are discussed with the external contacts. But interregional network projects are also carried out, in some cases together with researchers from the regional university.

Community profiling

The study groups' projects address current regional problems. The results may be reported on in regional media and official publications. The groups also undertake small commissions for citizens' groups and authorities.

An environment to stimulate the imagination has been built up around the Future Lab. The environment contains advanced technical equipment and is functionally suitable for project and group work.

REFERENCES:

Brennan, Matthew, 1989: Människan och miljön. Handledning för miljöundervisningen [Man and the Environment. Manual of Environmental Instruction]. Utbildningsförlaget, Stockholm.

Egidius, Henry, 1978: Riktningar i modern pedagogik [Trends in Modern Education Theory]. Natur och kultur, Stockholm.

Freinet, Celestin, 1975: För folkets skola (For the School of the People]. Wahlström och Widstrand, Stockholm.

Freire, Paolo, 1970: Pedagogy of the Oppressed. Herder and Herder, New York.

Illich, Ivan, 1973: Samhälle utan skola (Deschooling Society]. Wahlström och Widstrand, Stockholm.

Imsen, Gunn, 1992: Elevens värld, Introduktion i pedagogisk psykologi [The world of the pupil. Introduction to educational psychology]. Studentlitteratur, Lund.

Isling, Åke, 1984: Grundskola för allmänmänsklig kompetens [Elementary School of General Human Skills]. Sober förlags AB, Stockholm.

Jerlang, Espen, 1989: Utvecklingspsykologiska teorier [Theories of Developmental Psychology]. Esselte studium, Stockholm.

Johansson, Henning, 1985: Elevernas eller skolans kultur och samhälle (The School's Culture and Society, or the Pupils?). Pedagogiska institutionen, Umeå universitet.

Kallos, Daniel, 1978: Den nya pedagogiken (The New Pedagogy]. Wahlström och Widstrand, Stockholm.

Marklund, Sixten, 1987: Skolsverige 1950-1975. Del 5 Läroplaner [Sweden's Schools 1950-1975. Part 5: Curricula]. Utbildningsförlaget, Stockholm.

Persson, Sven, 1993: Flexibel skolstart för sexåringar [Flexible School Start for Six-Year-Olds]. Studentlitteratur, Lund.

Rousseau, Jean-Jacques, 1964: Émile ou de l'Éducation. Paris.

Sandström, Carl-Ivar, 1989: Utbildningens idéhistoria [History of Educational Theory] (3. uppl.). Svensk facklitteratur AB, Vällingby.

Skola för bildning. [School for Education]. SOU 1992:94. Allmänna förlaget, Stockholm, 1993.

Svedberg, Lars, Zaar, Monica, 1989: Boken om pedagogerna [The Book of Educators]. Utbildningsförlaget, Stockholm.

THE ECO-SCHOOL IN HÄRNÖSAND
Annex to the Johannesberg School
Architect: Krister Wiberg

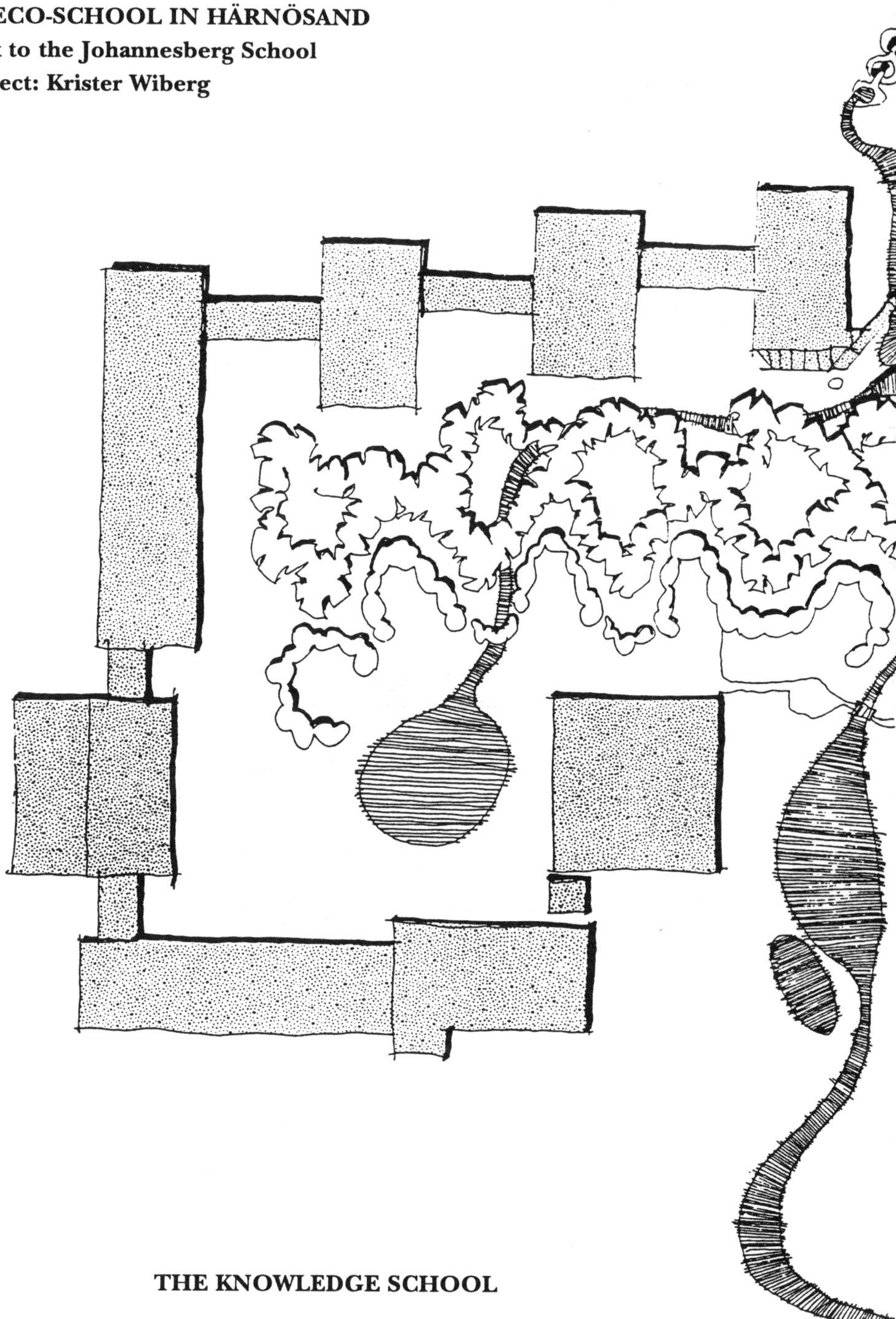

THE KNOWLEDGE SCHOOL

THE ECO-SCHOOL

ECO-SCHOOL IN HÄRNÖSAND
Structural plan

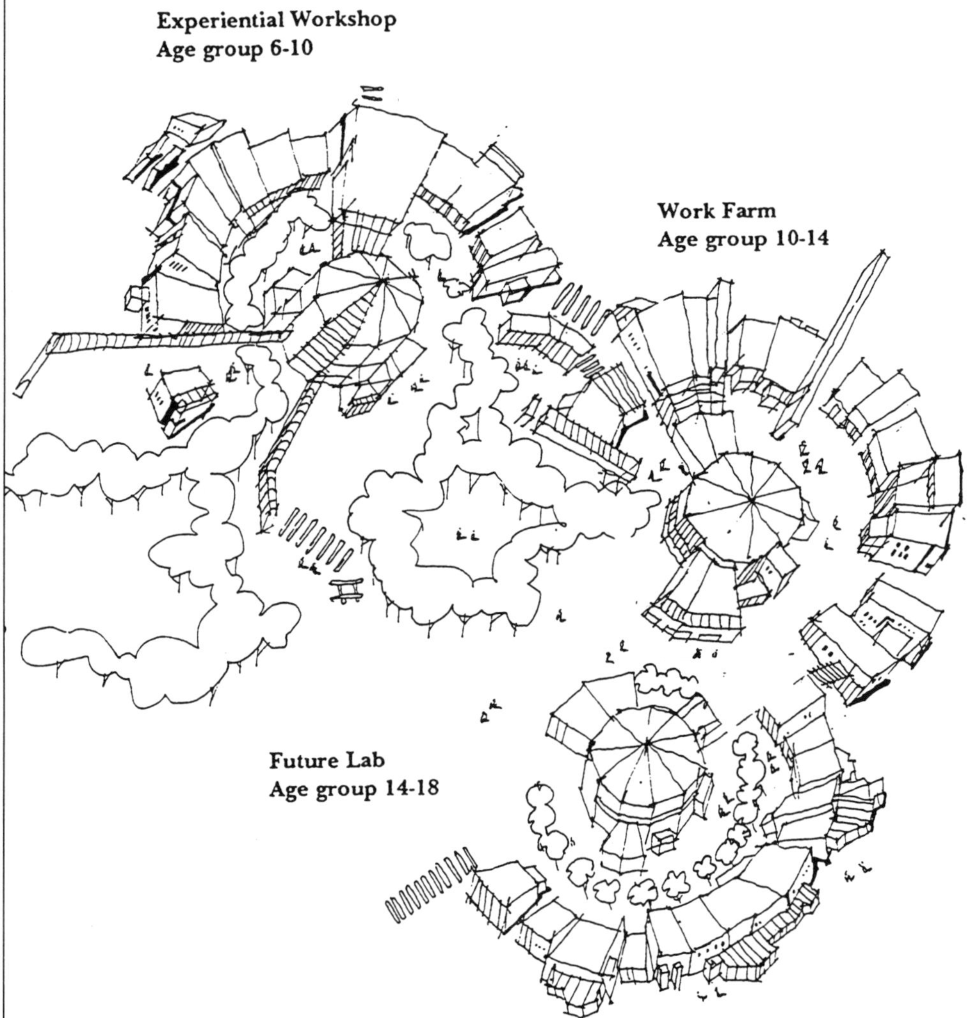

ECO-SCHOOL IN HÄRNÖSAND.
Functional plan 1: Exterior functions

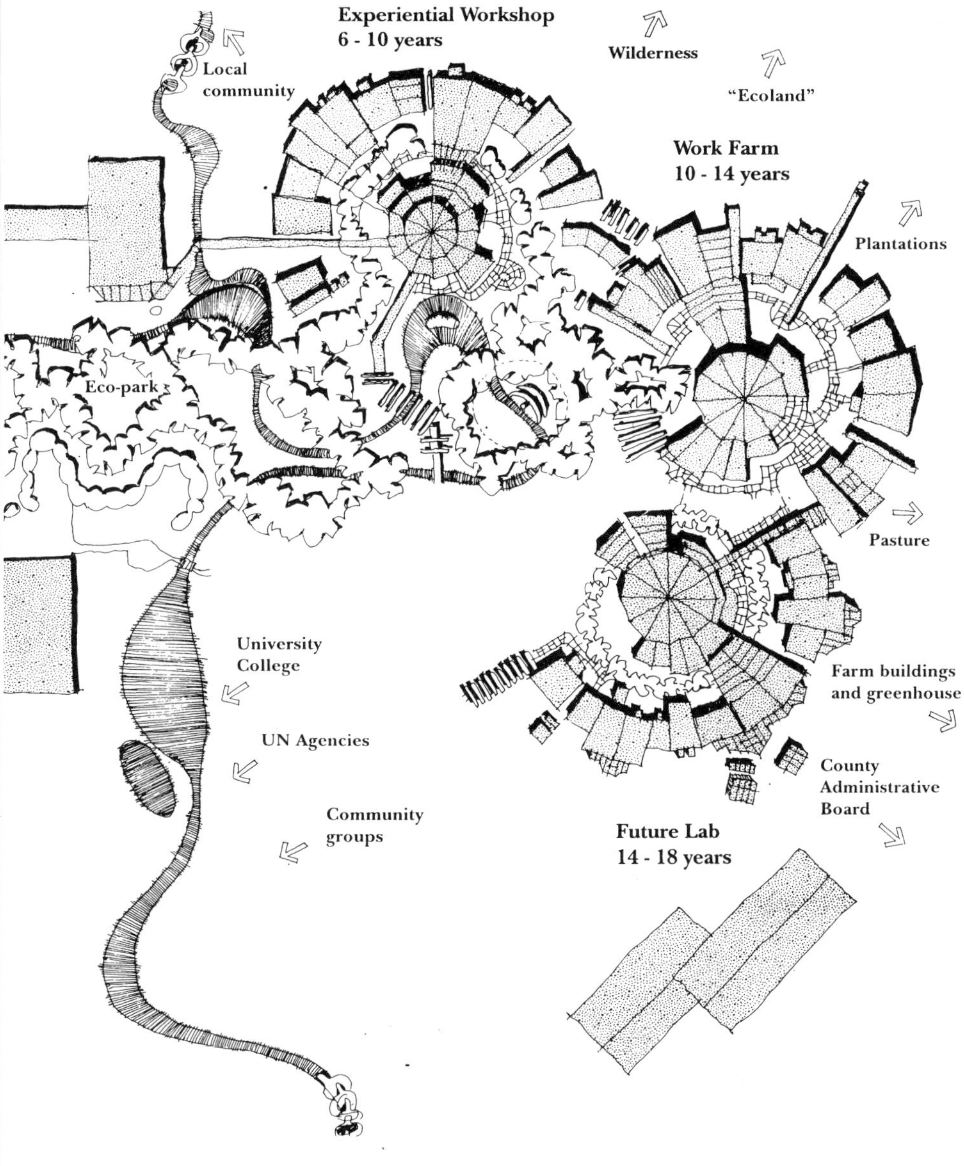

ECO-SCHOOL IN HÄRNÖSAND.
Functional plan 2: Interior functions

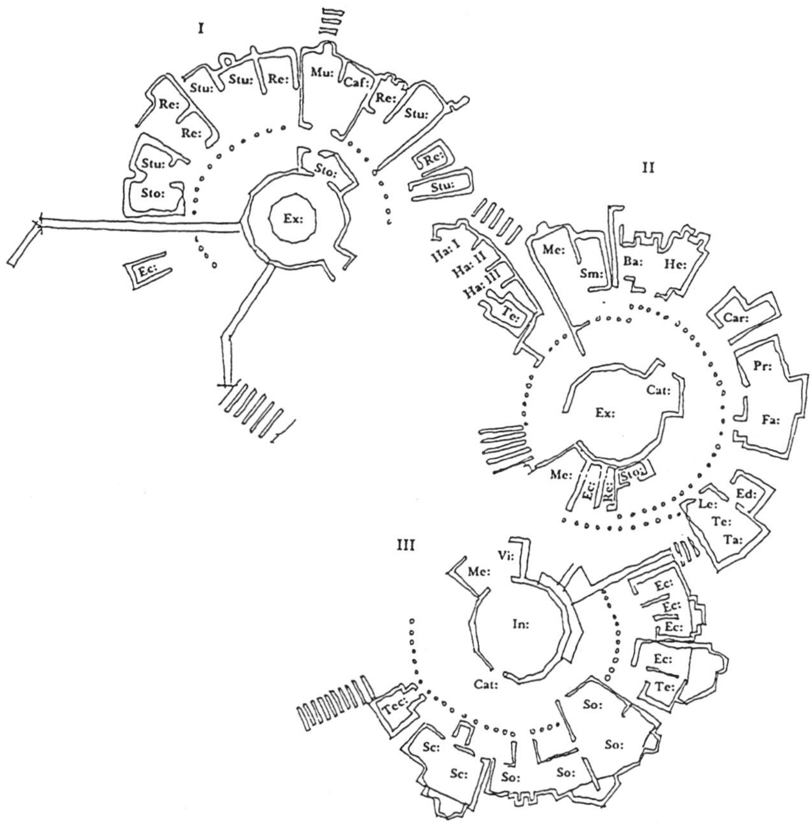

I. Experiential Workshop

Ec: Ecoforum
Sto: Store
Stu: Studio
Re: Recreation
Mu: Music and theatre studio
Caf: Cafeteria and assembly room
Ex: Exhibition hall

II. Work Farm

Ha: Handicrafts workshop
Te: Teachers' room
Me: Mechanical workshop
Sm: Smithy
Ba: Bakery
He: Health and care
Car: Carpentry shop
Pr: Print shop
Fa: Factory
Ed: Editorial and writing room
Le: Lecture-room
Te: Textile workshop
Ta: Tailors
Me: Meeting forum
Ec: Eco forum
Re: Recreation
Sto: Store
Ex: Exhibition and sales
Cat: Catering

III. Future Lab

Ec: Eco laboratory
Te: Teachers room
So: Socal laboratory
Sc: Science laboratory
Tec: Technical and system support
Cat: Catering
In: Information and contact centre
Me: Meeting forum
Vi: Video conference

Gullänget Town-Neighbourhood

An Experiment in Socio-Ecological Planning

A COMMUNITARIAN VIEW

Communitarian ideals occupy a central position in the regional development project which is run by the Umeå University Group in Västernorrland County. This is a fact which is especially relevant to underline in a paper which, like the one following here, is treating territorial planning in the municipality from an ecological angle.

Communitarian ideals imply, as the informed reader knows, that primary social groups are seen as the basic life cells of society, and that the interactive behaviour within these cells is regarded as society's life-giving essence. But in a radical view more than this is implicated: it means also that primary social groups and group values are regarded as the very basis of the overall societal organization. This is, in actual fact, an *ecological* idea which stands in opposition to mainstream definitions of "sociality" today.

Between the social sciences the community ideal has already provoked a century of discussion of the humanitarian concept of *Gemeinschaft,* which has - in a disdainful or sentimental tone - been contrasted with the anonymous *Gesellschaft* of the modern mass society. The well-known proposer of this distinction was the German sociologist Ferdinand Tönnies. Communitarianism as a modern phenomenon originally arose as a protest against the appearence of the anonymous Gesellschaft, which accompanied the growth of industrial society and the new urbanism in the West. It has retained its radical aura throughout modern history until the present day, within various political fringe movements and alternative currents of thought.

The intellectual history of the idea is interesting reading. It can be traced from the utopian socialism of the early nineteenth century (Fourier), through a phase of social philanthropy (Owen), and then via anarchism (Kropotkin) to the alternative communities of the 1960s and 70s. What is common to all these doctrines is that they stand up decidedly for the ideals of the primary population group, whether this is embodied in a rural village or an enclave of a teeming metropolis. Within their domain the community idealists have persistently cultivated the multifarious values of close community life, all the way from the theory of daily toil to the practice of human rights and freedoms.

In recent decades we have experienced that the community ideal has penetrated political thought and public planning in many places. In the 70s and 80s in particular, greater civic influence and local roots became popular slogans of both central and municipal planners. This is a trend which sensitively reflects both many citizens' belief that modern society is veering towards a technocratic dictatorship and the civil servants' growing need to rediscover the citizen behind the screen of official paperwork. In many quarters the community ideal has been incorporated in social legislation, leading to a reorientation of social services from welfare distribution towards participation and local self-reliance. Social scientists have been taking part in many projects with a bearing on the development of democracy and self-management in the community.

But this contemporary trend in social policy normaly remains deeply rooted in a tradition of officialdom, and depends on the latter's benevolence for its existence. It represents what has been called *a top-down attitude* to the primary life world. When the Umeå group sets out for a discussion in the community tradition it does so from a different starting point, more in line with the radical view mentioned in the introductory note. Its point of departure is a vision of society where the primary group ideal determines the entire social organization and defines its boundaries. The creation of national and even worldwide social networks in which communitarian ideals still provide the lifegiving spark is the ultimate dream of the socio-ecological approach. This is a utopia which emanates from *a bottom-up perspective* in which the dynamics of primary life is also the power source of the system world; it is the antithesis of the hierarchic dystopia which has shaped the contemporary techno-society.

This change of perspective gives rise to an entirely new strategical situation. In a society where the community ideal has gained power over collective consciousness, the civil communities must also assume responsibility for overall social politics. This means that they must themselves construct some form of public platform, although such a solution has long represented the incarnation of the repression they are trying to throw off. How can a public system level be created which does not at the same time threaten the freedom and well-being of the life-world? Solving this problem becomes the real test of communitarian thinking, if it does not wish to stop at an anarchistic level which forever confines it to the status of a fringe movement.

Trying to define a *public sector on the basis of communitarian ideals* presented the Umeå University Group - in the middle of its Västernorrland project - with an irresistible challenge. In the course of the social sector study we decided to tackle the problem. In our theoretical scheme, the primary task in order to achieve a viable solution in the field would be to establish a socio-ecological basic view, founded on delegation of responsibility "upwards" instead of "downwards". The public sector schould see itself as a responsive service agency of the primary networks, rather than as a superordinate authority. In order not to steal the initiative from the civilian life-world, it would be desirable to limit the institutions of the system level to the very essential. A classification of community institutions must therefore be undertaken from a socio-ecological standpoint. Certain public functions may be indispensable and can be taken over in their entirety (e.g. school and health care), whereas others need to be dropped. It is important also to point out areas which occupy a key function in a mobilization phase, in which the whole community is to be liberated from the prevailing state of technocratic repression.

PLANNING THE SOCIAL SPACE

One area which is probably going to occupy a prominent position in an alternative socio-ecological public sector is *physical community planning*. Physical community planning covers several important domains of practice and responsibility; one of them is the organized social space of the municipality. In modern society social planning often crystallizes at the point of intersection between the town planning and housing construction sectors. It is ordinarily located in an intermediate zone in the town plan and includes dwellings and residential areas. A tentative study to clarify *the scope of socio-ecological planning in the intermediate social zone* seemed to be a crucial task for the Umeå project.

The intermediate zone and its residential environment has for a long time attracted the interest of building researchers and housing sociologists. It has been a favored area of many institutes for building research in the industrialized world. However, in the socio-ecological perspective of the Umeå group the heart of the matter was not the individual dwellings, but the social potential of the residential area seen as a whole. The emphasis is displaced hereby radically from the housing modules to the external interactive environment.

Examining how the often monotonous housing of the intermediate zone may be transformed into a dynamic social life space appeared as the momentous challenge to the Umeå researchers.

The constraints of finance and time affecting the project meant that the subject had to be dealt with in outline. A suitable object for closer study presented itself at the centre of the project activities, namely in the Gullänget district of Örnsköldsvik. Gullänget is situated on the periphery of the town and forms a suburban environment which grew up in a fairly unrestricted way between the 30s and the 70s. It contains a mixture of residential and industrial buildings which has gradually become more dense. The area can be described as a district of the municipality, and covers a territory of approx. 250 ha. The population of the district in 1993 totalled 4.170 permanent residents.

It should be mentioned that Gullänget has been subject to considerable municipal interest in recent years. A new detailed development plan for the area has been drawn up and for several years there have been efforts to mobilize the district's social resources. A number of municipal departments took part in the latter project, in particular - initially - the recreational services department. Eventually these efforts also engendered a social science R&D project.

The Gullänget R&D Project represents a type of social research which was initiated in many parts of Sweden and Europe during the 80s. It started with the formation of a political democracy group at municipal executive committee level (1985). The group was briefed to examine methods of increasing civic participation in municipal matters, and a report was duly submitted. At a later stage the democracy group was transformed into a research and development group. One of this group's proposals was to start a project on democratic development in Gullänget. The actual project was placed in the hands of Matts Mattsson, secretary of the "National Association of Recreation Centres" (Föreningen Sveriges Fritidsgårdar). This research project, which has offshoots at both Gothenburg and Umeå universities, at the time of writing is nearing its conclusion.

The Umeå Group made contact with the actors of the Gullänget Project and suggested that an environmental renewal plan be prepared for the district, based on the socio-ecological ideas of the group. The suggestion was favourably received by the project leader, and also by a couple of leading architects in the municipality (Arne Wistedt, Kjell Mattsson and, at a later stage, Per-Eddie Bjuggstam). A discussion group was formed, to which were attached representatives of the municipal planning unit (Thomas Lundgren), the Gullänget Project (Gunnar Holmberg) and the local residents (Karin Engblom). A number of meetings took place at Örnsköldsvik Town Hall in the spring of 1993. At these a specific body of knowledge developed from analyses of the Gullänget Project, the local interests affected and the socio-ecological ideas. A survey of the existing residential area and its environmental characteristics was carried out. Several study visits also took place.

The next step was to design an alternative area plan for the district. There was a rush of ideas for functional and environmental innovations to enhance the existing surroundings. Eventually, the fragmentary ideas were consolidated into a *programme of socio-ecological environmental renewal.* The architects produced plans and perspective drawings for the entire area, and a scheme for the architectural renewal of Gullänget took shape. A preliminary version of it was presented at the University Group's social conference.

In this report we will present an updated version of the Gullänget renewal plan. The updating was carried out for the exhibition entitled "Gullänget Takes Off" (Gullänget lyfter), which was held in conjunction with the Örnsköldsvik Centenary in the summer of 1994. It must be stressed that the renewal scenario is still merely an outline. The outline is partly based on current trends in social planning but also presents a number of new socio-ecological ideas which we would like to offer for public discussion.

In order to facilitate an evaluation of the plans, we shall start with taking a look at the urban planning tradition which forms the historical background to our proposal.

THE COMMUNITARIAN PLANNING TRADITION

Community ideas have at intervals played an important part in developments in architecture and urban planning throughout the 20th century. Camillo Sitte's analysis of the social aesthetics of the medieval town and Ebenezer Howard's Garden City vision at the turn of the century was a start which has never ceased to intrigue the planners and architects of different epochs. The Garden Cities of the 1910s, the Grosshof blocks of the 1920s and the neighbourhood ideas and community centres of the 1940s are planning archetypes which all have their roots in a communitarian tradition. The British New Towns of the 1950s were also in essence a humanist protest against the remorseless growth of the contemporary conurbations - in this case the metropolis of London - and the loss of social identity which resulted from this growth. The New Towns became a popular town planning ideal which rapidly spread around the world.

In town planning, however, even more than in other societal areas, it is wise to distinguish between theory and reality. The observant historian of architecture will quickly realize that communitarian care has often served as an advertisement for a functional planning the true aim of which is continued growth of the conurbations and confirmation of the anonymous mass society. The readiness of the professional corps to stick a social label on their contribution has lent legitimacy to a development which has often pointed in exactly the opposite - the antisocial - direction. Modern social planning opportunism shows its true face in the colossal high-rise suburbs of the 60s and 70s, built using purely industrial production methods. In their human and environmental poverty these areas denote the high-water mark of the technocratic planning epoch and at the same time the point at which its inhumanity became unmistakable to all.

The anti-humanism in the urban environments of Late Functionalism eventually aroused widespread protest among the ranks of the architects, and its deterrent examples led to a new flourishing of communitarian ideas. The

70s and 80s are a period when a more credible social ethic appears and leaves many traces in home-environment, neighbourhood and district planning. A totally different scale is the first attribute of the new projects, a scale which places the individual and the neighbourhood in the centre of the community plan. High-density/low-rise is the motto of a whole generation of social planning experiments. The emphasis of the experiments may lie, as with the architect Ralph Erskine in Sweden, in the careful organization of social functions in close consultation with residents. In other cases - as, for example, in several projects of the Danish "Vandkunsten" group - architects have placed the accent on the design of the residential area as an elegant aesthetic setting.

Housing and building research organizations quickly adopted the new ideas. They assisted in urban renewal projects in which proximity, self-administration, care and environmental quality were passwords. Feminist research groups were formed to consider such themes as "the new daily life", and formulated radical schemes for the total home environment. Many architects became involved in social processes which occurred spontaneously in the community. They helped to build sheltered and collective accommodation for special groups or housing communes for experiments in social living. The collective unit for which plans were made could be a neighbourhood community in a renovated inner city environment or a village collective far out in the countryside.

Interest in the community perspective unfortunately has ebbed appreciably among architects in recent years. This is partly due to the susceptibility of the profession to changes in fashion, which do not always logically reflect the deeper needs of society. From a socio-ecological standpoint there is also cause for critical comment on the architect-created alternatives mentioned above. Neither in theory nor in practice was much concern shown for the relationship of the new social environments to the surrounding community environment. The neighbourhood often appears as a social oasis in an urban or provincial setting which may be a spiritual or material desert. But perhaps this criticism is a little harsh. These were the circumstances in which the socially interested architects had to operate. Otherwise the sketches would have stayed on the drawing board.

Of the new architectural trends that followed the communitarian wave, Post-Modernism is the most important one in its impact. It represents a big-city architecture with almost nineteenth-century aesthetic ideals. Certainly the results have often been fascinating from an artistic point of view. But there are socially more progressive ideas in the ecological planning impetus which developed also in the 80s, and which still possess momentum in the mid-90s. Introducing principles of sound, natural living on the terms of the bioenvironment and the natural setting is the main item on the agenda of ecological architecture. It has not as yet developed into a major trend such as Post-Modernism, but it is being cultivated assiduously by a growing number of advocates, who have obstinately ignored all rumours of its imminent demise. This trend has got increased momentum by the growing political interest in the introduction of ecological environmental norms in town planning in recent years.

We find the roots of the new planning ecology in the domain of environmental ecology, and it is from there, too, that the main watchwords have been borrowed. Recycling, reuse of organic materials, renewable sources of energy, pro-environmental technical systems and biodynamic cultivation are codewords which have inspired the environmental projects of the architects and the urban planners. The last ten years have seen the growth of an abundance of ideas and schemes in architecture, neighbourhood design and urban building which have been launched with the prefix eco- as a brand label. To the uninitiated, the sight of solar collectors on roofs, fully enclosing glass facades on south sides, greenhouses and gardens, carefully designed composting places, perhaps a windpower generator or a waterstairs, states that ecological experimentation is in progress. It may involve an isolated building that the architect has shaped into a circular system, or whole urban renewal projects such as “Green City” at Slagelse in Denmark. Block renewal in the heart of the big city and eco-villages in pastoral settings are found in the repertoire.

Ecological experimentation in architecture and social planning is important from the point of view of local and global environment protection, and we may expect it to receive increasing support. But after following the trend for a number of years one cannot avoid noticing that the ideas often follow a rather ritualized pattern. The ecocomplements sometimes acquire an irritating air of token action, and it is hard to see how they contribute to the solution of the real environmental crisis or to the promotion of a healthy life in the community. Similarly, one may wonder why ecological interest never extends to regarding man himself and his social setting as an ecological system.

Despite this it is important not to underestimate the ecological experiments. Even in their limited symbolic role they reflect an alternative stance with regard to both planning and community life, which is worth remembering because it may contain the seeds of a new general planning ethic. Clearly this planning ethic also includes communitarian ideals, even when this is not stated explicitly. This is very evident from the areas of application which are chosen. But it is necessary for the social circumstances to be clarified more than hitherto, as ecoidealism can quickly be reduced to just one more rhetoric among the others serving to give legitimacy to the next wave of technocratic social control. This wave is in fact already building up.

THE SOCIAL CYCLE

The community tradition in social planning has been an important guideline in the proposals sketched by the Gullänget group. At the same time a number of eloquent examples have drawn our attention to social shortcomings that we wish to avoid. They have helped us to articulate an autonomous operational profile. It is appropriate to try to describe this profile before presenting the outline material.

As has already been implied, the Umeå Group's model differs (to some extent) from the present communitarian tradition by virtue of its *socio-ecological* orientation. This means above all that the neighbourhood zone is brought into a total vision for the municipality and that a bottom-up perspective is applied in overall planning. The intermediate social zone is surrounded by zones where productive and cultural life functions dominate. Together these three zones form the basis of the cycle of local existence. Before we embark on zonal planning in the individual case it is important to consider the functional differences which exist between the zones, and also to reflect on the interaction that takes place between them.

As hinted at already, in the Umeå model the main task of the intermediate zone is not to contain housing but to form a framework for the social life functions. This means that greater demands are made on the social area plan than the mere siting of residential accommodation and its supplementary services. The area plan has to be provided with social community functions very different in nature and scale from the family modules. Their task is to add a meaningful social content to the civic life of the district. Inspired by sources as varied as the Anglo-Saxon settlement movement of the 19th century, the Danish architect Jan Gehl's book "Livet mellem husene" (Life among the Houses) from the 70s and the Nordic women's research group "Det nye hverdagslivet" (The New Daily Life), the Umeå project has formed a model for social district planning. In our presentation we have for pedagogical reasons used the metaphor of the ecological cycle. We describe the forms of community life in the local neighbourhood in terms of a *social cycle* (Fig. 1).

The cycle of the social district can most simply be regarded as a socio-dynamic system, where a number of basic functions in the local neighbourhood form points of action in a circular chain of events. The logical starting-point in the system is in the functions which are most elementary to the continued existence of the species: the loving care of children and young, care of the sick and old, the many themes and rituals of collective relationships. Here, the human foundation is laid for community life, and here the warm empathic ties are made which are the basis of the social learning process, collective co-operation and finally a democratic social order. Socio-cultural functions with a more specific task are later added to these collective primary functions - such as the aesthetic conventions of local culture, the internal information system and the area's health unit. A central unit in the cycle is formed by the *Neighbourhood Forum,* which is the direct democratic parliament of the area; it is represented by an executive

Fig.1. **THE SOCIAL CYCLE at neighbourhood level**

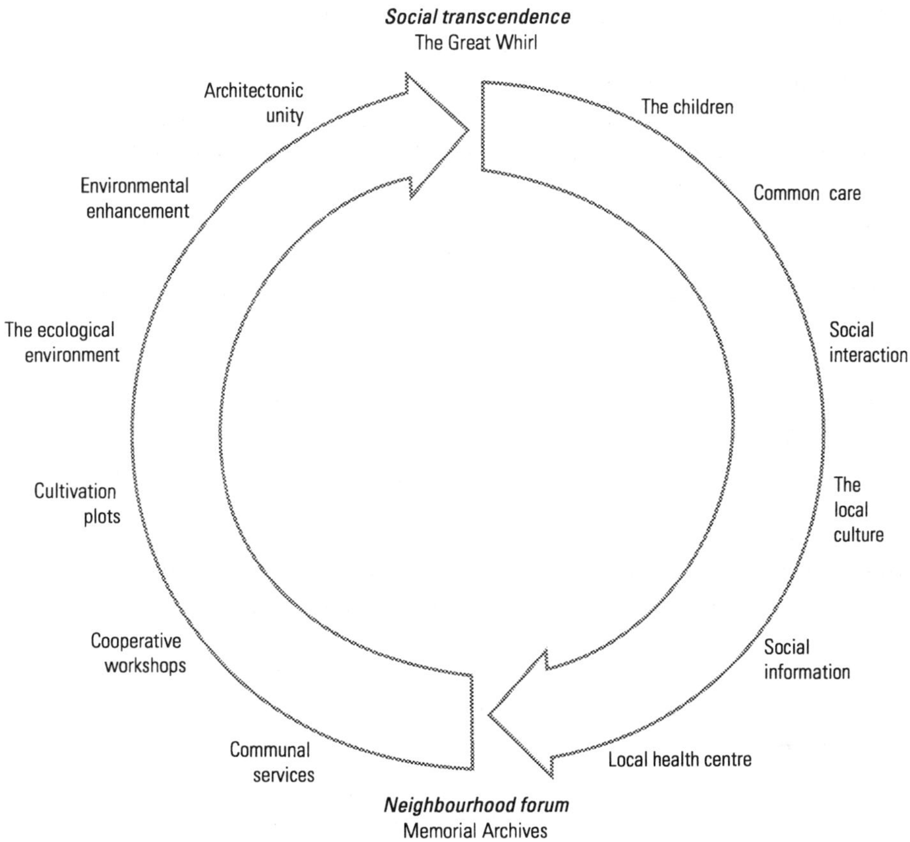

committee with the task of carrying out public decisions and handling contacts with other levels of society.

Beyond the Neighbourhood Forum the social cycle continues with collective functions of a more operative character. These are necessary in order to uphold a reasonable standard in the communal life of the area. Here supplementary social services of several kinds can be found, such as area caretaking, sanitation, messenger services and postal services, but also co-operative workshops for repair work. The activity can be described as an exchange of services between the inhabitants of the area. All these are activities that the community jealously defends, since they play a decisive functional role which is important for the feeling of social meaningfulness. They represent a value which the collective is unwilling to hand over to contracted or higher organs. The civic activities assume a more natural and resource oriented character further on in the social cycle, where we find themes such as lot cultivation and maybe animal husbandry on a limited scale and for local benefit, as well as care of the local ecological environment. The emphasis on a well-functioning interplay between natural ecological and social ecological life processes is an important aspect of social pedagogics.

Caring for the aesthetic environment in the neighbourhood is also an important aspect of social pedagogics, partly where different details and temporary arrangements are concerned, but above all in the total architectonic plan. Through these initiatives, the area becomes enriched with an aesthetic dimension, which is of great importance in giving intensity and qualitative overtones to the neighbourhood. The empathic, elevated aura hereby created finally forms the inspiration for the phase which concludes the cycle in the Umeå model: the empathic transcendence, which is also the phase of comprehensive social identification. After a terminological quandary, we have given this point in the cycle the name *The Great Whirl*, the terminus where the striving for community has its definitive emotional discharge in varying expressions of euphoric joy and friendship. These expressions can be achieved in many ways and are logically followed by a phase of recovery, which in its turn awakens the need for a new beginning.

The social cycle can, however, also be described in a different way. A reliable image shows the cycle's activities as a tension-filled interaction between two poles: the well-organized Neighbourhood Forum with its calculated directives and the collective source of inspiration of The Great Whirl. Between these two poles the segments of the circle flex their bows, the first encompassing the genuinely humanitarian functions, the second outer and area-related tasks. In both segments the interactive dynamics appear as an interplay between input and feedback reactions, where the incentive can originate from either of the two social pools.

Considerable gains in the socio-ecological pedagogy can be achieved if the efforts to implement ecosocial functions are made in a well-structured manner in the physical area plan. The basic social functions should be brought together in social action points and meeting places which also stand out architecturally in the neighbourhood area. This will cause them to function more effectively as gathering points for large or small groups among the local population. They assume a visible role in the everyday life of the community and thus give the district a new identity which is communitarian in the true sense. At the same time the new adjuncts inject a pulse of life into the zone which is often lacking in traditional residential areas. The largest gain from the socio-ecological interventions is that they transform the whole area into an environment for dynamic communication and thus into a space for vital civic contact.

The local district can, in this way, be transformed into an existential setting for man's personal life - if and when he wishes to use it - from the cradle to the grave. The district also serves the whole community as a social safety net which warrents a dignified life to the individual and a social communication based on reciprocity and trust. With this, the local network forms a good soil for the growth of a genuinely, and not merely formally, democratic consciousness. The Umeå group suggests that the social archetypes in the urban plan be given a new name. We call them *town-neighbourhoods.*

PROFILE OF THE GULLÄNGET PROJECT

The town-neighbourhood in the Umeå vision is based on an overall view of the intermediate zone. But the practical steps taken then differ according to whether the task involves projective or renewal planning, a densely populated urban area or a more rural setting. In our specific context the argument has been based on the needs of renewal planning. This means that the fundamental infrastructure in the form of buildings, communication systems and technical networks is already in place. In such cases the interventions acquire the character of selective initiatives, which are dimensioned according to the space available for socio-ecological supplementation. The objective is as far as possible to strengthen the collective identification with the place and to give it experiential expression.

Another limitation to the actual study is caused by the short time available in connection with the Umeå conference in Örnsköldsvik. The renewal plan for the Gullänget district presented in this paper should be seen as a rough draft, and is primarily meant as a declaration of intent.

Within the given frames the Umeå group and its associated architects have chosen to propose environment-enhancing efforts for the Gullänget district on three different levels. The first covers the architectural plan of the whole local area and concerns the addition of *aesthetic environment supplements*. Shaping the Gullänget area into a visual and experiential unit was regarded as the primary task in an overall plan intended to encourage social identification. The main aesthetic contributions suggested by the group are four in number: 1) *accentuating the centre of the district,* 2) *drawing a boundary around the district,* 3) *adding a number of entrances to it and* 4) *creating homogeneity in the conglomeration of buildings.* A basic requirement of a good experiential identification is that the area is reasonable in scale and therefore remains visually comprehensible to the individual inhabitant.

Architectural design is seldom used as a systematic socio-aesthetic instrument in modern professional area planning. Naturally this is not to imply that the aesthetic side is neglected and ignored. But the aesthetic solution remains largely the secret of the architect and his personal signature. Still, behind the mystique or, possibly, cultural conventions, many architects intuitively keep socio-aesthetic goals in mind. In order to further the artistic quality of the social environment, however, intuitive aesthetic awareness should be lifted to the level of conscious knowledge - in this case meaning knowledge of common aesthetic needs.

We have called the second group of socio-ecological improvements in Gullänget *social function supplements.* Here, in accordance with the model, we have in mind environments for humanitarian care, social communication and co-operative activities of a varied nature, which are managed by local groups and provide a communal service to the residents. With the limited effort which has been possible, we believe that the best results are reached by assembling the civic community functions at a limited location in the district, preferably near its physical centre. We found a natural place in this case next to the Gullänget Civic Centre called "Gnistan" (The Spark). The proposed social environmental supplements can actually be seen as an expansion of this already existing, popular citizen centre.

The functional complements through which we are trying to strengthen the civic centre surroundings are made up firstly of a *Community Square* for formal and informal meetings, but also for the marketing of local crafts and cultivated produce. In a nearby arcade a number of *Co-operative Workshops* have been quartered, mainly intended as an internally organized basic service for the residents of the area. A prominent location has also been given to a *Youth IT Centre* to be administered by the young people of the area and to be used as a social meeting-point. Further, a self administering *Centre of Local History* has been inserted which houses an archive of personal history and space for temporary exhibitions. Finally, with a view of the Högland Lake, a Summer Forum for the area has been sketched with a name symbolic to the town: *Eagle's Wing.* The Wing houses arenas for concerts and local theatre groups, a dance restaurant and rooms for meetings. With these varied and wide-ranging features we have created an architectural ensemble which may serve as a magnetic point and a social activation factor for community life in Gullänget.

The third level of environmental improvements is named *ecological environment renewal.* The easily grasped purpose here is to instil an ecological consciousness in the district residents and to make the preservation of the local environment a common concern. In recent years, there has appeared a whole range of applications for social planning in this field. In our socio-ecological plan we have made cautious use of these props. We have deliberately avoided some of the current ritualistic patterns and propose applications which either have a real function in the local environment or possess a clearly instructive mission.

A main point in the propositions which we have sketched under this heading is formed by the *Eco-Estate.* The Eco-Estate is an experimental town farm run according to bio-ecological principles, with renewable sources of energy, recirculation and toxic-free cultivation as guiding programme points. An *Ecological Environment Laboratory* has also been linked to the farm, with the aim of carrying out continuous environmental monitoring in Gullänget. Both institutions are managed by local efforts and play an important role in youth pedagogics. The Eco-Farm is directly annexed to the Änget school in Gullänget. The theme of renewable energy has also been taken up in another project, which aims to set up a *wind-generated power station by the Högland Lake.* The Lake has for a time constituted a nature reservation with high priority in Gullänget, the importance of which, both as an ecological study object and as a recreational area for local inhabitants the study group wishes to underline.

Another project where aesthetic values are mixed with the ecological message is *The Glass Palace.* The Glass Palace is a large-scale greenhouse, which has been formed into a subtropical winter garden and is located in the centre of Gullänget's housing establishment. In the bitter winter of northern Sweden The Glass Palace (together with The Paradise Baths in Örnsköldsvik) constitutes a welcome place for contemplation and recreation in balsamic, climatic surroundings. Poetic additions are the butterfly house with its exotic fruits (figs, peaches, bananas) and plants (cypresses, palms, etc.).

PLANNING RESPONSIBILITY

Who is responsible for ensuring that the socio-ecological area plan is implemented in accordance with the guidelines sketched here? Is it the population of the district or the local authority? We remind the reader of the theoretical issues dealt with in the first section of this paper. In the communitarian bottom-up society the situation is rather different from today's top-down variety. The institutional level attends to the genuine interests of the citizen in a much more conscientious manner. We therefore entrust the drafting and implementation of the town neighbourhood plan to the level of the communitarian authority. It is here that the necessary knowledge and expertise is to be found, and such an admission leaves no stain on our social honour. However, close consultations of the citizens of the town neighbourhood is also a matter of course, and should be deepened concurrently with the growth of local conscience and responsibility. The everyday management of the district, and also limited renewal projects, should on the other hand remain in the hands of the local population.

These brief notes on the subject of socio-ecological environmental renewal will have to suffice here. The observant reader will already have realized how many things are lacking before the alternative Gullänget plan can be described as a comprehensive renewal plan. These include, for example, a proper demographic analysis. They also include environmental studies in and immediately around the dwellings. There has not been room for these subjects in this limited project.

Before proceeding to a presentation of the architects' drawings, we only wish to add a brief comment on Gullänget as an area plan and a planning object. Gullänget as a social environment is a relatively variegated area, with a heterogeneous mixture of industry and housing on various scales and from different periods. The fragmented impression is underlined by the fact that routes for through traffic cut broad swathes across the area. Creating a unit and a feeling of local identity in this area may in reality be an insuperable task. At any rate without much more radical intervention than that proposed here.

In the course of working on the plans we have nevertheless enjoyed ventilating flights of fancy and testing new ideas. Perhaps these can still help to stimulate the inhabitants of the area and the municipal planners to make an effort which will transform Gullänget from a dreary fringe belt into a stimulating living environment.

REFERENCES:

Arendt, H., 1969: The Human Condition. University of Chicago Press, Chicago.

Berg, Per (ed.), 1993: Biologi och bosättning. Naturanpassning i samhällsbyggandet [Biology and Settlement. Adaptation to Nature in Community Planning]. Natur och kultur, Stockholm.

Bjur, H., et al., 1984: Ett ekologiskt synsätt i översiktlig planering (An Ecological Approach to General Planning). BFR och EFEM arkitektkontor, Göteborg.

Croft, S., and Beresford, P., 1992: "The Politics of Participation", Critical Social Policy 35. Longman, Harlow.

Dahlgren, L., 1984: Samhällsplanering och lokalsamhälle [Community Planning and the Local Community]. Research Reports no. 79. Sociologiska institutionen, Univers. Umeå.

Etzioni, A., 1993: The Spirit of Community: Rights, Responsibilities and the Communitarian Agenda. Crown, New York.

Galtung, J. et al., 1980: Self-Reliance - A Strategy for Development. Institute Universitaire d'Études du Développement, Geneva.

Ge rum för framtiden (Make Room for the Future). Statens råd för byggnadsforskning 1992, Stockholm.

Gustafsson, G. (ed.), 1993: Demokrati i förändring. Svenska kommuner och landsting i det lokala välfärdssamhället [Changing Democracy. Swedish Municipalities and County Councils in the Local Welfare Society]. Publica, Stockholm.

Hall, P., 1988: Cities of Tomorrow. Blackwell, Oxford.

Hall, T. (ed.), 1991: Perspektiv på planering [Aspects of Planning], Urban Studies. HSFR, Stockholm.

Holmberg, P., 1976: Socialpolitisk teori och praktik [Sociopolitical Theory and Practice]. Prisma, Stockholm.

Hultberg, R., 1991: För den goda stadens skull [For the Sake of the Good City]. Tiden, Stockholm.

Håkansson, P.A., and Nyström, L., 1991: Miljö för livet [Environment for Life]. Allmänna förlaget, Stockholm.

Illich, I., 1982: Det gemensamma samhället [The Common Society]. Federativs förlag, Stockholm.

Johannisson, B., Mårtensson, B. (eds.), 1985: Lokala gemenskaper och kooperation - vägledning för lokala mobiliseringsprojekt (Local Communities and Co-operation - Guidance for Local Mobilization Projects]. LTs förlag, Stockholm.

Kanter, R.M., 1972: Commitment and Communes. Communes and Utopias in Sociological Perspective. Harvard University Press, London.

Kretslopp. Basen för en hållbar stadsutveckling. [Recycling. The Foundation of Sustainable Urban Development]. SOU 1992:42. Rapport från miljövårdsberedningen, Stockholm.

Lacroix, B., 1981: L'Utopie Communautaire. PUF, Paris.

Lundahl, G. (ed.), 1991: Den naturliga staden. Bygga, bo och leva ekologiskt [The Natural City. Build and Live Ecologically]. Arkitekturmuseet, Stockholm.

Malbert, B., 1992: Ekologiska utgångspunkter för planering och byggande [Ecological Assumptions for Planning and Building]. Statens råd för byggnadsforskning, Stockholm.

Mattsson, M., Kindahl, G. (eds.), 1993: Folkhemmet på lokalplanet - en studie av samhällsarbete i Gullänget [The Folkhem on the Local Level - a Study of Community Work in Gullänget]. Statens råd för byggnadsforskning, Stockholm

Mårtensson, B., Alfredsson, B., Dahlgren, L. and Grahm, L., 1988: Det hotade lokalsamhället [The Local Community Under Threat]. Statens råd för byggnadsforskning, Stockholm.

Oresjö, E., 1993: Förnya bostadsförnyelsen [Renew Housing Renewal]. Statens råd för byggnadsforskning, Stockholm.

Rexroth, K., 1974: Communalism: From Its Origins to the Twentieth Century. New York.

Ronnby, A., 1991: I skogarnas land (In the Land of the Forests]. Statens råd för byggnadsforskning, Stockholm.

Råberg, P., 1972: Funktionalistiskt genombrott [Functionalist Breakthrough]. Norstedts, Stockholm.

Sundh, K., and Turunen, P. (eds.), 1992: Social mobilisering [Social Mobilization]. Publica, Stockholm.

Svensson, R., 1984: Social planering i praktiken (Social Planning in Practice). Socialstyrelsen, Stockholm.

Vedel-Petersen, F., Jantzen, E.B. and Ranten, K., 1988: Bofælleskaber. En exempelsamling [Housing Collectives. A Collection of Examples]. Statens Byggeforskningsinstitut, Hørsholm.

Veier til det nye hverdagslivet (Roads to the New Everyday Life), 1987. Forskergruppen for det nye hverdagslivet. Nordisk Ministerråd, København.

Werner, C., 1991: Staden som livsrum [The City as Life Space]. Statens råd för byggnadsforskning, Stockholm.

Åström, K., 1993: Stadsplanering i Sverige [Town Planning in Sweden]. Byggförlaget, Stockholm.

ENVIRONMENTAL RENEWAL IN GULLÄNGET.
A socio-ecological plan for renewal.
District plan

Architects:
Per-Eddie Bjuggstam
Kjell Mattsson, FFNS Ångermanland
Arne Wistedt, Arking AB

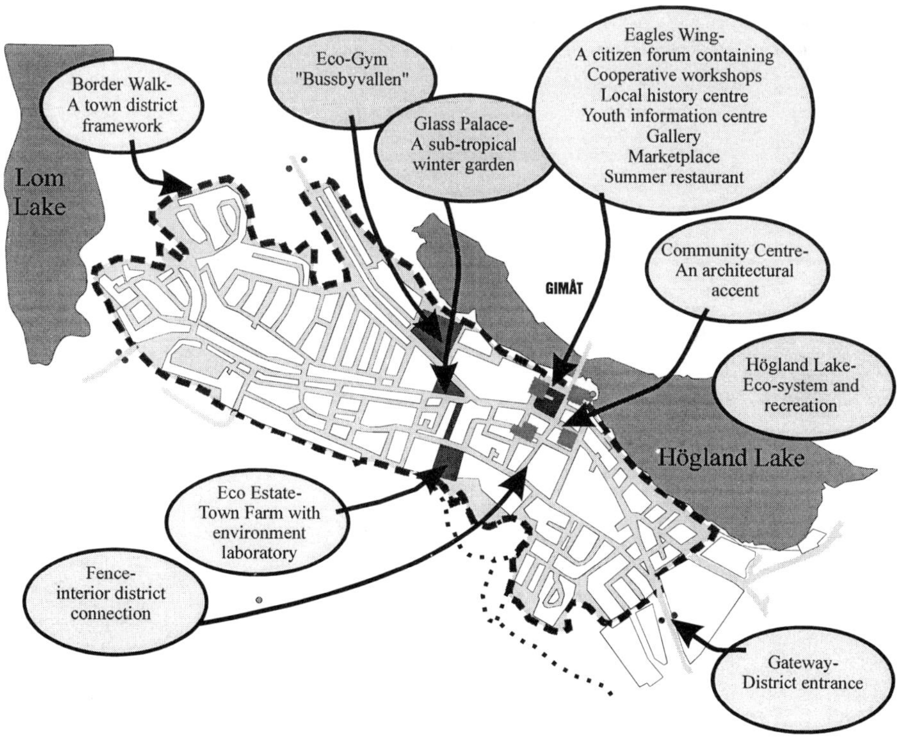

I: AESTHETIC ENVIRONMENT RENEWAL

GATEWAY

Entrance to Gullänget town district. The entrance columns in steel contribute a welcoming accent.

BORDER

A town district framework for Gullänget. A border walk symbolically marks off the area.

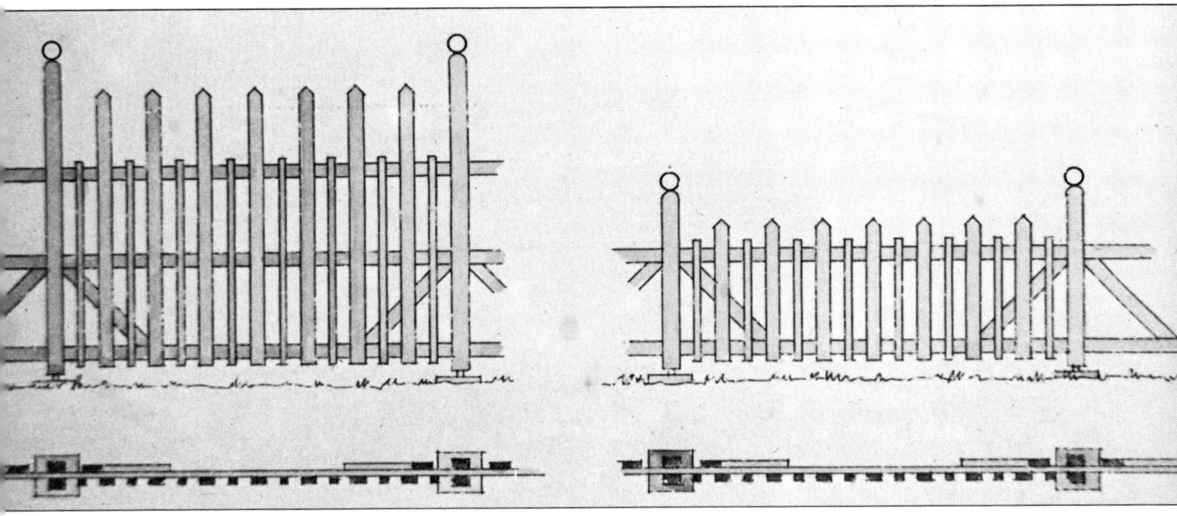

Fence, sections and details.

FENCE
A unifying element which reinforces the architectural entity of the area.

COMMUNITY CENTRE

An environmental accent which strengthens the local identity.

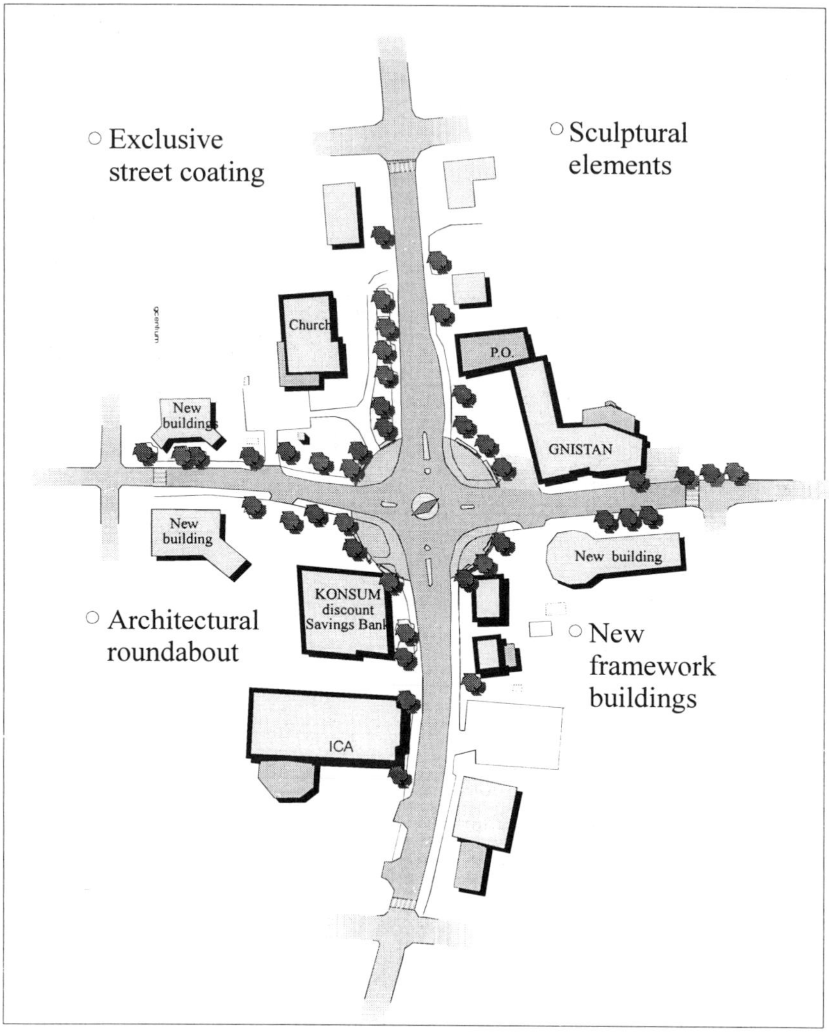

Central crossroads are laid out as a main square.

- **Architectural roundabout**
- **Sculptural elements**
- **Exclusive street coating**
- **New framework buildings**

II: SOCIAL ENVIRONMENT RENEWAL

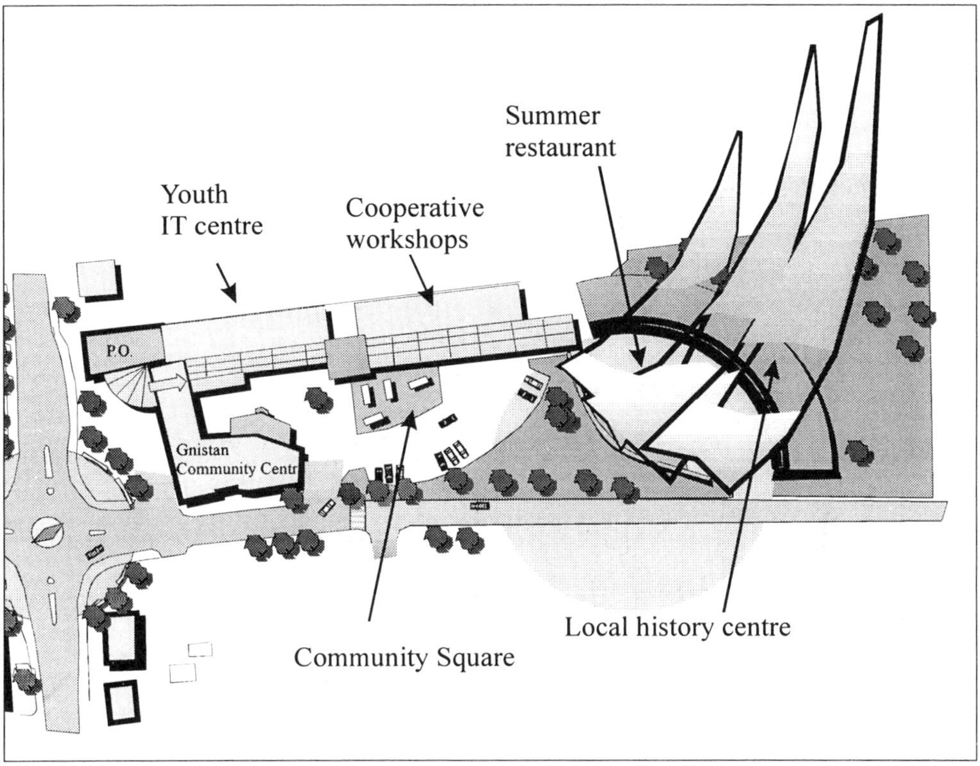

EAGLES WING. Citizen centre in Gullänget.
A concentration of cooperative functions strengthens the social identity of the town district.

III: ECOLOGICAL ENVIRONMENT RENEWAL

ECO ESTATE

Town farming with non-toxic cultivation and ecological livestock husbandry. Environment laboratory for local environment control and ecological education.

GLASS PALACE

Sub-tropical winter garden, butterfly house with exotic fruits (peaches, figs, bananas) and plants (cypresses, palms). Exterior garden rich in species.

HÖGLAND LAKE
A local ecological reservoir.

WIND POWER PLANT AT HÖGLAND LAKE
Renewable local energy sources for a sustained community life are utilized. Technical plants, produced at the settlement, are used.

Megamedia Centre Bothnia

An IT Vision for Västernorrland

TECHNICAL BACKGROUND

"Communications technology and information processing technology are coalescing and forming a unit within information technology (IT). IT facilitates the provision of information, communication and collaboration irrespective of geographical distance and thus creates a new infrastructure. All social activities are affected by this development. The dynamics are greatest in the area where communication and information processing overlap. *The gap between what is technically possible and what is actually in use is growing*" (our italics).

This is an extract from the programme of the Swedish SiREN Project, written in April 1993 at the behest of the Ministry of Industry and Commerce. The quotation offers a concise picture of the dynamic state of the field of information technology in the middle of the 90s.

To start with, it gives us a pointer to the self-image of the project creators on the IT front. Despite the dramatic implications for social change presented in the field, their picture of the world remains ineradicably technological. Secondly, it illustrates the radical change in the practical purposes of information technology which has taken place since the breakthrough in the 60s and 70s. Completely new fields of application have opened up alongside the monolithic systems of the databases and the electronic gadgetry. In the mid 90s these new tendencies are realized in academic pilot projects and government development programmes all over the world.

The following trends are typical of the research and development work of the 90s:

• In microelectronics and photoelectronics, the tendency continues to be towards smaller formats, shorter time constants and increased packing density. These permit steadily smaller and more flexible applications.

• The development of mobile networks with a higher transmission capacity and more powerful microcomputers (e.g. pocket models) is increasing mobility of the users. This development is backed up by sophisticated geostationary satellite networks.

• The changeover from analog to digital technology permits integration of text, speech, image and animated or other information and also the transmission of larger quantities of data. Multimedia concepts are emerging in consequence of the new technological possibilities.

• The distribution networks are being improved by refined fibre optics and the spread of broadband networks for two-way communication. New transmission systems using coaxial cables (ATM) facilitate high-speed communication of general information. All over the world, public data networks are being constructed. Considerable effort is made to identify new applications.

• The opportunities for interactive information retrieval and data processing are steadily increasing, thanks to new generations of high-performance computers. Receiver technology and terminal equipment are also being improved as 2D interaction is supplemented by 3D reproduction via high-definition TV, cybernetic image manipulation and real-time holography. Screen, microphone and earphones are being placed in spectacles, and computer gloves and pens are replacing physical keyboard, mouse etc.

• The cost level is falling in large sectors as a result of system rationalization and mass use of the new technology. Software is becoming increasingly reliable and program production wider-ranging and more imaginative.

ORGANIZATIONAL DEVELOPMENTS

We have become used to regarding the constant stream of IT innovations as a process of mythical proportions. We seldom feel dubious about the role of the new electronic world in our working environments and homes, or question its claim to be in the vanguard of societal development. Words of warning from early critics of technology, such as Jacques Ellul and Lewis Mumford, have made no considerable impression on either the public or the decision-makers.

Nevertheless, the establishment of information technology in society is urged on by very definite interests. IT products are launched smartly and spectacularly on the wide consumer market, and in more sophisticated forms pervade industry and commerce and the public sector. The range of offerings extends from technical innovations such as CD-ROM disks and neural networks to mundane applications such as the flood of electronic games. It includes portable processors in paperback (and soon in wristwatch) format, and large-scale alternatives such as the "paperless office" (or the "intelligent house"). With the aid of new graphics programs we can ourselves produce virtual realities on our cyberscreen and climb easily into the electronic dream world.

The technopolitical IT decisions taken at national level, together with the infrastructural measures in which the decisions result, are an important motivating force for development. With the exponential growth of IT and media installations the need of comprehensive co-ordination, networks and normative systems is also growing. In large parts of the world, governments are now working on systems development for IT applications. In many projects network organization is a leitmotif. Discussion is very concrete in nature and is conducted in terms of broadband solutions and multimedia transmission.

The USA is leading the way, having begun under the Clinton Administration the construction of a nation-wide information structure (NII), popularly known as "the electronic highway". A new transmission system with an expanded capacity (Integrated Services Digital Network - ISDN) has been created to meet the very diverse requirements of users of the network. A Japanese equivalent, placing a stronger emphasis on technological product development, is the research programme "Real World Computing". Australia is going in for high-speed networks in the "Experimental Broadband Network" project.

In Europe, too, there are a series of large IT projects with a similar profile. We find them both in the national research programmes of a number of countries and within the EU system. RACE, ESPRIT, EUREKA and DRIVE are some of the more interesting pan-European ventures. EBONE is a collaboration project between about 25 regional networks in Europe. Much experience has been gained during the last ten years in university and research-based IT networks. The best-known of them is INTERNET, which can point to global coverage. Two related networks in the Nordic countries are SUNET and NORDUNET. Another initiative, at present still in its infancy, is the Swedish SiREN project.

The lasting impression remains that the mainspring of the IT revolution is the innovative power generated by the actual technology itself and its cadres of electronics experts. However, harnessing of this power is very much the preserve of the leading companies in the field. Their names are already well embedded in the subliminal consciousness of many people: Microsoft, Intel, IBM, Apple, Sony, Matsushita, Philips, Compaq... There is also a definite relationship between creative development and the industry's sales income and market expectations. Thus the circle is completed.

The current trend in the IT world is towards mergers between the leading telecommunications companies and the big companies in the computer, media and entertainment sector to produce veritable global electronics giants. This development reflects the fusion of advanced communications methods into multimedia schemes which is at present taking place and which is ultimately based on the definitive breakthrough of digital technology.

DOES I.T. HAVE A SOCIAL POTENTIAL?

Year by year we can follow the continuous colonization of the system and life worlds by information technology and the media. The alluring vision of the information society paraded in the 70s will be realized - if current trends continue - much sooner than anyone expected. The "global village" of the media philosopher Marshal McLuhan is definitely on the way to becoming a reality.

But will the concept of the *information society* survive? Is it an adequate term for describing the high-speed networks of multilateral services which are emerging as the new infrastructure of the global society? The designation appears outdated from several standpoints. It has acquired its legitimacy primarily in the global view implied by the international market system, in which the information society represents the top step in the climb of the economy from rural livelihoods and local crafts, through industry and trade, to a global money and credit market. The transactions there are based on the lightning-fast flows of information made possible by modern computer technology.

But this picture of the world, autocratic in its general application, has today acquired a number of challengers. Depending on the future paradigm chosen by the younger generation, classic information technology may give rise to either an *electronics society* (the technological dream) or a *communications society* (the humanist dream). Possibly the ideological discussion of the immediate future will take the form of a dialogue between these three paradigms of earlier and later date.

It is the humanist dream which is at the heart of considerations in the actual paper. But to proceed in that direction a further question has to be raised: Does the new information technology really possess a social and humanist potential? How, in this case, has this potential hitherto been realized, and what forecasts can be made concerning its future?

These are questions which are obviously of interest to anyone who has noticed that there is a world outside the technology villages and the labyrinths of technocratic power. To the researchers in the Umeå regional project they appeared important enough to merit a special study, of which this paper is the result. This study, too, was carried out in conjunction with the social development project discussed in this section of the book. The project managers set up a study group in the usual manner to examine the social aspect of IT in relation to the regional problems of Västernorrland.

The core of the actual group consisted of a number of regional experts.

The regional economy unit at the County Administrative Board in Härnösand was chosen as the venue for a series of informal discussions, which took place during the spring of 1993. Imaginative proposals were made by several participants, in particular the county administration's IT expert Lennart Blomgren, K.G. Svensson from the Communications Science Centre of the Mid-Sweden University, Sundsvall, and architect Hans Gillgren, also of Sundsvall. At the social policy conference in Örnsköldsvik Lennart Blomgren presented an account of the county administrative board's plans with regard to IT in the county's development.

This article reports on the ensuing dialogue on IT as a social medium. Certain differences in the visions of the regional future soon became apparent. The county administrative board's main interest at present is in a decentralization of the IT network, i.e. its potential in rural community development. The project group fully shares this interest, but wished in particular to discuss an even more radical social environmental alternative. Both visions were given concrete expression in sketches of alternative IT environments. In the final editing of the book, perspectives were further broadened by consultation with a number of centres of expertise. The exchange of thoughts with the following experts and organizations was particularly rewarding: Stefan Larsson (Telia-Mitt, Sundsvall), Mats Brunell (SiREN Project, SICS, Kista), Olle Findahl (Media and Communications Science, Umeå), and Teldoc, Stockholm.

The architect Abelardo Gonzalez (Malmö) played an especially prominent part in the project work. Gonzalez is the creator of the integrated vision of a regional media centre which is presented in this paper as a synthesis of the accumulated thoughts of the study group.

SOCIETY'S VIEW OF I.T.

The first question in an analysis of IT as a social factor is of course whether information technology as such possesses qualities which support humanist and social development aims. Two of the visionaries of the information culture, Alvin Toffler and Yoneji Masuda, assert in their well-known studies ("The Third Wave" and "The Information Society", both 1980) that IT has inherent decentralist features which will lead to a decentralized society. Therein, according to the Umeå researchers, lies a humanist potential.

The decentralization argument has since been put forward many times on various grounds. In the last ten years, for example, regional researchers have made much of the idea. The obvious assumption is that information technology creates quick-and-easy networks which function unimpeded by the sluggishness of the physical infrastructure. This permits a fragmented geographical social structure. However, other researchers claim that information technology has not led to any unambiguous move towards decentralization, but has instead contributed to the process of centralization already in progress.

At this point the comment might perhaps be made that the question ought to be put differently. IT networks create a new infrastructural level, and at this higher level new power structures are formed, with their own centres and peripheries. These often coincide with the terrestrial infrastructure, but this is by no means inevitable. The IT-borne power networks are largely invisible and their sociogeographical effects still remain difficult to assess.

The gist of this discussion is that the new information technology at first glance appears socially and humanly indifferent. In the contemporary phase of civilization it has become an instrument for strengthening the existing structures, but it would be premature to discount the possibility of quite different, perhaps humanist, applications. Nevertheless, these observations apply only for as long as IT is regarded as an instrumental service-culture alongside the central life culture. When IT itself becomes a part of the central culture, the situation at once becomes more complex. This is the point that we are now approaching, and very fast indeed.

A second introductory question is how far modern society, through its political and public sector, has helped to guide the development of information technology in a social and humanist direction. The question is quite easy to answer. In the present technological climate the social idealism of the power elite is notoriously trumped by the wish to adapt to prevailing trends. The wielders of power give generous priority today to an extension of information technology. This happens principally as a concession to the economic and technological pressure groups, or to their value system, which politicians have internalized. The predominant impression is that this internalization takes place without any regard for the possible human and social implications.

However, political standpoints on the IT question are seldom based on ideological conviction; they are inclined rather to be opportunistic. One important cause of the lack of social commitment on the part of the technocratic elite is probably the absence of effective lobbying groups in the social sphere. Academic humanist and social research ought reasonably to constitute one such well equipped pressure group upon politicians. Despite this, social support from the university world is conspicuous by its absence. Creative contributions to a futurological debate on IT's humanitarian potential seem, almost by definition, to fall outside the bounds of contemporary human science.

One modest exception is books on topical issues from a human ecology viewpoint, in which a morose criticism of the dark side of the IT society is the main message and a prestigious academic tone conceals the absence of enlightened alternatives. Whenever humanist academics treat IT questions more closely, they do so on behalf of the authorities and in order to create the defensive role of technology assessor. Those interested will find characteristic examples in the bibliography.

SOCIAL EXPERIMENTS WITH I.T.

These harsh generalized comments do not overlook the occurrence of social experiments using information technology. Grass roots movements in a number of countries have, with varying degrees of success, built up IT networks (e.g. the British "Community Information Project" and the Danish "Folkedata").

Projects which might generously be labelled social have also been started by the governments of a number of countries. During the 80s, in particular, there was extensive activity, in which social scientists also took part. However, the social experiments using IT show considerable difference in character. One type was represented by the French government's MINITEL venture. (There were similar projects in Germany and Italy.) The project was initiated in the early 1980s and involved providing large groups of the population with simple teledata equipment for seeking various social services. The aim was, as it was put, to increase the computer literacy of the population and equip them for the coming IT age. The perspective was clearly that of the central bureaucracy and the social philosophy that of the mass society. The project cannot be labelled social in the humanist sense of the word which we use here. Rather it implied the opposite.

Another type of IT project with a social profile had a regional policy bearing and was designed to promote social mobilization in disadvantaged local communities. Often the communities concerned were of moderate size. The Nordic countries played a prominent part in this context, for example by means of a series of short-term experiments carried out in Denmark, Norway, Sweden and Finland. Qualified researchers took part in several of them, and there was detailed documentation (see bibliography).

Most of these projects (including the Norwegian and Swedish ones) were primarily intended to contribute to commercial and industrial development. Nonetheless they are interesting from a social-humanist perspective, owing to the interwoven character of the processes of life in the small-scale community. One example is the Rogaland Research Project "Telematikk for lokal og regional utvikling" (Information Technology for Local and Regional Development) which embraced eight Norwegian communities, evenly distributed around the country.

The Danish IT experiments place a clearer emphasis on genuine social values. As a part of a state research programme, about fifteen experiments involving IT in the local community took place in the mid-80s. The characteristic of the Danish experiments was the prominent place given to the citizens' networks as generators of progress. Creating a user-friendly information technology and shaping IT systems as an instrument for social communication in the community became one of the main themes of the experiments.

Within the framework of the Danish research project the idea of local information technology centres arose. The idea found concrete shape in the telehouse scheme, and in this form it spread into the other Nordic countries. As an instrument of local mobilization in rural areas and in sparsely populated parts of the country the telehouse has since been introduced into a broader European discussion through various channels, including the OECD and the Council of Europe.

The telehouse may be described as a compact information technology centre, created for daily use by citizens, companies and associations in the local community. Its areas of application are personal contact, dissemination of information, and provision of social services of every conceivable kind. Adult education, municipal information, business service, tourist and hotel bookings, membership registers, long-distance study and library and database services are among the tasks that are usually mentioned. Its advocates dream of the telehouse taking the role of an "electronic community centre" in the political and cultural life of the locality.

The standard technical equipment of the telehouse is a number of personal computer terminals, telefax, teletex, interactive video plus printing and copying equipment. At the end of the 80s there were about sixty telehouses in the rural areas of the Nordic countries, relatively evenly distributed between the four countries. Even if the boom of the 80s was followed by a sharp levelling off, a lot of the telehouses are still in use and serving their purposes.

The Danish social scientist Lars Qvortrup has long been a well-known expert on local experiments in information technology. He is also the author of a number of books on the subject, including "Det levende eller det døde samfund. Nye veje i informationsteknologiens organisering" (1988) [Living or Dead Community. New Routes in the Organization of Information Technology]. Qvortrup recommends a community-friendly local IT development. He levels criticism at government research projects (mainly the Danish ones), because despite alleged local and social intentions they embody an authoritarian technological approach.

The Danish research programme was a remote-controlled experiment designed to identify new areas of application for the technology of the IT culture. Social experiments with IT should be the opposite, states Qvortrup, namely an analysis of how social primary networks and local values can be strengthened with the aid of the new technology and enabled to survive in the long term in an advanced industrial society.

Not that Qvortrup lays the blame for the limited success of social IT experiments only on the lack of imagination of the commissioning governments. The situation which has arisen is also due to the indifference of university research. To set up meaningful trials of social information technology at all requires a cross-boundary research in the humanities and social sciences which finds expression in creative experiments. What society needs, in order to maintain a balanced development, is a revolution in social scientific thinking which manifests itself in real *social inventions.*

A BROADER VIEW OF I.T.'S SOCIAL POTENTIAL

The Umeå IT study group shares Qvortrup's view of the social IT experiments of the 80s and the premises of social experimentation. What the acute problems of society demand from the sphere of human sciences is certainly not technology assessment but social innovations in the IT field.

But the Umeå group does not alltogether share Qvortrup's opinion as to the field of application of the IT proposals. We believe that more progressive prospects for social IT-installations exist in other sectors of modern community life. Recognizing these requires that our view of communitarian involvement in society is liberalized. A liberalized view in this case means acceptance of *a radically widened action sphere to civic and communitarian participation.* It is the socio-ecological studies of the Umeå group and the model of the ecological life-region which has opened our eyes to the new prospects. What, then, can be taught from the life-region model?

A first assumption of the Umeå model is that harmonious civic community life requires a well-organized spatial framing, a permanent habitat. In this respect the local IT experiments of the 80s appear as a progressive beginning. A second hypothesis in the model is however that civic organization in the habitat should also have a bearing on a global sphere of action. The local level of basic sensory life functions should be overlaid by a global level of more abstract cognitive functions. The establishing of a global action level goes back on an evolutionary process, which can be described as a leap in human consciousness development from a sensory (local) to a cognitive (global) strategy of action. (The closer arguments for the model are presented elsewhere in the book.)

In the opinion of the study group, the most interesting social IT challenges are not within the local sphere, as normally claimed in communitarian debate, but in the global communications sector. The leap in man's mind from local to global awareness has caused, since long, new cosmopolitan structures to spring up all over the human society. Entirely new instruments for mastery of the global scale are being designed in many fields. Social communication is among the areas which have, for various reasons, drifted into the backwaters of this epic movement towards a global paradigm.

For example, it is becoming increasingly apparent that the fine-mesh transport networks for physical goods which today span the planet are a primitive instrument, at least for social communication and information exchange. And the far-flung metropolises created by man in his quest for global vantage points may function symbolically as planetary communication nodes, but are hardly a brilliant solution in social practice. The individual more often than not gets isolated in an urban ghetto, and the access to the global scene remains a privilege to the few.

Our civilization, in actual fact, still lacks a relevant social communication structure on the global level, as an equivalent counterpart to the civic networks of local life.

It is high time that we begin to realize clearly to which extent technological progress has altered the preconditions of civic communication. Regular telephone links and modern media (radio and television), during the first part of the 20s century, signalled the arrival of a global communications structure at the cognitive level of modern man. Still, within their respective domains the new media suffered for a long time from serious limitations (the telephone from its narrowness and the ether media from their one-way direction). For these reasons they were long stuck with the roles of technical substitute and pleasurable indulgence respectively.

The most recent developments in IT have altered this situation dramatically. The smart solutions of microelectronics, the expansion of digital transmission technology, the fibre optic broadband networks and the multimedia schemes have created the preconditions for a civic network medium with global coverage, high capacity and previously unknown transmission speed. By the fusion of the informative and suggestive media, the mediated message can be brought to life and given an authenticity which has up till now been impossible to achieve.

As a consequence, a system for multidirectional communication with an enormously broad human interface with the global environment is now emerging. Maybe the most interesting thing is that this is not only a concise but also a warm medium, which combines rational presentation with sensory and emotive informatics. At the same time the whole concept remains firmly anchored on the abstract-imaginative level of cognitive man and is thus extremely flexible and easy to handle.

Many debaters with a humanitarian profile express anxiety about the expansive power displayed by IT, and fear that in the hands of authorities seduced by technology this will lead to undesirable computerization of every corner of the living world (see for example L. Ingelstam and L. Sturesson: "Brus över landet" [A Noise Over the Land], 1993). The research team of the Umeå University Group does not share the resigned pessimism of these critics. In its eyes the current integration has, on the contrary, created a huge potential for a humanist reorientation of society.

The advanced transmission systems increasingly assume the contours of a medium for broad civic contact and communication between local communities all over the world. A social innovation of evolutionary proportions is actually taking place before our very eyes.

SOCIAL I.T. AS A GLOBAL MEDIUM

It is primarily the discovery of this intrinsic potential which arouses great expectations in the Umeå Group. The applications which are being proposed by today's leading electronics companies hardly achieve this. They still bear the clear imprint of the modern electronics market and of the expected demand from the business world, the public institutions and private consumers.

To respond to the new needs of the civilian social sector, as they are defined in this article, the IT applications must undergo adjustment and modification, and entirely new applications must be invented. Easily accessible databases for interregional civic information are one large-scale social application, and international networks of scientific knowledge banks are another. Systems for selective global news service are a third alternative; in this case experimental activity has already begun.

Moreover, the system's transmission functions need to be profiled to invite the public to take part in global friendship networks and electronic public meetings, cultural debate and politics. Equipment for program production over the whole audiovisual spectrum has to be made more easily available and educational users' programs need to be designed. One important wish is for the reception functions to be given a versatile multimedia shape, which breathes life into the messages from the info-universe.

Furthermore, the functional realignment must be accompanied by new structural thinking. A feature of contemporary IT culture is the extreme individualization of the operating functions. A computer screen in every cottage and in every corner of the office seems to be the goal when regional planners and media researchers discuss IT from a participatory perspective. Such a terminal-fixated philosophy has hitherto guided product development in the majority of the electronics companies.

In a discussion of IT as a civic medium of the future, the need becomes apparent for applications which have already found an initial environmental and public shape - which are spatial. One reason for the demand for *spatial applications* is of course the desire for the widest possible interface with the global transmitters; this greatly stimulates the feeling of involvement in a global communication culture. A functionally and visually well designed media environment also possesses an attractiveness which tempts citizens to use it more frequently.

Another very important desideratum is for the spatial IT environment to function as a *social forum* in the local community and the home region. The global communication net is incomplete if the exchange stops at a discussion between two individuals who are tied to their VDUs. The network becomes social only when the external information is made the subject of internal debate in the home region. The civic IT vision must leave room for local citizens reactions to the network message, in small groups or in plenum. Only via collective evaluation in special sessions does the individual find his global social identity and have it regularly confirmed.

All these considerations made the study roup realize the multimedia technology's capacity to contribute to a social communications structure with a global signature. But at the same time our reflections were accurate paths in a stepwise concretization of the ideas. They increased our impatience in front of the decisive question: what would be the functional form of a multimedia application erected along these lines? Which visual and environmental appearance would it take in the regional landscape where it was supposed to be realized?

A PLANETARY MEDIA APPLICATION

The Umeå group and its regional team decided to accept the challenge implicated in the question. It embarked on a journey whose final aim was to deliver a preliminary draft for an IT-based citizens communications centre with a cosmopolitan scope.

Concretizing such a complex idea was certainly not an easy task. It involved considerable demand on innovative thinking, and several traditional views in the IT field had to be opposed. Not least the wish to create a social terminal environment was a provocation to conventional schemes. The large scale necessary for a well-equipped plant also presented a number of obstacles, but at the same time was a strong enticement to creative imagination. This certainly was a great chance to shape a symbolic environment for citizens identifying with the global dimension of modern life. At the same time it could give increased prestige both to the IT-medium and the regional community as a social living space.

But the task also included a set of very close-to-earth questions concerning the regional placing and organization of the plant. When the study group ventured into the project these issues were the ones first brought into focus. Let us, therefore, start our presentation of the utopian draft at this point.

In the present case we had the County of Västernorrland in mind.

We imagined that the communications centre would be the result of a joint local authority investment. This would make it possible to finance more qualified IT equipment than individual municipalities of average size can normally afford. We also wished to launch the idea of a shared social asset which might serve as a unifying and identity-reinforcing factor in the region.

In a good intermunicipal spirit the facility ought to be sited in the geographical centre of the territory. Convenient public access via the regional transport net was a precondition for such a location. By its scale and position, the installation should indicate a regional identity and at the same time state its place in a universal civilization. What was a suitable name for such a media centre? "Multimedia centre" seemed far too weak. The scale and the spatial profile had added a new dimension. We opted for the designation *Megamedia Centre.*

How well did the group's ideas fit in with the existing and planned IT networks of Västernorrland? A telecommunications map and a phone call to the Swedish telephone company in Sundsvall ("Telia-Mitt") revealed that the extension of an optical fibre network is in progress, and proceeding at a rate of 120 km per year. The network is being prepared for ATM services, giving a large capacity for multimedia transmission. The system adheres to the European standard laid down by the Commission of the European Communities.

Within a few years Västernorrland will be well connected with the electronic network which is now being cabled out over the Nordic countries, the continent of Europe and large parts of the globe. The attached map of the optonetwork shows that the place chosen for a regional media centre allows convenient access to the net (Fig. 1).

In order to take the final step towards a visualized functional environment, a qualified architect had to be engaged in the project. Our positive experience on an earlier occasion led us to turn to the office of Abelardo Gonzalez in Malmö.

After joint theoretical discussions the task of creating the architectural framework was entrusted to the architect. His unique assignment was to mould a functional synthesis which also could serve as a captivating symbolic framework for the new communications structure. We report below on Abelardo's visualization of the theoretical drafts. A tight budget has limited the effort to a summary sketch. This may have the advantage of leaving a lot of room for the interested reader to develop ideas of a global communications structure in his own imagination. This is of course just as it should be in an area where, still, creative chaos rules.

MEGAMEDIA CENTRE BOTHNIA

The site chosen for *Megamedia Centre Bothnia* - as the establishment was named - is on Svanön, an island in the estuary of the Ångerman River (Fig.2). The Centre has been placed on a cliff, near the shoreline to the east. The Centre is of substantial size, and visible from a fair distance.

Access roads link the Centre with the mainland and the E 4 highway, which runs close by. This is an idealized portrayal, and therefore the general plan does not take account of other schemes which might affect the island. Certain amenities (e.g. for overnight lodging) are a necessary adjunct if the facilities are to meet reasonable user requirements.

In the sketches presented, the architectural totality of the installation and its visual expression may speak for themselves. The large scale and the constructivist allusions announce the special area of use and emphasize the cosmopolitan significance. The building neatly contains the proposed media functions, but can easily be extended for new applications. In itself it represents a communicative structure for enjoyable excursions into a planetary megazone.

We present below a summary of the functional environments which the Centre contains. The technical systems are only referred to in passing.

NewsWorld

The foyer of the Megamedia Centre is occupied largely by NewsWorld, a planetary news centre for the common public. NewsWorld is divided into sections for the different continents. Each section contains a large high-definition TV screen for general news services, together with a number of video terminals for individual news retrieval from different countries and regions. The flow of regional and national news can be followed in a special section.

Sections within NewsWorld also have text TV and printer facilities. News analyses and weekly digests are distributed on screen or in printouts. As an experiment a section for interactive news debate has been introduced. As far as the media content is concerned, high standards are required. NewsWorld strives for a high information value in its reporting.

KnowBase

KnowBase is a virtual library. By this is meant a knowledge base whose services consist largely in providing documents from other databases. Via the electronic network the library is linked to knowledge bases (or a selection of them) all over the world. Special orders of various kinds are executed at high speed. KnowBase also disposes over its own database, in which knowledge documents from the home region are available to the public and other virtual libraries.

The user is guided through the knowledge banks and subject fields by electronic and manual catalogues. Special library staff assist with complex searches and give introductory courses. The searches are undertaken at screen terminals which may also function as study positions.

The system uses multimedia microcomputers. For the storage of the volumes of data there are traditional hard disks or CD-ROM diskettes. Equipment is also supplied for printout of the documents. KnowBase consists of the following three sections: *Reception* (with reference system, library service and manual library), *Public section, Research section.*

InterFace

This facility houses an electronic communications environment for global dialogue between individuals and groups. The thematic content of the network is determined by the users and extends from pure hobbies to genuine research. The network is addressed to interested citizens of all ages, but can also serve as a working environment for scientists in the region. Within the global structure, group research into special projects can take place.

The computer environment is composed of small-scale modules grouped in sets. Each module is provided with a personal computer terminal for multimedia communication and supplementary services. A manned support and information centre is included. The functional and technical organization of the proposal is to a considerable extent inspired by INTERNET.

ConferenceNet

ConferenceNet is also a multimedia environment, in this case for bilateral and multilateral group conferences between different parts of the country or the world. The most important technical prerequisite of the conference system is access to a broadband network with ISDN capability.

The local conference environment is equipped for on-line communication by video, sound, text and data. The unifying medium is a large, high-definition screen. The peripherals include, in addition to loudspeakers and microphones, equipment for the exchange of written and pictorial documents. The conference environment differs from traditional video studios by virtue of its lobby character. This solution is made possible by access to mobile camera technology.

ConferenceNet includes conference rooms, maybe five or ten in number. Some of them have technical equipment for more modest requirements. The environment is designed to meet the need of the public sector, regional commerce and industry, and other local interests for group communication with other parts of the world.

WorldScreen

WorldScreen is an IT-communications mega-environment of the same standard and status as concert halls, theatres and congress halls. WorldScreen is used for real-time attendance at global premieres and top events in theatre, music, sport and entertainment taking place anywhere in the world. At the same time it is a forum for inter-regional conferences, up to world level, in a number of configurations. The facility can accommodate a maximum of 3,000 people, but can be dimensioned for smaller events. The projection screen in omnimax format and the outstanding camera equipment permit optimum experience of planetary presence in two-way communication.

WorldScreen is the region's great public platform towards the world. It is an environment which fulfils, with maximum realism, the vision of the local community's participation in a global culture with regional simultaneity and equality between centre and periphery. But the dreams to be realized are not those of a global monoculture. Every region makes its unique contribution to the overall picture. Equipment and premises for regional programme production are a part of the vision.

CyberCosmos

An arena has been created for simulation of virtual worlds of experience of a more or less imaginative nature. The arena can be used for aesthetic experiments which cover the whole spectrum of experience of the individual, but is also a contribution to the creative and sensory parts of youth teaching. The enclosed virtual space (cyberspace) can also serve as a background for public events on a limited scale (dancing, jazz, entertainment, parties).

The technical equipment in CyberCosmos consists of a projector and control panel placed centrally in the studio. From this panel the projection of image and sound is controlled on the electronic screen which is formed by the enclosing walls, ceiling and floor of the room. Cross-boundary aesthetic sessions are acted out in CyberCosmos, with the artist as conductor and the public as inescapably involved participants in the transformations of the virtual work of art.

The realization of this vision requires current cyberspace technology (special spectacles) to be supplemented by a projective cyberspace technology. A development in this direction has begun.

ServicUnit

Central administration.

Central database.

A common database serves as a memory and processing unit for the different IT environments in the Megamedia Centre. It also constitutes a storage place for the selected body of regional information.

Programme editing.

Editorial and production units are needed for the required programme and news production. A joint programme unit might possibly be set up for the purpose and also serve as experimental and rehearsal studio.

Technical department.

A unit is of course needed for regular maintenance of the technical equipment, new acquisitions and technical development work.

REFERENCES:

Benedikt, Michael (ed.), 1991: Cyberspace. First Steps. The MIT Press, Cambridge, Mass.

Churchman, C. West, 1984: The Systems Approach, (rev. ed.). Dell Publishing Co., New York.

Cronberg, Tarja, 1990: Fremtidsforsøg [Future Experiments]. Akademisk Forlag, København.

Findahl, Olle, 1986: Svensken - en världsmedborgare i framtidens massmediavärld?? [The Swede - World Citizen in the Mass Media World of the Future??]. Wahlgrenska stiftelsens rapportserie, Malmö.

Forester, Tom, 1980: The Microelectronics Revolution. Basil Blackwell, Oxford.

Habermas, Jürgen, 1981: Theorie des kommunikativen Handelns. Suhrkamp Verlag, Frankfurt am Main.

Hesseldahl, Peter, 1993: Computerdrømmen og de økologiske realiteter [The Computer Dream and the Ecological Realities]. Munksgaard, København.

Hetland, P., Knutzen, P., Meissner R. and Olsen, O.E., 1989: Nært, men likevel fjernt. Telematikk og lokal utvikling (Near and Yet Far. Information Technology and Local Development). Rogalandsforskning, Stavanger.

Ingelstam, L., Sturesson, L. (eds.), 1993: Brus över landet. Om informationsflödet, kunskapen och människan (A Noise Over the Country. On the Information Flow, Knowledge and People]. Carlsson Bokförlag, Stockholm.

Ivanov, Kristo, 1991: Pragmatic and Cognitive Constructivism: Steps Towards a Humanistic Computing Science. Report UMADP-WPIPCS 34, Umeå.

Kommunerna inför de nya medierna [The Municipalities and the New Media]. Kommunförbundet 1985, Stockholm.

Krueger, Myron, 1991: Artificial Reality II. Addison-Wesley, Reading, Mass.

Lundberg, Eric, 1993: ESPRIT, EUREKA och RACE - tre paneuropeiska IT-satsningar [ESPRIT, EUREKA and RACE - Three PanEuropean IT Ventures]. Teldok, Stockholm.

Machlup, F. and Mansfield, U. (eds.), 1983: The Study of Information. Interdisciplinary Messages. John Wiley & Sons, New York.

Masuda, Yoneji, 1981: The Information Society as Post-Industrial Society. World Future Society, Washington.

Qvortrup, Lars, 1988: Det levende eller det døde samfund. Nye veje i informationsteknologiens organisering [Living or Dead Community. New Routes in the Organization of Information Technology]. Blytmans Forlag, København.

Ramos, E. and Schroeder, A., 1994: Contemporary Data Communications. MacMillan Publishing Company, New York.

Rosenberg, Richard, 1992: The Social Impact of Computers. Academic Press, Inc., San Diego.

SiREN. Ett initiativ till nationell samverkan inom IT-området [SiREN. A National

Co-operation Initiative in the IT Field]. Förstudie, utförd på uppdrag av Näringsdepartementet. PM 1993, Stockholm.

Skot-Hansen, D., Duelund, P., 1990: Lokale kulturlivsformer og ny informationsteknologi [Local Cultural Life Forms and New Information Technology]. Institut for Kultursociologi, København.

Svanberg, Lasse, 1991: Stålsparven. Om 90-talets medier och om "informationssamhället" [The Steel Sparrow. On Media of the 90s and the "Information Society"]. Bokförlaget Prisma, Stockholm.

Søby Kristensen, Peter, 1978: Kommunikation [Communication]. Nordisk Forlag A/S, København.

Tannenbaum, Andrew S., 1989: Computer Networks. Prentice-Hall International, London.

Tengström, E., 1987: Myten om informationssamhället (The Myth of the Information Society). Rabén & Sjögren, Stockholm.

Thyssen, Ole, 1993: Nutiden. Det overfyldte rum [The Present. The Overfilled Space]. Gyldendal, København.

Toffler, Alvin, 1981: The Third Wave. Dan Brooks, London.

Trauth, E.M., Kahn, B.K. and Warden, F., 1991: Information Literacy. An Introduction to Information Systems. MacMillan, New York.

Wiener, Norbert, 1963: Menneske og automat. Kybernetikken og samfundet [Man and Automation. Cybernetics and Society]. Gyldendal, København.

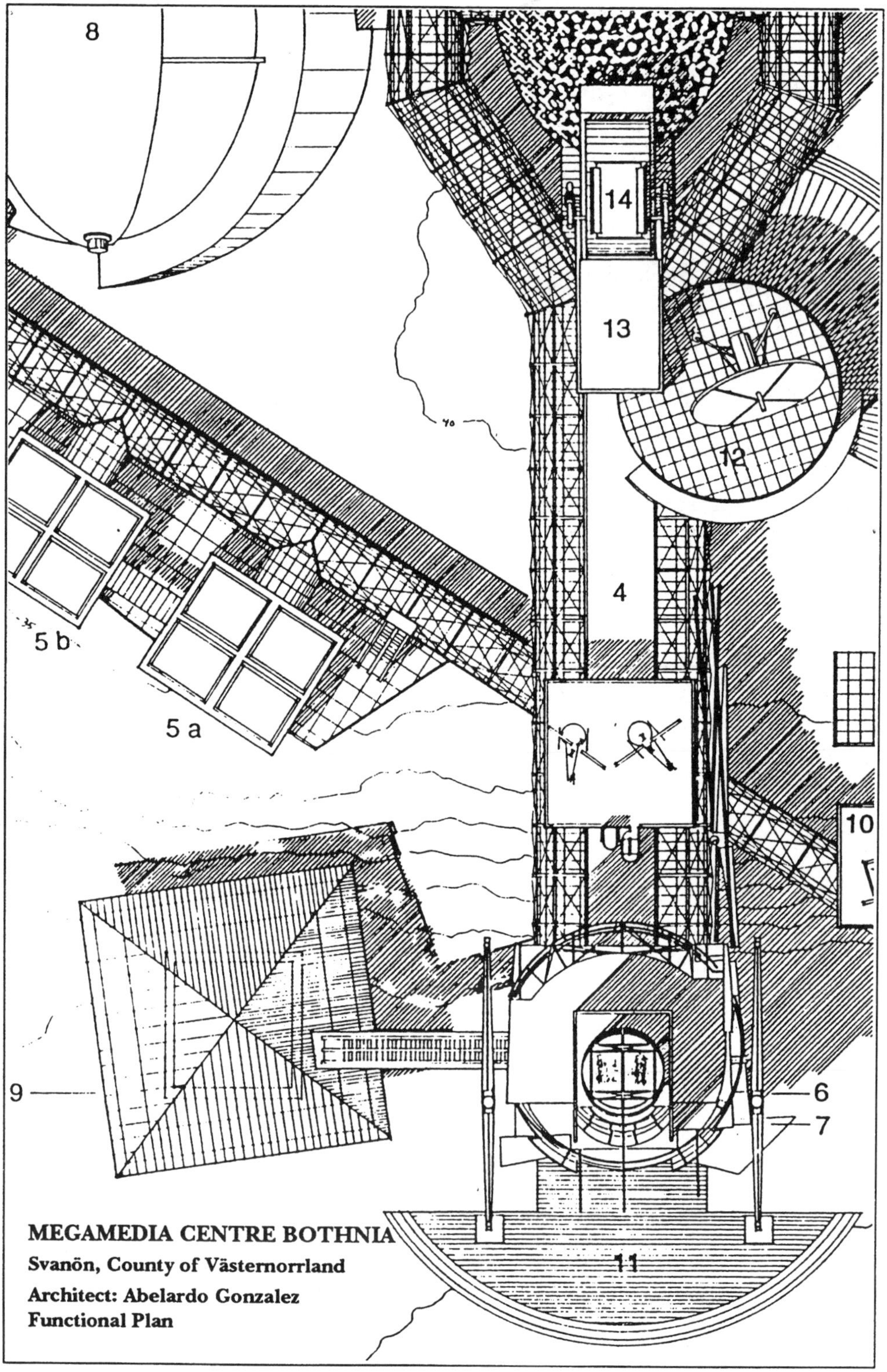

MEGAMEDIA CENTRE BOTHNIA
Svanön, County of Västernorrland
Architect: Abelardo Gonzalez
Functional Plan

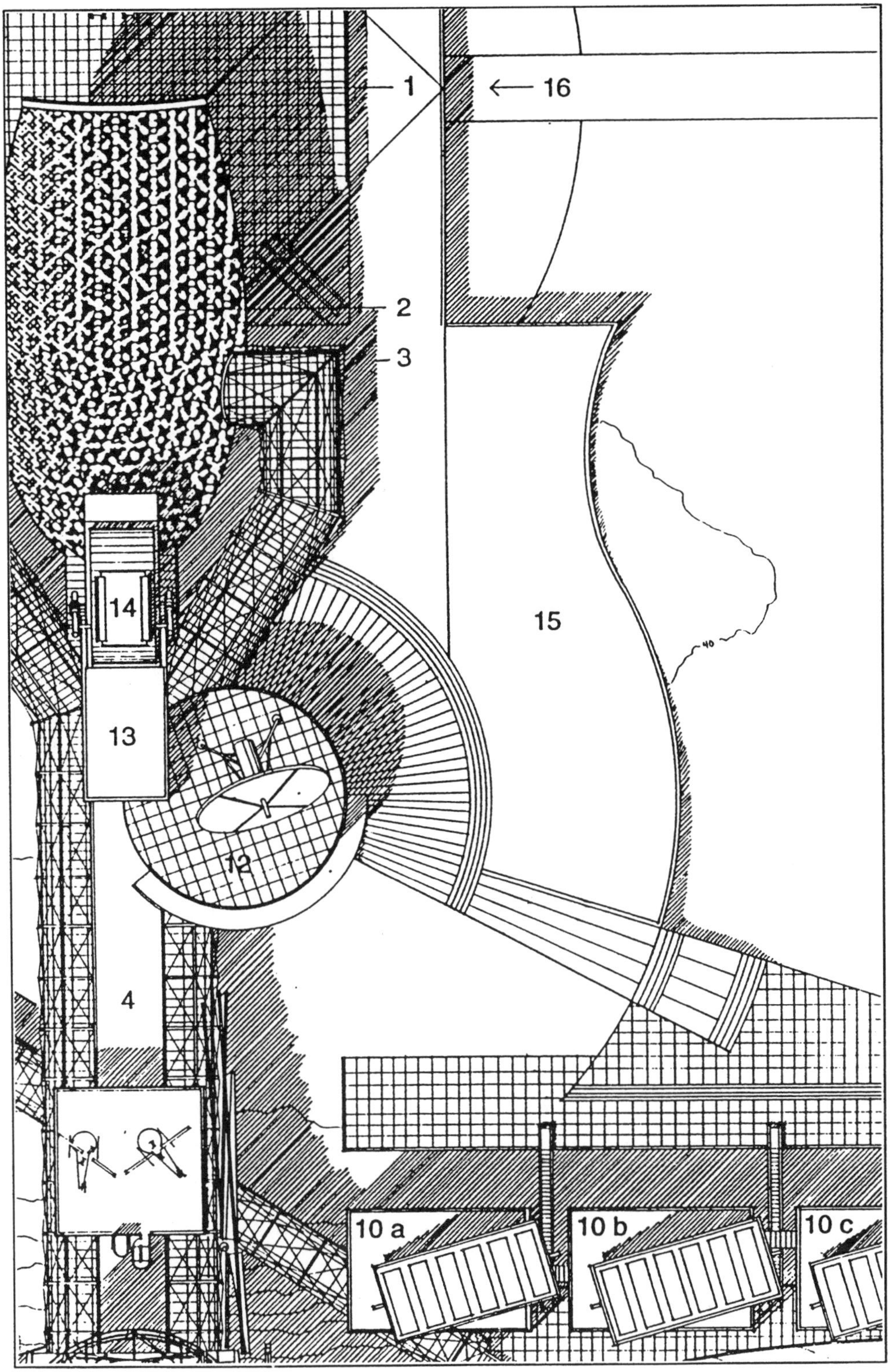
1
16
2
3
15
14
13
12
4
10 a
10 b
10 c

MEGAMEDIA CENTRE BOTHNIA

Functional plan. Fig. pp. 380 - 381

1. **Entrance concourse**
2. **Arrival hall**
3. **NewsWorld**
4. **Central Street**
5. **KnowBase**
   - a **Reception/service**
   - b **General department**
   - c **Research department**
6. **InterFace**
7. **ConferenceNet**
8. **WorldScreen (1.500 persons)**
9. **CyberCosmos**
10. **ServiceUnit**
    - a. **Central administration/technical department**
    - b. **Programme editorial department**
    - c. **Database**
11. **Water arena**
12. **Restaurant**
13. **Café**
14. **Observatory**
15. **Pond**
16. **Parking**
17. **Auditorium (300 persons)**
18. **Cloakrooms/toilets**
19. **Service street, deliveries/stores**
20. **Conveyor belt**

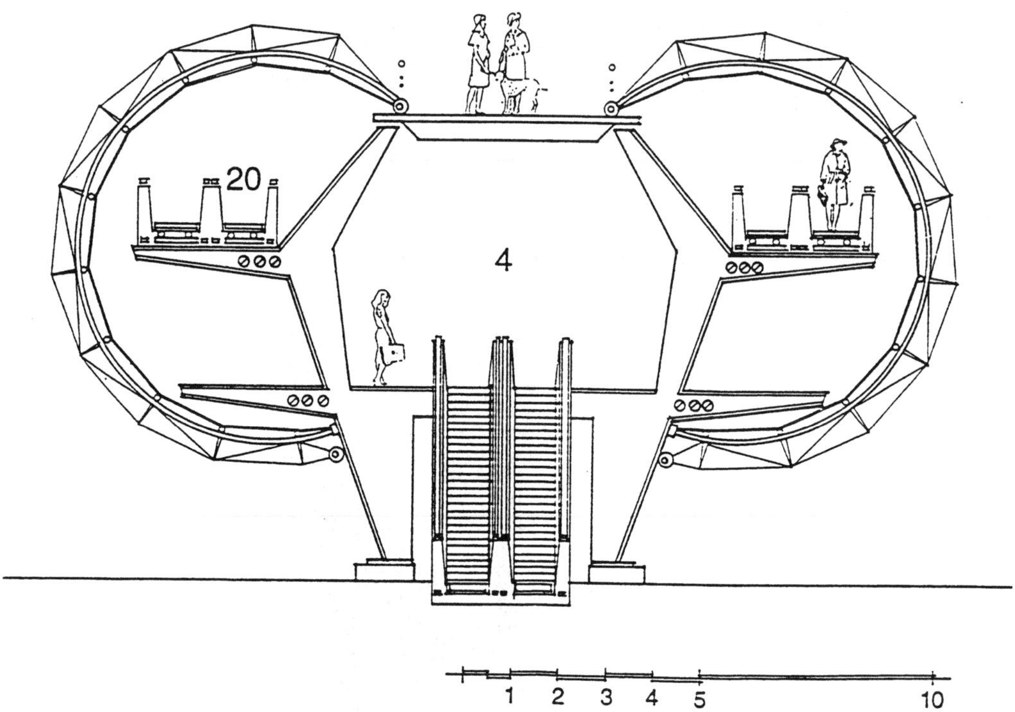

MegaMedia Centre "Bothnia"
Section of public corridor

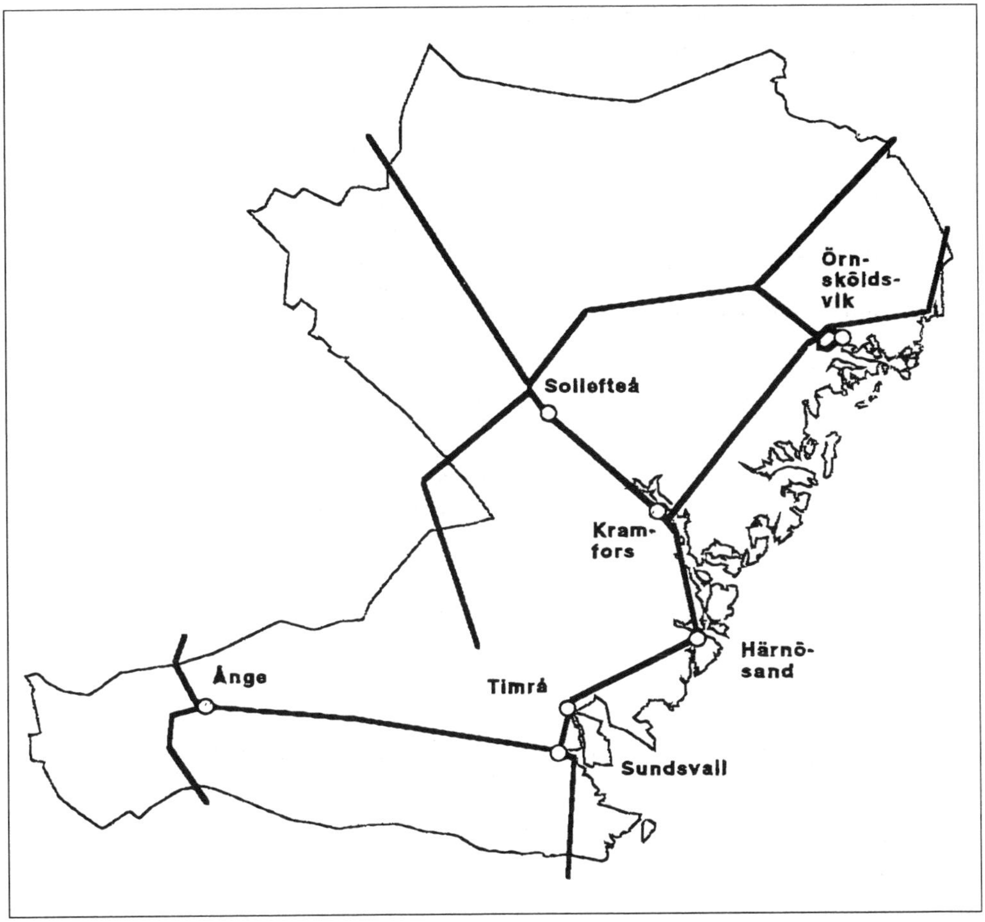

Fig.1. **Fibre Optics Network in Västnorrland County**

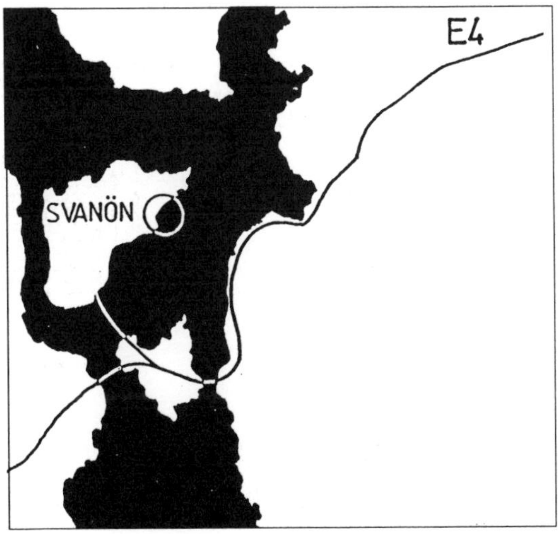

Fig.2. **Megamedia Centre Bothnia Site plan**

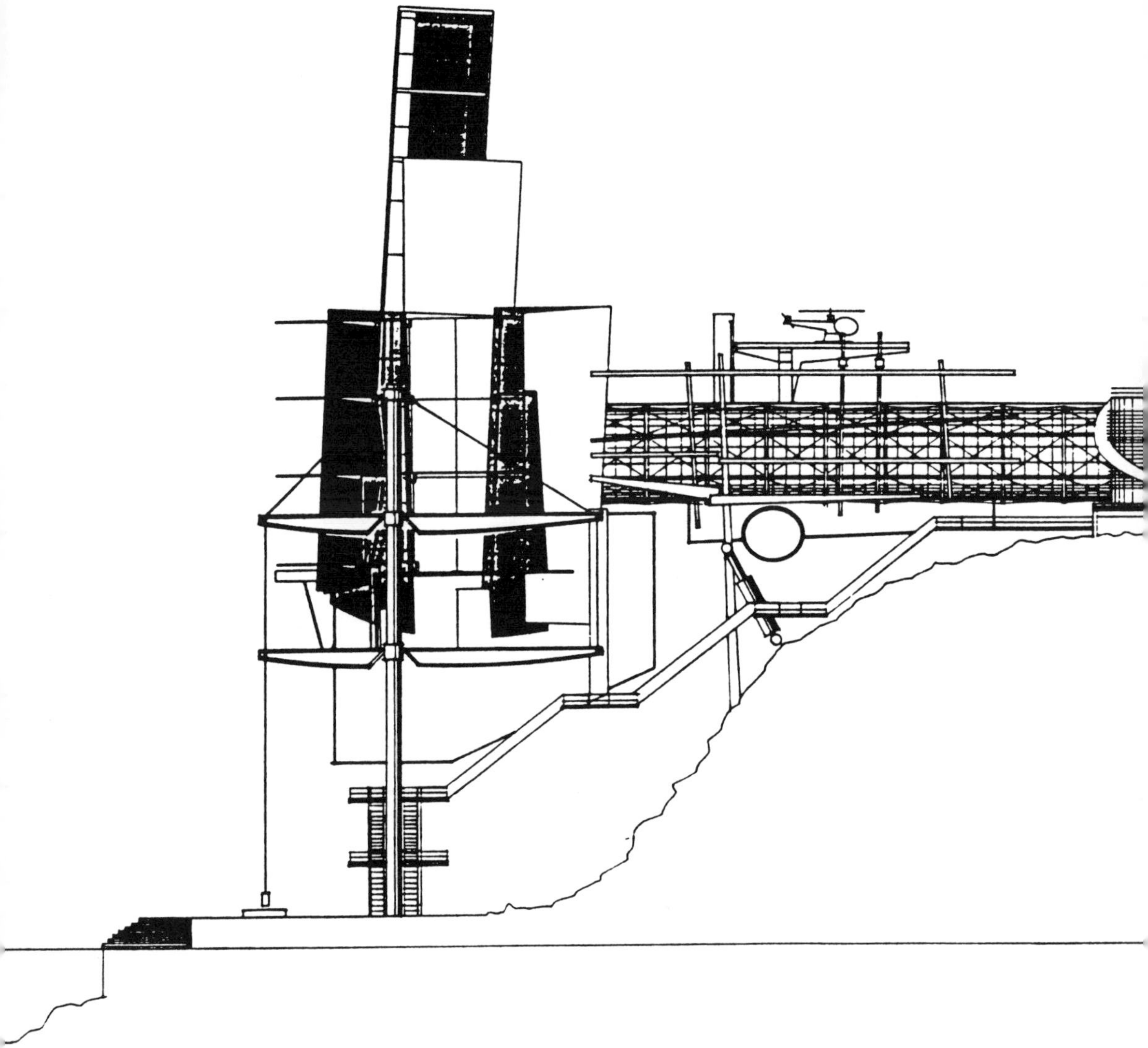

MEGAMEDIA CENTRE BOTHNIA
North elevation

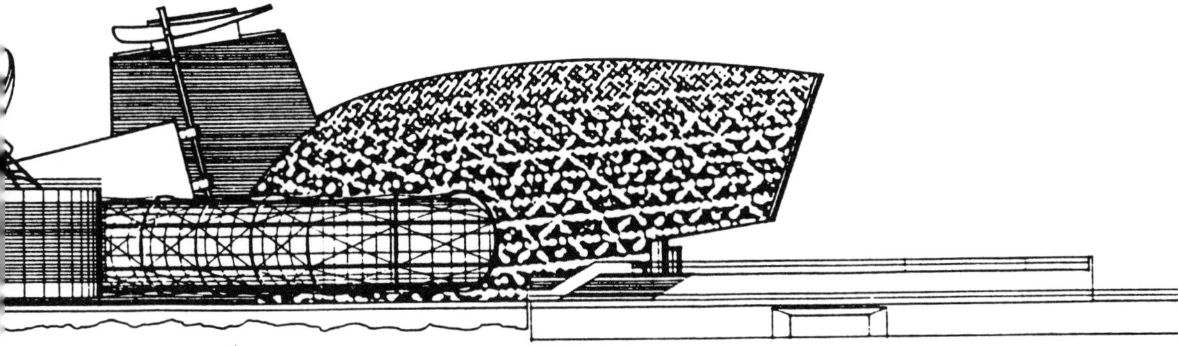

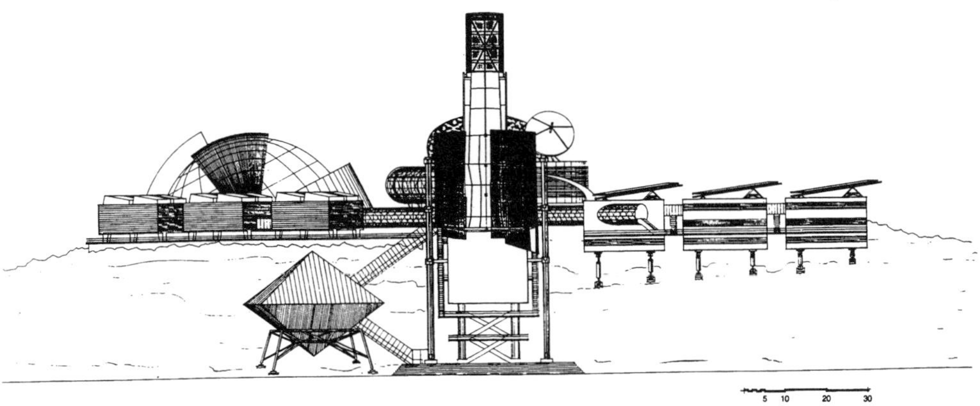

MegaMedia Centre Bothnia
West elevation

Public Health Is Social Welfare

Legacy of the 1991 WHO Conference in Sundsvall

Public health agencies, during the last decade, have made a major contribution to the debate concerning sustainable social development. One leading actor has been the World Health Organization (WHO), which has launched a new, socio-ecological, concept of public health. The significance of this is that the health risks caused by careless social environmental policy and planning are pointed out. A more offensive public health policy is called for, which makes the promotion of human well-being in the social setting its supreme goal. This is an attitude which comes very close to the ideas of the Umeå University Group discussed in the book.

By a coincidence, which appears almost predestined, the Umeå Group made contact with the WHO initiative during its social R&D project in Västernorrland. The purpose of this final article in the series of social reports is to give an account of the promising consequences which this unexpected meeting brought about.

THE WHO WORLD CONFERENCE IN SUNDSVALL

Long-term initiatives to improve the general health of the population have come to play an increasingly important part in futurological discussions in recent years. A broadening of the public health agenda has become a practical concern alongside traditional measures in many municipalities. Growing professional attention is now being given to a wide range of societal factors, ranging from work environments, residential planning and leisure activities, to social services, drug policies, unemployment and social insurance. Within the municipalities public health committees are being considered and at national level institutes of public health are set up.

The UN's World Health Organization has played a prominent part in the promotion of a new public health during the 80s. By way of a series of international conferences a platform for radical professional discussion was created. A special innovative role has been assumed by the WHO's European Office and its "Lifestyles and Health" Department. The Department has put

"Angelica Silvestris". Photo by Margareta Klingberg

forward a new socio-ecological health model, which implies a considerable sharpening of the health objectives.

According to WHO officials a number of serious health risks characteristic of our modern society are not being alleviated by sectoral means such as tobacco and alcohol policies. Many of today's health problems have their roots in lifestyle and environmental factors which are typical of the whole societal structure and its "normal" mechanisms. The elimination of these health-impairing factors requires structural modification in the very basis of society.

WHO has not limited its active interest in the matter to administrative and inspirational initiatives. It has also developed a set of strategical criteria for the application of the socio-ecological health scheme in community work. The new WHO policy for environmental health promotion includes the following characteristics:

- A holistic view of social life and community processes.
- A positive definition of the concept of health: to create preconditions for public well-being is the fundamental task. Health in itself is not a goal, but should be seen as a means to a good life.
- Public health policy work should be locally oriented and identifiable at citizens level.
- Professionals have to take responsibility for the implementation of the local public health programmes.
- The health promoting policy should be integrated within a general social policy vision.

In the summer of 1991 WHO held a world conference in Sundsvall (in Västernorrland County) titled "Supportive Environments for Health". The Västernorrland County Council made a tremendous organizational contribution to this event. The agenda of the conference included the following six topics:

1. *Housing and health*
2. *Social support*
3. *Food and health*
4. *Importance of education to health*
5. *Work and health*
6. *Communications and health*

For all six themes preparatory material was distributed in the form of briefing booklets. The WHO Conference also issued an impressive number of professional and more popular publications. Its general summary was entitled "The Sundsvall Statement on Supportive Environments for Health".

A parallel national conference took place at the same time under the title “Folkhälsoarbete i samverkan” (Working Together for Public Health). The reason for this arrangement was to give a larger number of participants from Sweden the opportunity to take part in the Sundsvall event. The subjects of the national conference were largely based on the same themes as the international conference.

THE UMEÅ PROJECT ENTERS THE STAGE

On the occasion of the Umeå Futurology Group’s social conference in Örnsköldsvik two years had passed since the Sundsvall manifestation. This gave spontaneous cause to reflect over what traces the Sundsvall conference had managed to leave in the county. But even more it prompted a resumption of the discussion on the supportive environment scheme, perhaps in part from a widened - sociopolitical - angle. How should the many radical ideas be put into practice in future public health work, but also in overall social politics in a region like Västernorrland?

Therefore, when the Municipality of Örnsköldsvik and the Future Studies Group at Umeå prepared the conference “Social Identity and Life Region”, the new public health was chosen as a theme for one of the group studies precedingthe conference. Because of the short time available, the brief was limited to *the role of the public sector* in a future socio-ecological health policy. A dozen or so representatives of public activities at municipal and regional level took part (cultural and recreational services, environmental and social services, youth centres, county council and county administrative board). Per Råberg of the Umeå Group acted as chairman.

MEMBERS OF THE STUDY GROUP

Birgitta Bergfelt, Director of Cultural and Recreational Services
Dr Göran Carlsson, County Council
Stig Ernestål, Director of Social Services
Dr Christer Gunnarsson, Örnsköldsvik
Christina Hammarström, Social Worker, Youth Centre
Gunnar Holmberg, Recreational Services Adviser
Marianne Lundström, Research Secretary, Social Services
Birgit Mollberg, Social Services Director, County Administrative Board
Lars Gunnar Olsson, Departmental Manager, Environmental Department
Ella Sohlberg, Health Planner, County Council
Per Råberg, Chairman, Umeå Future Studies Group

This article starts with summarizing the results of the workshop discussions. Five separate themes were chosen to form the basis of the group’s study.

It should immediately be noted that this account is of a very general and somewhat improvised nature. Its practical purpose is possibly twofold: to help keep the ideological debate on radical public health alive in the county, and to encourage municipalities and other authorities to take practical measures designed to promote ecological public health in the region.

Theme 1

The Sundsvall Conference of 1991 produced a number of radical suggestions for future public health work. Which influence have these ideas so far had on health work in the county? Is there any criticism as to a) the introduction of the health promotion programme and b) the local follow-up?

Group's comment:

A number of reports and publications on the improvement of public health work have been published but they have not yet yielded many results on the practical level. The start of public health activities in the county may be traced back to the Västernorrland County Council's initiative, the "Satsa på hjärtat" (Heart First) project in 1987, which involved forming cross-sectoral working teams in the municipalities.

These have since developed into what may be called public health teams. Quite successful campaigns for public health information have been arranged. But large sections of the population have not yet grasped the message. There is even a risk that current health information methods increase the gap between groups in society.

Some other initiatives also deserve to be mentioned. The Municipality of Sundsvall has included public health aspects in its objectives, and has run the "Sunda Sundsvall" (Healthy Sundsvall) project. The Municipality of Sollefteå has initiated a public health committee, which is attached to the environmental board. And in Kramfors the municipality has adopted a health policy programme, for which overall responsibility rests with the municipal board. The Municipality of Örnsköldsvik is currently considering how public health work should be organized. And finally, the County Council is working on a public health programme which will set up goals for future health work.

Additionally it should be mentioned that The Swedish Association of Local Authorities has linked up some 25 municipalities in different parts of Sweden in a "Network for Healthy Municipalities".

Theme 2

The "Sundsvall Statement for Environmental Health" and the "Public Health Declaration" lay down guidelines for a socio-ecological public health programme. In what way do these guidelines differ from traditional principles of preventive health work? Is there reason to regard the radical WHO initiative as a special policy field and to give this field its own name?

Group's comment:
Parts of what is now called "the new public health work" were previously a part of the public sector and its various branches. When the Social Services Act was drafted from the directives laid down in the late 1960s, public health in the modern definition hardly existed. But the objectives set out in the initial paragraph of the Act are very consistent with the public health ideas:

> *"On the basis of democracy and solidarity the community's social services shall promote people's*
> - *financial and social security*
> - *equality of living conditions*
> - *active participation in society*
>
> *Taking account of the individual's responsibility for his and others' social situation, the social services shall concentrate on releasing and developing the innate resources of individuals and groups.*
>
> *Activities shall be based on respect for people's right of self-determination and integrity."*

The Social Services Act also emphasizes efforts on different levels - structure-oriented, generally oriented and individually oriented. Structure-oriented efforts are those intended to promote a good social environment, e.g. by participation in social planning.

The WHO principles involve a deeper study of the causes of general health problems and have to be applied using other approaches and methods. There is reason to make a distinction between traditional and socio-ecological public health work in order to spotlight the new problems.

Theme 3
The socio-ecological public health programme places great emphasis on participatory and action-oriented strategies. But does the radical programme also contain a creative role for the public sector alongside other agencies in a coherent policy for the promotion of public health? How do we define - generally spoken - the new duties of the public sector?

Group's comment:
The public sector has an important role as defender of the long-term social development goals of the community and as protector of their social foundation. Within this sphere there are major creative tasks alongside the purely administrative ones. The public care sector can never be replaced by total privatization in an ecological health policy.

However, promotion of public health needs to be built into the training of architects, planners, social workers etc., if a common attitude to supportive environmental work is to emerge. Also a more detailed discussion of humanist values is needed.

What constitutes our feeling of welfare? How do we adjust to a changing society - if adjustment is the solution? Young people's situation and unemployment can be taken as a starting point for a discussion on crisis and crisis abolition.

Theme 4

What specified areas of work can be distinguished in a practical socio-ecological health policy for the regions? In which sectoral organs of the public authorities should responsibility for practical efforts be placed?

Group's comment:

Practical fields of ecological health include lifestyle-promoting efforts who strengthen the population's regional and social identity, but also land-use planning from a social perspective. Another important duty may be to acquire new human and social knowledge and to support research efforts.

An actual problem, however, is that the detailed local knowledge of social workers is not being properly utilized. There is also a risk of relevant health knowledge being lost in the general mass media offerings of world events. Whether we are ready to accept the implications of the knowledge we have or acquire is also a problem.

Within the public authorities there are a number of sectors who include work relevant to general public health, even if hitherto we have not always defined it as such. The new public health work should be based on cross-sectoral co-operation, with bridges being laid between different departments.

Theme 5

Does socio-ecological health work as a field of skills belong within traditional public health administration or should responsibility be shared with other sectors? And further: should the socio-ecological perspective in municipal administration result in a general social policy in order to give satisfactory coverage?

Group's comment:

Suprasectoral goals and initiatives are needed, and the municipal authorities have the crucial task of setting up these goals and co-ordinating the work. A public health (or welfare) committee under the municipal board may provide the solution to this problem of co-ordination. One of its primary tasks would be to formulate the aims of an ecological health and welfare policy.

In a public welfare committee social and humanistic values might, it is hoped, receive greater emphasis than in the existing administrative units and in sectorial policy. However, social ecological values must be implemented at an early stage in planning, when the guidelines of social policy are being

PUBLIC HEALTH DECLARATION

adopted by participants in the conference "Folkhälsoarbete i samverkan" (Working Together for Public Health) at Sundsvall, 10 -12 June 1991.

We note that inequalities in health are increasing in Sweden.

1. AIMS

The aim of public health work is to contribute to a sustainable long-term societal development in which the life environment gives all people the possibility of maintaining a good standard of health.

2. BASIC PRECONDITIONS FOR GOOD PUBLIC HEALTH

- *peace and freedom*
- *an ecosystem in balance*
- *equality*
- *necessary material preconditions*
- *social relations in the family, at work and in leisure time which lead to personal development*
- *daily life with a meaningful content*
- *social justice and solidarity*
- *influence and participation*
- *individual responsibility*
- *equality between the sexes*
- *opportunity of basic education*
- *right to meaningful work*

3. PUBLIC HEALTH WORK MEANS

- *that decisions in all sectors of society will take public health considerations into account*
- *that supportive environments will be promoted*
- *that efforts will be made to promote health and prevent illness*

Public health work is based on

- *democratic principles*
- *influence and active participation by the population*
- *good knowledge of the population, society and health*
- *use of local community resources*
- *cross-sectoral co-operation*

(Extract from the declaration)

formulated.

Citizen participation is a keyword in the development of supportive environments.

CHAIRMAN'S COMMENT

How useful are the views which the health policy study group managed to formulate in its brief seminars? Can they be of benefit in regional public health work in Västernorrland or more generally? We must remember that the Umeå University project operated for a limited period and that no special preparations were expected of those taking part. In reality the exercise was simply a reminder of the great social importance of the public health sector.

On the other hand, all those participating were professionals working in the health and social service sector, holding positions of responsibility with the municipal and regional authorities. Even spontaneous contributions from this group must be regarded as significant.

As the Umeå Group tries to summarize the discussion from its academic viewpoint, three observations, in particular, deserve attention.

The first is the positive attitude of the study group to WHO's radical public health programme and the message of the Sundsvall Conference. Health for all, an environmental overall view and a participatory perspective were points to which everybody subscribed. The interviews reinforced the impression of true social compassion which radiates from the public health declaration of the parallel Swedish conference (see quotation).

The second impression is the willingness to translate the radical WHO goals into practical health policy. One concrete example from the county is the public health programme of the Municipality of Kramfors, which was delivered by the County Council hot off the presses to the conference in Örnsköldsvik. Another is the work which was done in Örnsköldsvik municipality after the conference on the organizing of a broadly based political public health council. (It is said that the social conference of the Umeå Project was also an activating factor.)

Our third observation is that there is nevertheless a certain irresolution with regard to the implementation of the radical ideas as a practical policy of the local authorities. This is an impression which, I think, is quite well endorsed by the foregoing summary. The proposals are inclined to hark back to a well-known stock of health and social policy methods. They reflect a tradition focused on risk prevention and individual lifestyle correction. The health issues are seen as tasks for a sectoral policy, the aim of which is the restoration of the individual to the good society. A holistic grasp which signals a deeper social and ecological awareness of crisis is still lacking. The questions of participation are merely touched upon.

The omission is not confined to the actual group or to the public attitude to health in Västernorrland. This is easily confirmed by a cursory glance at the avalanche of public health reports emanating from local authorities, public health institutes and government committees in recent years. Acceptance of the radical WHO ideals is professed in many documents. But practical action is in most cases taken under the auspices of the established public sector, without any acknowledgment of the fact that this, too, is often questioned today.

The new interest in our living environment is shown in numerous initiatives in the housing area, in schools, in social services, in the cultural and health sector and in recreational programs - all the familiar sectors of municipal activity. Not very much is seen, up to now, of the critical detachment and holistic thinking of the exemplary model in these initiatives. On the other hand they stake out far- reaching new areas of work for public sector agencies.

It is, in actual fact, a remarkable transformation that the radical programme for health promotion undergoes on its descent from ideological programme level at WHO headquarters to local implementation level. Or, in a nutshell, from idea to action. How is this striking discrepancy to be interpreted?

Is the explanation the banal one that the sectoral interests of the health authorities have got the better of the WHO ideal? Or has the radical message been too difficult to assimilate and to translate into plans of action? Or is perhaps the Umeå Group reading too much into the new message? To find the right explanation we have to produce a more penetrating analysis. Let us therefore look more closely at WHO's radical public health programme as it now appears.

POLICY TRADITION IN PUBLIC HEALTH

WHO's health promotion programme arises from a tradition of health policy thinking which has its source in the nineteenth century and which runs parallel with the modernization of industrial society. The agenda displays the same mixture of protest and observance of tradition that we find in many radical trends in modern times.

It is a process which begins with *the environmental-hygiene period.* The characteristic of this period is great passion concerning the health problems of industrial society as reflected in social wretchedness, pitiful housing conditions and appalling urban environments. Administrative measures and health policy legislation, and also traffic planning and improved sanitation, are important instruments. To the agents of a radical health ideology today, this epoch, with its holistic view and its environmental orientation, is in certain respects a model.

The theme of the second phase of public health work is that of *preventive medicine*. Medical care, backed up by increasingly more advanced technology, becomes central and diagnostics is individualized. A preventive health care, mainly concerned with epidemiological matters, takes over. Professionalization and technical modernization are trendsetting ideas during this phase, which extends from about 1900 into the 1950s. In practical public health work it still plays an important role.

The third policy stage primarily covers the 60s and 70s and is usually called *the behaviour-modification phase*. Interest turns from medical diagnosis to risk factors in the social environment and in the personal lifestyle of the individual. The work environment appears on the agenda and campaigns are mounted against alcoholism and smoking. A pedagogic attitude appears, and a finger is wagged at the individual, who is himself regarded as ultimately responsible for his personal health. Society is exonerated from responsibility, but may assist with advice and care.

As a protest against the individual and medicinal fixation of established public health policy, a movement for *public health promotion* appears around 1980. This movement launches a positive health concept: the new goal of health policy is active promotion of not only the physical health of the individual but also *his overall well-being* in society.

The triggering factor behind the initiative is a new global environmental consciousness, coupled to a radical social commitment. That even the welfare state can be a health hazard is a new discovery which turns out to be supported by a wealth of empirical findings.

A UN body, the World Health Organization, is the main instigator of the new movement. The first major contribution is the health conference in Alma Ata in 1978, which produced the influential "Health for All" document. This follows up the earlier interest in environmental factors, but expands it into an overall view of society and the ecological environment. With the broader perspective new forces need to be mobilized in order to realize the health goals: social policy must also step in and all general policy areas have to share responsibility.

THE POLICY OF HEALTH PROMOTION

This is an exceptionally bold programme. Even if in the professional practice it adds up to a well-practised routine, it breaks so radically with the policy tradition of health work that many must see it as a provocation. It is also apparent that the new trend has drawn inspiration from radical movements at the time of its inception, particularly the youth movement of the 60s and the 70s with its perceptive political and ecological criticism of an alienated technological society.

But if we examine the programme more closely we find that beneath the rhetorical overtones there is a very exact medical diagnosis of an emerging series of new and acute health problems. An entirely new type of psychic insufficiency, which cannot be related either to the so-called welfare diseases which occupied the previous generation or to medical epidemiology, has appeared on a broad front. It may be labellled *psychosocial sickliness* and it produces a whole spectrum of pathological symptoms, which cannot be explained or cured with the aid of the old textbooks.

This phenomenon has acquired the name of the *third wave* in modern public health development (Fig. 1). Its public debut is seen as coinciding with the appearance of post-industrial society. According to some researchers, the new psychosomatic disorders are connected with generally growing doubt about society's existential foundations. Children and young people appear to be particularly hard hit. The disorders are not due to isolated risk factors in the social environment. They seem to have their roots in the very systemic structure of our society.

It is a diagnosis which appears scientifically very reasonable. It is corroborated by the professional methodologies of both individual-related medicine and lifestyle analysis. The main curative strategy for coping with the new ill health is also intellectually irreproachable: it is to create a living environment and a social climate which safeguard the well-being of the citizenry and the individual.

But it is also evident that in the new approach an important link is missing between diagnosis and programme: a carefully fashioned operational strategy, whereby the ambitious goals can be realized in the existing society and with clinical precision. It is this link which is lacking in Västernorrland's radical health programme, and it is also missing from the model, the WHO "Health for All" programme with its 38 points. This gap forms a serious obstacle to the realization of the radical programme as intended.

Nevertheless, extensive practice has developed in the field, based on professional energy and human inspiration, and much has been achieved. Only one question continues to worry those who recall the inspiring words of a number of big health conferences: Are we digging deep enough? Are the remedial measures adequate reactions to the clinical diagnosis? May the sociomedical system crisis which has been so skilfully exposed be too imminent to be dealt with solely with current sectoral methods?

The new movement for health promotion has quickly achieved great success and won broad professional backing. But if at heart it cherishes a dream of realizing its great vision, it must sooner or later review its operational strategies. It must continue to map out the problem areas of "the third wave" until a strategy has been devised which will really be capable of sweeping away the pathological patterns in our civilization. These are patterns which are appearing more clearly in global society day by day, and which it is no longer possible for anyone to ignore.

SEARCHING FOR AN OPERATIONAL STRATEGY

On what lines, then, should the work of devising a penetrative operational strategy for ecological public health proceed? Let us conclude by following this train of thought and seeing where it takes us. Such work plainly leads in several directions and far beyond the boundaries of medicine and health care. But where should we begin?

In our opinion, the development of the radical WHO programme itself gives important indications of where to search. The three international health conferences in Ottawa in 1986, Adelaide in 1988 and Sundsvall in 1991 produced a series of interesting specifications of the original "Health for All" scheme. In the Charter for Health Promotion of the Ottawa Conference the positive new programme was presented in all its socio-cultural expanse. Health is a resource for daily social, cultural and economic life which is achieved by a mobilization of the vigour of the population. In Adelaide the programme was further clarified, with the political action level being emphasized and designated a social responsibility. "Healthy Public Policy" was the unifying slogan.

The big innovation at the Sundsvall congress in 1991 was the direct linkage of the radical health trend and the ecological environmental movement. This was a linking of ideas which opened the way to entirely new policy initiatives of global scope. But even more important were the ideological and intellectual implications. In a statement of aims written in 1989 (see bibliography), Ilona Kickbusch, leading theoretician and policymaker at WHO, says:

"The major difference between the old and the new public health lies in the new ecological risks we are faced with worldwide and the challenge of developing a global public health."

And later she continues, now focusing directly on the social health dimension:

"An ecological paradigm of public health is not only a plea to pay more attention to new environmental problems and to create strong administrative structures in doing so, although that would be a reasonable first step. Ultimately, we must reconsider health in terms of an ecosystems approach." This basic idea, with all the reflections that it provokes, possesses a tremendous factual energy. It actually points straight to an operational strategy for radical health work.

Reaching this goal, however, requires yet another leap of imagination, or perhaps two. First we must shift the focus of the ecological system analysis from the external global to the internal social environment. Also we must consider human society as an ecosystem - and an active one.

We can learn from ecological science how ecosystems in nature function at different organizational levels, which will make us better fit to observe the patterns of interaction in our own life system and habitat. We can also learn how whole populations and ecosystems may be affected by pathologies which devastate their societies and the ecological resource base.

From this interesting reading we can finally learn to regard the environmental crisis not as something external but as a symptom, and the most serious one among several, of the internal system crisis at the social core of our species. The conclusion is self-evident: to fend off the environmental crisis the social system crisis must be overcome. This is also what Ilona Kickbusch says.

But we still have not got to the heart of the problem. We reach this only when in our calculations we take account of the force which keeps the whole system working and by which it acts: the living participants, ecological man. Who is ecological man? What are his distinguishing characteristics? How has he developed and how does he create his ecological system, his society? In which respects is he unique and in which does he resemble other species in the ecological cycle?

From the answers to these central (and scientific) questions we will obtain knowledge not only of man's ecological nature but of the ecological social environment which constitutes his natural element. And with this knowledge we can see more clearly the anomalies and pathologies in our own social system, and avert the systemic threats to the planet and to man.

This should - as we see it - be the focal point of the discussion of future ecological public health, which will start when the message of the Sundsvall Conference has lodged itself in people's consciousness.

TOWARDS A RADICAL SYNTHESIS

This focal point can be approached in a number of ways. The Umeå University Group has also been working in this direction for a long time. It has been doing so at a considerably more modest level, within the framework of social ecology research conducted by its Swedish and international network. Our inspirational starting point was the research into the quality and values of life which flourished in the 60s and 70s, and which reflected a new awareness of the human alienation discernable in the technocratic welfare state.

One theme of this early phase was the study of human needs, with

Abraham Maslow as one of the foreground figures. A seminal work in this direction was "Att ha, att vara, att älska" [Having, Being, Loving], by the Finnish sociologist Erik Allardt. The concentration on existential questions seemed fresh and attractive to hungry souls at this juncture, but scientific results were sparse.

In the Umeå project, which started about 1980, a systematic ecological approach was adopted from the outset, but with the focus on man: Designing models of man's ecological nature and interaction with his life surroundings was set up as the research objective. The intention was then to test these models in actual social experiments. If the results justified it, the models would serve as foundations for (or contributions to) a qualitative (social and humanist) methodology in social planning. The models could also be used as a corrective in human-ecological assessment of existing social environments.

This work has now brought us to the point where with a certain assurance we are launching ecological models of society in a global debate on development. The project "The Ecological Life-Region" is one example of this. In these models it is the local and regional core of the eco-social totality that is mainly considered. Yet, allowing the regional life cells to cooperate to form communicative networks up to continental and planetary level is the overall vision. The environmental impact of the new social model is evident.

It was no coincidence that the Umeå Futurology Group invited Ilona Kickbusch from WHO to give the opening lecture at its social conference. This was done knowing that the eco-social focal point was in fact also the meeting point of the radical public health programme and socio-ecological welfare research. The whole health-promotion scheme can be used without major changes as a political programme for a humanist social order with its roots in ecological doctrines.

From a humanistic standpoint, the scientific anchorage offered by the linking of public health to medical diagnostics opens promising perspectives. Human feelings of welfare and wellbeing may perhaps become medical facts, with all the prestige that this entails in our epoch.

Conversely we imagine that the human-ecology models of the Futurology Group can - with certain modifications - form a basis for the operative strategies which an ecological health policy needs in order to realize its grand vision. Or that it may serve at least as one of the sources of inspiration for an updated mainstream public health strategy. However, further consideration of these possibilities must be left to those active in the health-policy field.

After all the arguments, the impression remains that the two initiatives mentioned are in essence closely related. In both cases the objective is humanitarian, that of creating a society with greater human dignity. In both cases the frame of reference is an ecological systems perspective, which takes the whole global ecosphere as its principal witness. Similarly there is in both a demand for objectivity, the persistence of which entitles it to be considered scientific.

One experience which both types of project will probably go through sooner or later is that by intrinsic logic they tend to reach unmanageable proportions. This insight arouses an interest in seeking partners and creating alliances. We are not thinking primarily of temporary constellations, but of coalitions on a paradigmatic level. Participants from a number of key areas ought to form chains, in which each link is a functioning unit but at the same time interacts with the others, scientifically or politically. All would gain inner strength from this interaction, but above all society would benefit. Perhaps a paradigmatic coalition could be created from these four links:

- A global environmental defence on a scientific basis;
- A social-policy programme founded on socio-ecological and participatory ideals and global in scope;
- A radical public health policy which reveals and averts the socio-ecological health risks in the local and regional community;
- Basic research in social ecology which charts out man's ecological nature and habitat.

The question is whether a socially and ecologically viable society can arise from the interaction of these new fields of force. A society where perhaps the ideas of health and quality of life have entered into an alliance and become synonymous.

That is for the citizens to decide.

REFERENCES:

Allardt, E., 1977: Att ha, att vara, att älska [Having, Being, Loving]. Argos, Lund.

Andersson, R. and Kickbusch, I. (eds.), 1990: Health Promotion. A Resource Book. WHO Regional Office for Europe, Copenhagen (unpublished).

Dahlgren, G., 1991: Strategier för folkhälsa - några framtidsperspektiv. I: Folkhälsans villkor (Strategies for People's Health - Some Future Perspectives. I: The Conditions of Public Health). Folkhälsogruppens skriftserie nr 9:1991. Stockholm.

Dean, K. and Hancock, T., 1992: Supportive Environments for Health. WHO Regional Office for Europe, Copenhagen.

Diderichsen, F., Östlin, P., Dahlgren, G. and Hogstedt, C. (eds.), 1991: Klass och ohälsa - en antologi om orsaker till den ojämlika hälsan [Class and Ill Health - An Anthology of Causes of Unequal Health]. Tiden, Folksam, Stockholm.

Folkhälsorapport Norrland (Public Health Report, Norrland). Samverkansnämnden, Norra sjukvårdsregionen. 1992, Sundsvall.

Folkhälsorapport Örebro län [Public Health Report, Örebro County]. Samhällsmedicinska enheten, Örebro läns landsting. 1993, Örebro.

Hela folkets hälsa. En nationell strategi (All the People's Health. A National Strategy]. Folkhälsogruppens rapport nr 8. Allmänna förlaget 1991, Stockholm.

Kickbusch, I.: "Back to the Future. Moving Public Health into the 90s". In: Wolczuk, P. et al. (eds.), 1989: Proceedings of a National Symposium on Health Promotion, British Columbia Ministry of Health, Victoria, B.C.

Kickbusch, I., 1989: Good Planets Are Hard to Find. WHO Healthy Cities Papers No. 5, Copenhagen.

Kickbusch, I., 1990: A Strategy for Health Promotion. WHO Regional Office for Europe, Copenhagen.

McMichael, A.J., 1993: Planetary Overload, Global Environmental Change and the Health of the Human Species. Cambridge University Press, Cambridge.

Milio, N., 1981: Promoting Health Through Public Policy. F.A. Davis, Philadelphia.

Svenska kommunförbundet, 1991: Kommunerna och folkhälsan [The Municipalities and Public Health]. Stockholm.

The Sundsvall Conference on Supportive Environments, 1992: Playing for Time, Creating Supportive Environments for Health. People's Health Västernorrland, Sundsvall.

The Sundsvall Conference on Supportive Environments, 1993: We Can Do It! The Sundsvall Handbook. People's Health Västernorrland, Sundsvall.

The World Bank, 1993: World Development Report 1993. Investing in Health. Oxford University Press, New York.

WHO Regional Office for Europe, 1985: Targets for Health for All. Copenhagen.

WHO Regional Office for Europe, 1986: Health Promotion Concepts and Principles in Action: A Policy Framework. Copenhagen.

WHO Regional Office for Europe, 1990: World Health Organization Healthy Cities Project: A Project Becomes a Movement. Review of Progress 1987 to 1990. Copenhagen.

WHO Regional Office for Europe, 1991: Targets for Health for All. The Health Policy for Europe. Summary of the Updated Edition. Copenhagen.

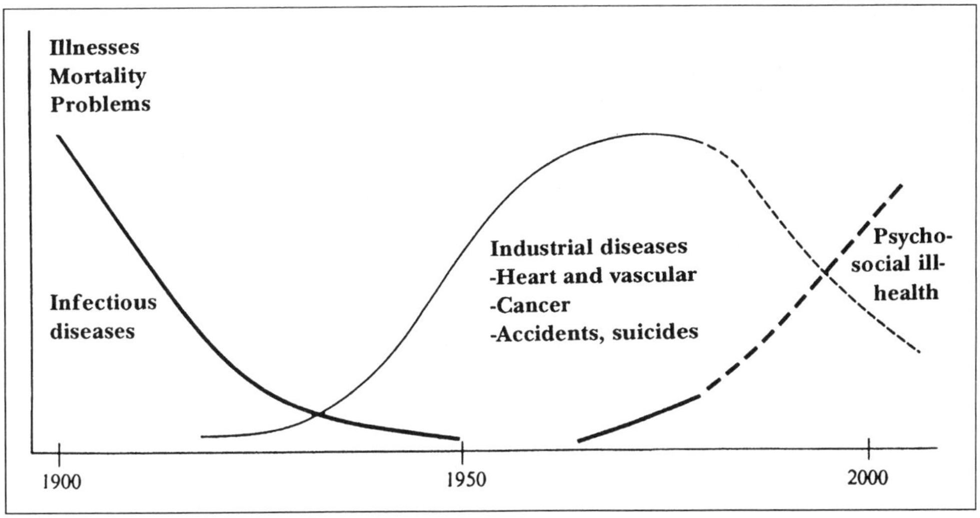

Fig.1. **The three waves of public health development in the 1900's.**

V EPILOGUE

"Angelica Silvestris". Photo by Margareta Klingberg

Summary

FINAL ACCOUNT

It is time for a final summary and, perhaps, for one last look into the future.

In this book we have attempted to outline an alternative strategy for regional development. It is a strategy which differs from the routines of traditional regional policy in its greater interest in the individual region's potential and visions. The region is not seen primarily as a contributor to national economic growth, but is regarded as the living space of a local population, which is the guardian of a social tradition and a cultural heritage.

One of the main themes of the alternative strategy is the idea of self-reliance, which contrasts sharply with the regional philanthropy often advocated by central authorities. In the alternative vision the local population itself is the main authority, bearing responsibility for the management of resources and for taking initiatives to promote the common good. From these ideological starting points the book has been discussing three aspects of a regional social structure: *the planning of the living space, the cultural policy and the social dimension.* We have coined a phrase to designate this type of region: *the Life-Region.*

The intention of the book, however, is not to argue for an individualistic regionalism, which cuts itself off from the outside world and self-sufficiently defends its own interests. The Life-Region principally maintains an open and cooperative - not a confined and competitive - attitude to its inter-regional surroundings. The same is true of the life-region´s attitude towards its inner social structure. In its political programme it stands up for communitarian ideals and agitates for a reconstruction of the basic networks of close human relationships.

The background to these studies is the social disintegration which is taking place in contemporary society, and which constitutes a growing threat both to human well-being and the global ecological balance. The task that we have set ourselves is to try - after thorough critical analysis - to outline an overall model for a socially sustainable community. This is a task which necessitates working with a very light but at the same time accurate hand.

We have described our approach as a *socio-ecological* one. This means that the relationship between the individual citizen and the community is considered as a functional system. From the scrupulous analysis of this interactive system, suggestions for new structural solutions can be deduced. The alternative approach to regional policy advocated has resulted in a number of unexpected proposals. Perhaps they can help to strengthen local identity and improve the quality of life in the regions.

When the socio-ecological scheme of the Life-Region is uncompromisingly implemented in geopolitical reality, the result will be *networks of life-regions*. The networks will branch out across continents and may finally embrace the whole planet. Each region in such a network will be a socio-cultural unit and as such form a closed ecological system. At the same time each region will be functionally an open system, economically, socially and culturally interacting with the other life-regions in the planetary net. The intriguing process of social and territorial organization reflects basic principles of the functioning of the biological life systems on our planet. These are principles which guarantee, in an evolutionary perspective, a balanced coexistence between the species of the biosphere.

The essential intent of the socio-ecological strategy is, therefore, to try and readapt human social organization to the natural cyclical laws of the biological life systems. With this, the utopia of the life-region offers not only a radical solution to the community crisis of our time, but also an action strategy for dealing with the global environmental crisis. These are all ideas which are in harmony with the new awareness of the deeper origins of the ecocrisis which we find in the documentation of the UN Environmental Conference in Rio de Janeiro (1992) and in several humanitarian UN undertakings in recent years.

How veracious is the outline of a society in inner and outer balance which we launch here? Is the vision of the ecological region, after all, merely a rhetorical chimera doomed to exist in the shadows of intellectual history?

This is not at all as certain as might at first be imagined regarding the physical infrastructure of vast extent which surrounds modern society. This whole unwieldy apparatus, with its historical stratifications, is showing clear signs of disintegration. It is already having obvious trouble defending its own central institutions. Nor will it be capable much longer of offering particularly attractive prospects of a good, secure and healthy life to its citizens. In the light of the possibility of an alternative form of social organization, today's geopolitical dogma, with its division of society into metropolises and provincial zones, appears ever more clearly an antiquated and outworn model. In the long term it ought perhaps, if at all possible, to be abandoned on the grounds of its qualitative and functional - not to mention ecological - shortcomings.

The Life-Region, with its combination of ecological awareness and insistence on human quality, is a solution which may turn out to be much more attractive to the socially and ecologically enlightened men and women in the mature global information society. If only the alternative models deliver what they seem to promise, it is easy to envisage strategies which lead to their realization, and to point to areas where they can form bridgeheads towards a viable future.

Such bridgeheads are of course the provincial regions all over the world, today notoriously impoverished by the central forces of power, but latently possessing a series of the qualities which the socio-ecological vision describes. The Västernorrland region, which has played an active part in the current research and development project, is a good example.

While we are waiting for greater changes, the vision of the Life-Region can also be used to inspire limited sectoral contributions in all parts of the community where social improvement and cultural renewal are needed. Supporting the social resilience of threatened provincial and rural regions is a task which appears as acute in the industrial nations of the North as in the developing countries of the South. There are also major sectoral planning tasks waiting to be tackled in many heavily burdened metropolitan regions all over the world.

In order, however, to realize a sustainable social vision, more than alternative community models is required. It is necessary to identify players who are able to carry out the assignment, and who can apply the commitment which is necessary for a successful outcome. To attach all hopes to future grass-root forces is probably naive with the hardening social climate which is spreading. It is also absolutely imperative to apply better structured efforts. What is needed, first and foremost, is radical readjustments of priorities at the highest political level, which place social development at the centre of overall politics.

The new social and human knowledge which is now emerging from alternative scientific environments, will enable a strong and radicalized social policy to be obtained. This is a policy which possesses the necessary human and ecological competence to take up the struggle for the long-term development objectives with the biased economism and technologism in current political thinking.

THE VIEW AHEAD

This book has not been planned as an integrated whole and its formal arrangement does not give an altogether harmonious impression. This is because the contents are the result of an R&D project which has been in progress for several years, and the course of which has not been entirely predictable. The whole process which the book has undergone is reflected in its partially improvised design.

In the book, a number of authors and study groups present their scientific and personal viewpoints on the selected topics. The basic strategy, which is visible throughout the whole book, has been to show openness to differing opinions within the framework of the chosen theme. Such openness is of course the only possible attitude in a dialogue project where researchers and regional representatives come face to face.

While this has to be said, it is apparent that running through the book there is still a logical, unifying theme. This theme is the socio-ecological theory of the Umeå University Group, which runs all through the text and which may perhaps acquire a more definite import towards the end.

A radical regionalist vision of society naturally extends into many more areas than those described in this book. The most important are *the political and administrative sectors* and what I shall here term *the social economy*. The fact that these topics are not dealt with at greater length in the book is a failing. But tackling these two perspectives in as ambitious a manner as the earlier themes is beyond the current capacity of the project. As far as the political theme is concerned, we hope to be able to return to the subject before very long. The economic theme is so fundamental to the discussion of the eco-society that it should at least be touched upon in this book.

We therefore round of these studies and the whole book with a discussion of the possibility of *a socio-economic strategy* for sustainable regional development. And to underline once again the serious intent we raise one last issue: the scope for *a science of social ecology*, by means of which the alternative approach may be underpinned theoretically, pedagogically and in community work.

Economy in the Viable Region

THE ECONOMIC CRISIS

It is reasonable to start considerations of a socio-ecological economy by examining the status of the prevailing economic system. Following the fall of the Soviet state and the planned economy, the economic world scene is now dominated without serious reservation by market capitalism. This situation represents the logical point of departure for our study and for its opening question.

What, from our actual point of view, are the deficiencies and limitations of today's version of the market economy? And, reversedly, what is its potential for realizing the regional vision that we are pursuing here? From the answers to these questions we will be able to begin formulating the alternatives which we require for our particular purposes.

Before we begin our argument, however, one thing needs to be pointed out: the economic discussion that we wish to conduct does not concern only the industrial countries of the West. Our examination of economic alternatives is addressed to a much broader circle of interests, including the countries of the East and the South. But industrial capitalism of western type still dominates the international economy, especially as a model. An examination of the precarious state of the North Atlantic economies is therefore also salutary from the point of view of the general global debate, not least in a discussion of the role of the economy in a vision of sustainable global progress.

The first thing that inevitably strikes the eye when we take a look at the current world economy is the long series of symptoms of crisis. Ever since the oil scares of the 70s the leading industrial countries in the West have found themselves in a protracted - latent or manifest - state of crisis. This is a pattern of development in which unforeseen recessions have given way to sudden upswings, and where the result is an economic curve which is strikingly different from the rising diagonal of the neoclassical vision of growth.

All agree that the oil crisis alone cannot be blamed for the problematic course of events - that it was merely a precipitating factor. We are dealing with a crisis the roots of which lie considerably deeper among the mechanisms of the contemporary market economy.

The differing symptoms of the western world's economic malaise are well known to everyone since a long time. For decades they have been analysed in a voluminous economic literature and debated by the columnists of the daily press and the weeklies. Every day the inexhaustible material is presented to the public in graphs, development curves and economic forecasts. Media analyses point to continued inflation, rising unemployment, a slowing rate of growth, industrial decline, budget deficits, financial crises, currency outflow and a menacingly growing national debt. Sophisticated expositions, which the majority of us can follow only imperfectly, describe how falling profits in the competitive sector lead to poor savings and then a decline in investment, and how the downward spiral is aggravated by a cost crisis in which rising wages and increasing expenditure in the public sector are negative factors.

In contrast to the abstractions of the economic commentators, popular reportage offers a different, more easily grasped picture: in uncanny panoramas we perceive how whole communities come to grief, how vast regions are deindustrialized, how once-leading industrial countries with high standards of living sink back into insignificance.

The menacing reality can no longer be repudiated; it intrudes into our homes and shakes our lives to their foundations. Social insurance, public health, accommodation, social amenities for children, young people, the sick and the elderly: the whole basic security system is being demolished before our eyes, with no real alternative being presented. The vision of the democratic welfare state, of the all-embracing "people's home" (as the Swedes put it) dissolves like a mirage - just when we have invested our assets and our human capital in the vision.

Many feel that they are ruined, and not only financially, but in their belief in society and the future. At the same time the ordinary citizen's option to develop alternative forms of livelihood and security has been almost squeezed out. The capsizing welfare project pitches the faithful citizen into a sea of despair and desperation.

Then, suddenly, the tone changes. All the newspapers are writing of a turning tide. Once again the banks are making record profits, financiers are investing in new companies, exports are rising in leading sectors. The deepest pessimism becomes rosy-spectacled optimism overnight. Brighter times are coming, a fall in unemployment is predicted, many entrepreneurs have cause for joy. The race can continue as before (if only the international interest trend can be brought under control...).

But all those who have had to ride the economic roller coaster for decades are already asking nervously when the next downturn is coming, finding it difficult to see the sudden excitement as much more than another kind of crisis symptom. They very reasonably wonder if the time has not come for really serious considerations on the economy.

ECONOMIC POLICY

Government economists and their legions of expert advisers ponder on strategies for coping with the economic crises and their destructive effects. The historical knowledge of two centuries of industrialization forms an important part of their professional expertise. The experience of decades of unbroken progress since the Second World War is another cornerstone of economic authority. A set of economic and financial policy strategies, and various philosophies concerning the state's interference in industry, form the tools with which attempts are made to tackle the issues.

Depending on which aspect - investment, foreign trade, production, regional policy, inflation etc. - is of current concern, a particular political strategy is chosen.

Devaluation has long been one of the trump cards of economic policy. By means of repeated devaluations, efforts have been made to compensate for the high cost level of domestic industry and increase export revenue, with a view to stimulating investment and further industrial growth. Policies of restraint (e.g. cutbacks in public expenditure or increased taxation) have been urged in the hope of dampening domestic purchasing power and thus improving the balance of trade, reducing the budget deficit and stemming inflation. At other times more Keynesian policies of stimulating growth and employment by increasing public expenditure have been tried.

Redistribution policy is another instrument. Tax reforms and other redistribution policy measures (higher wealth tax, transaction tax on share purchases, wage-earners' funds and renewal funds) have in certain cases been introduced in order to influence wage structure.

Financial and monetary policies are also used as a method of infusing new life into the economy and internationalizing capital and investment markets. Currency deregulation is one of the recipes. Also regulation and deregulation in many other areas (agricultural policy, the foodstuffs sector, the transport sector, credit facilities) have long been a part of the arsenal of economic policy.

Unfortunately, the actual course of economic development gives ample evidence that such interventions have achieved only limited results. Despite increasingly sophisticated strategies, the negative symptoms have quickly recurred and the economic crises have increased in severity.

One easily appreciated explanation is the notorious preoccupation of economic policy with the monetary level, where the direction, circulation and ramifications of flows, and, therefore, the actual effects of economic policy, are always difficult to anticipate or to influence. (The experts persist in telling us that we must learn to live with this state of affairs.)

Another explanation is the exaggerated faith of the experts in the idea that the failures of the market can be corrected with minor tinkering with various variables. The contemporary crisis has unfortunately proved much too deep and far-reaching for correction with the finely calibrated instruments of the economists.

A growing insight into these troublesome circumstances has led in many countries to the trial of more robust methods of handling the crisis. For the most part they represent a step forward for economic liberalism. Preventing inflation is given priority and unemployment is allowed to rise unchecked. Saving the productive economy at any price becomes an overriding objective. Increasingly drastic pruning of the public sector and the democratic social services takes place. A double-edged regional policy gains ground, which, while paying lip service to the survival of rural vitality, concentrates on strengthening a small number of cosmopolitan growth centres. It is hoped that the internationalization of the market will save the national economies, or, at least, their leading corporations.

For the sake of this pious hope there is a willingness to surrender power and control over the domestic factors of production, and even to interfere with the political foundations of popular sovereignty.

ECONOMIC ALTERNATIVES

Will a dose of radical liberalism restore the lost balance in the national economies, so that the overall growth of prosperity can continue? Nobody seems to know for sure, but there may be reason to doubt it.

The painful interventions hardly represent a renewal based on a deep theoretical awareness of crisis. They rather signify a coarsening of economic policy which reveals that desperation prevails at the seats of political power. The most unfortunate feature is the almost total absence of a discerning critical debate on the basic principles of the economy. Since the collapse of the socialist system there is a suspicious consensus on general questions in the leading camps. The copious expert commentaries in the daily and the specialist press examine the technicalities from every angle, and the political discussions between liberals and social democrats degenerate into a charade in which the symptoms of the economic crisis - unemployment, lack of growth, declining exports etc. - serve mainly as weapons for use in party-political jousts.

Both the specialist and the political debate are conducted in an abstract jargon and with a fragmentation of the issues which make it difficult for most people to discern the economic reality that is at stake. Often the rhetoric serves as pure disinformation and has a paralyzing effect on the will to start a broader general debate on the economic conditions of society.

Despite all these alarming symptoms, the neoclassical model continues to occupy the public stage. To obtain a more considered commentary on the current crisis, therefore, we have to dig deeper: into economics as a science, into our history and into the alternatives. In the margins of the economic debate, anyone who makes an effort will in fact find currents of thought showing greater critical acuity than those served up by the media. Three of these currents, in particular, deserve to be spotlighted here: those of *institutional economics, ecological economics* and t*he communitarian tradition (or grass-roots economics).*

Of these three schools, institutional economics is the oldest, with roots in the nineteenth century (its main advocate is Thorstein Veblen) and the American pragmatism of the early twentieth. It attaches particular importance to the institutional and social framework which surrounds the system of the market economy. The communitarian tradition also has deep roots, but derives in its modern form from the alternative debate of the 70s with focus on local mobilization and self-administration. Ecological economics may be seen as an offshoot from the same stem, but it is the one most recently established as a professional concern of economists. Its main theme is the problems associated with the sustaining of the world's natural environment and resources.

All three alternatives level serious objections at the model of neoclassical economics, and the interesting point is that none of them expounds its arguments in Marxist terminology. Also, their spokesmen all decline to conduct the discussion in the terms of neoclassical economics, or on its favoured ground: fiscal policy. The arguments generally point to deeper structural deficiences in the economic system than the leading lines of economic thought are normally prepared to consider.

The *ecological economists* put an unwavering finger on the very pulse of market economy dynamics: the idea of growth and of an eternal expansion of prosperity.

This is a thesis which even common sense declares unreasonable, they plead, but whose absurdity is fully revealed in the environmental destruction which has been brought about by modern industrial society. The ecological economists maintain that the viability of the natural ecosystem has now been stretched to - or even beyond - its limits and that there are strong indications of imminent collapse. As the global ecosphere makes up the resource base of human society and a precondition of social life, expansion must be arrested, not only by sectoral efforts but also by new ideas in economic policy.

The ecologists assert that the quantitative model of economic growth ought to be replaced as soon as possible by the term *economic development,* which has a qualitative connotation and means resource management designed to maximize general well-being. Ecological economics tries to wrest the theoretical weapons from the hands of current economic policy by rejecting its only commonly agreed prescription for overcoming the crisis in the long term. This amounts to a dramatization of the debate which ought to provoke a serious attempt to escape from the intellectual straitjacket of neoclassicism on the part of economic theoreticians.

The *communitarian tradition* attacks from the opposite flank. Its main criticism of the neoclassical model concerns the damage that it wreaks on the elemental levels of social life, i.e. in the local community and its social networks.

The rationalization of the production economy and the growth in scale which this has implied have transformed labour from a presumed value of life into a burden, and the everyday existence of the individual from a life of freedom to a pseudolife of powerlessness. With this, the motivation for prudent husbandry of resources and a vigorous work input has disappeared from working life.

The market economy is also blamed for causing an extreme segmentation of the local social household and for the rise of monoeconomies. The local need of goods and services has to be met by extensive trade with other specialized economies near and far. This is a supply system which requires a voluminous organizational and technical infrastructure, and which in its purer forms proves unexpectedly expensive. The socio-economic profit appears after a certain limit to decrease gradually, and may even turn into a loss by comparison with a system which knows how to use the advantages of dense local networks.

Against the background of these arguments, the grass-root economists recommend a revitalization of locally based production and distribution systems. The co-operative form of activity is described as one of a number of alternative instruments which can restore influence and initiative to the citizens. The quality of life thus regained is regarded in many cases as a higher value than a technological superstandard.

The arguments of the communitarians for an improvement of the local and informal sector also imply a serious questioning of the theory of the market economy: it is emphasized that the neoclassic strategy aims, in principle, at liberation from the local subsistence economy, which is from every point of view interpreted as limiting.

A third broadside against the market economy is discharged by *the institutional economists*. Here the market model's claims to represent a comprehensive economic theory is called in question. With its bias towards market functions the neoclassical theory gives a very misleading picture of the economy as a whole, say the institutionalists. The market is in reality always surrounded by an organizational and institutional structure, which is one of the preconditions of its functioning. Socioeconomically this structure must frequently be seen as a cost.

According to the institutional economists, the economic system should always be considered as an interplay of institutional and functional factors. The well-known Swedish economist Gunnar Myrdal has proposed a theoretical formula for describing this socioeconomic interaction in his "principle of interlocking dependencies within a process of cumulative causation".

Many institutionalists are economic historians. The study of the interaction between the institutional structure of society and the productive economy under capitalism reveals hidden logical connections and prompts interesting reflections. One of these is the idea that the gradual rationalization of the market economy requires new institutional structures, which at every stage involve an increase in the relative social cost. Douglas C. North, winner of the Nobel Prize for Economics in 1993, develops this line of reasoning at length in his book from 1990 "Institutions, Institutional Change and Economic Performance".

After this overview of alternative schools of economic thought the impatient reader probably raises the provoking question: How can it be that all this significant criticism was not a long time ago able to change the questionable course of events in the economy?

One first answer is that the insidious points made are seldom made the theme of discussion either in specialist or public economic debate. This is quite peculiar in the light of the current crisis, and probably a further sign of the depth of that crisis. But another important reason for the status quo of radical debate is the absence of comprehensive and operational alternatives. This is a fatal situation, which radical economics will have to overcome quickly in order to avoid dwindling into insignificance. What is needed now is both a more holistic and more offensive stance towards the main issues at stake.

Dogmatism should be opposed in classical economics but also in the alternative schools themselves. New economic models ought to be constructed which take into account all the restrictions imposed by the alternative economists, but which still embrace the global dynamism possessed by industrialized society and representing the position of modern man. In other words: models which also take appropriate advantage of the experience and competence of the market economy!

Discussion of an alternative theory of economy must continue, and with greater intensity, as the economic situation becomes more and more untenable in many countries and regions. It is also important to guide this discussion into practical applications. But it is not isolated and sectoral applications that should primarily be sought, leaving the established economic policy view unchallenged. What is really required now is alternative policy strategies, with the capacity to handle the overall economic questions more effectively than the contemporary policy schemes.

The new economic knowledge that society needs in order to realize such a vision must naturally be sought in various quarters. Four possible sources have already been discussed in this article. In the following part of this chapter I will describe a fifth: *the socio-ecological alternative.* This alternative has culled elements from all the approaches described, but integrates them into a whole which is encompassed by the theoretical perspectives outlined in this book.

However, to facilitate a proper - which here means undogmatic - understanding of this alternative we must make an historical digression. We have to cast our eyes back to the theoretical and historical starting points of contemporary economics.

HISTORICAL FOUNDATIONS OF ECONOMICS

The theoretical foundations of modern economics lie in the science of political economy. Political economy has long been regarded as sacrosanct by the political and intellectual elite, not least its academic branch. This respect has functioned as an effective shield against all kinds of more fundamental criticism. Today it definitely constitutes an obstacle to new thinking and creativity in the field, which in the last resort is to the disadvantage of society.

Underlying this respect there is the vague idea that political economy represents an objective and exact science, whose theoretical and applicable pronouncements are the pinnacle of the rationalist culture which many see as the distinguishing mark of our civilization. This is a myth which has been cultivated and repeatedly polished by the political economists themselves. Despite persistent criticism from many quarters, economists have been allowed by the spirit of the age to play the role of high priests of politics, to whose dictates the whole of society, including its social, cultural and ecological expertise, is expected to submit.

It is the political economists' penchant for building theoretical models at a highly abstract level that has done most to create this myth. At the heart of the science of economics is the (frequently visualized) idea of economic life as a monetary circulation system, in which streams of currency flow between market components in a continuous cycle. Into this cycle the profit factor constantly injects new incentives for growth, while the price mechanism creates guarantees for the maintenance of internal balance.

In the vision of the economy as a mathematical system, in which all the factors of life can be converted into money and all values are measurable, the rational society of technological utopianism comes as near as it possibly can to fulfilment. In its self-containment and its rigid regularity it resembles a

Newtonian mechanical universe.

But make no mistake! The authoritarian model is not pure fantasy. It is actually based on the realities of development dynamics and it is supported by powerful pragmatic arguments. The economic-cycle model summarizes the strivings of modern developmental forces to form a functional system embracing the whole of commercial life, characterized by great unity and yet also by great factor mobility. In this system the combined contribution of labour and capital can easily be translated into a common medium: money. The accumulation of money that the system permits makes it a power factor. It enables really large-scale investments to be made and thus allows economic growth with far-reaching effects.

The construction of a functioning monetary market economy is the great *practical* invention of modern economic thinking. To all appearances it represents a lasting value.

In order better to understand the meaning of the model, however, we must take a look at its historical genesis. This coincides with the breakthrough of a monetary market economy on a large scale. It is a process which is intimately connected with the Industrial Revolution in the eighteenth and nineteenth centuries. The Industrial Revolution opened up a vigorous product market of cosmopolitan extent. The preconditions of this development were a combination of scientific and technological progress and the emergence of an entirely new infrastructure, based on the enormous capacity of mechanical engineering. The factory system, an increasingly effective global transport network and the advent of a number of world centres of trade and finance were the factors which triggered society's leap forward into a functioning international civilization. These conditions also opened the way to a previously quite unseen growth in the material standard of living.

But the success of the monetary vision in the industrial epoch of the eighteenth and nineteenth centuries would not have occurred without the benevolent interest of the political authorities. The implementation of the major projects required the active support of state government. Following the positive outcome of this collaboration the political elite soon capitulated to the market as a model and elevated it to a *theoretical system.*

Private enterprise became the foundation of the whole national economy and of economic policy. With the assistance of experts from the emerging science of political economy, strategies were devised for directing and regulating the system and controlling its mechanisms. At the same time respect for the self-generating dynamics of the system served as a moderating factor. Strategies of restraint and, eventually, of expansion were formulated and their effect on the market cycle was tested. A conscious financial and monetary policy took shape.

The circulation model was soon enlarged by the incoming and outgoing flows of foreign trade and the welfare machinery of the public sector. The net effects of the factor and product markets were made the basis of national accounts which contained well-balanced columns for the national product and the national income. In this way the market economy model was calibrated to serve as a comprehensive socio-economic theory. This led to the classical designation of "political economy" as the name of the science.

This economic "theory" has functioned reasonably well for a long time as a support for the productive economy and as a regulator of relations between the productive and the public social sector. It has also served as an actual welfare-creating instrument in a broad popular sense. As long as the model has worked reasonably well, and consensus has been reached concerning the distribution of profit, no need to question the theoretical foundations or to examine them more closely has been felt. Not even, it would seem, by the academic experts.

CRISIS OF THE POLITICO-ECONOMIC MODEL

However, when the market-based economy functions as imperfectly as it does today, when its fluctuations threaten to devastate the whole welfare system and the smartest policy strategists are no longer capable of coping with the crises, it is high time to point out the fundamental inconsistency of the economic system *as a theoretical paradigm.*

And there is every reason to ask whether the crisis might not possibly be caused by the internal theoretical conflicts and sparked off by the inconsistencies which are intrinsic parts of the prevailing economic model.

In clearing the view historical knowledge comes to our aid. What economic history teaches us is that the market economy is basically a production economy, not a social economy. Little reflection is needed in order to realize that there must be a difference in kind between an economy geared primarily to productive interests and a social economy, i.e. an economy taking into account socio-economic processes, not least in an advanced technological society like ours. The central concern of the social economy should, quite naturally, be the care of society's institutional, physical and geographical infrastructure in the long term. If a deeper view is taken, the public social and cultural institutions should also be included.

The social economy is concerned with the management of society's historical capital, and not only for our own epoch, but also for future generations. Economic policies must pursue a long-term plan, which is certainly revised regularly and which should not be without spontaneous features, but which primarily possesses great consistency and purpose.

The basis of a purposeful social economy is a thorough knowledge of the available resources necessary for long-term survival, and good care to ensure that they are present in adequate quantities. It also includes a commercial policy which secures the local conditions in which regional industry can produce and ensures access to complementary products from the interregional market.

The situation with regard to the production economy is different. It is in the nature of things that the productive sector is guided by shorter-term goals than would be reasonable in a social economy. The dynamics of the production economy include more dramatic processes, running over shorter periods and with a more rapid pulse. Competition and improvisation, in some cases even disloyalty, are legitimate in the ethics of the business world, but hardly so in community life.

Sudden innovations which are followed by phases of stagnation, decline and closure are logical courses of events in the creative business sector. So are concentration of power, cutthroat competition, rationalization and automation of production. The more the range of goods is diversified and freedom of choice between different products is seen as a part of the quality of life, the clearer becomes the difference in kind between the systems of the production economy and the basic social economy.

When, therefore, the perfect production economy is dressed up as a social economy and is granted hegemony, problems may be expected to arise sooner or later. The problems consist of inherent conflicts of values and goals between the two perspectives, which cannot be fudged over. They are made up of differences in spheres of interest and in tempo between the social and the production cycles, which entail frequent faulty connections and severe synchronization problems.

They also consist of the obfuscation of large parts of the social economy which is necessary in order to preserve system logic, and the costly improvisations which must nevertheless subsequently be undertaken in order to plug all the leaks. And they are made up of the distorted and misleading picture of the economic situation which is presented by official national accounting. It is also evident that an economic policy based on an incomplete picture of the economy will to a large extent be groping in the dark.

Furthermore, the negative effects of the confusion of schemes are also severe to the production economy. It has brought with it a dangerous disregard for the societal basis which is also the necessary foundation of efficient production. Many entrepreneurs have learnt in recent years, by unpleasant personal experience, that a working productive economy requires a well-functioning social and institutional base. The long-term stability of the society is the necessary prerequisite for smooth production and a harmonious labour market. And a stable society is at the same time the ideal market for the competing products of industry and commerce. A flourishing business sector demands, in other words, a society with a well-balanced social economy.

Neither the institutional nor the production economy functions well with the total confusion of terms that exists today. Truly constructive co-operation between the complementary sectors is even farther away.

Is there, then, a connection between the current succession of economic crises and the systemic structural problems to which we have pointed here? These are questions which the researchers of the Umeå University Group feel deserve to be illuminated more in depth.

Unfortunately, the available official data have been collected with other intentions than to comment on this question, and this makes it difficult to give a reliable answer. One hypothesis which would be interesting to test is that in its uncontrolled growth the productive economy, by its own inner logic, gives rise to escalating social costs (of many kinds), which are never revealed or discussed. After a certain interval they nevertheless burden the social economy, and eventually rebound on the productive economy. Is it possible that the manic drive for expansion and growth in the global economy is in part a desperate flight from the soaring social costs that were originally caused by the enterprises themselves and their deliberate rationalizations?

Is it not also reasonable to think that the total penetration of society by the market, and society's devout adaptation to market forces, are creating a very expensive society? Expensive, considering the modest needs fulfilment at citizens' level which is achieved through the gigantic efforts; in an astonishing degree these efforts in fact are bypassing the basic needs of the citizens. Sooner or later the informal and local social needs make their presence felt anyhow and demand attention in a socio-economic environment which has been removed very far from the conditions of a normal community life. For example, modern economic development has destroyed the traditional social safety nets of the civil society. Through modern social policy these nets have then been reconstructed at the public levcel. This is not only a doubtful waste of money but also amounts to a lowering of the general living standard, as regarded from a qualitative perspective.

Perhaps the presence of the local production networks is also what differentiates a well-functioning from a dys-functional market economy. In that case, the future looks gloomy. The triumph of the global market over the local community is today almost total. But maybe it will be a Pyrrhic victory. Maybe socio-economic bankruptcy is also threatening the West.

A SOCIO-ECOLOGICAL ECONOMY

In the scenario which has been outlined here - may I just repeat this - we are not laying the blame for the failure at the door of the market economy as such. In its genuine role as a production economy the market system will presumably continue to play a prominent role in every thriving economy by virtue of its dynamic qualities. It is the lack of an adequate social economy, geared to fundamental societal needs, to which we are (here tentatively) pointing as the cause of the precarious situation into which the advanced technological society has drifted.

In this case a total capitulation to the production economy will not repair the traumatic situation. What is instead needed is a discussion of socio-economic basics. Such a discussion must be conducted within the scope and theory of social science in a broad sense, and be free from the dogmas which encumber today's economic debate. It ought in turn to penetrate the institutional basis of society, the basic conditions of social life and man's elementary interaction with his natural surroundings.

The task ahead is nothing less than to try to devise a social economy tailored to sustainable development, locally and globally alike. Naturally the discussion has to include the productive sources of livelihood, both the neighbouring ones which guarantee our local security and the cosmopolitan ones which may give us big profits.

The unobservant may perhaps suspect that all this implies some form of planned economy. This would be a misunderstanding. The capsized socialist economy was in its theoretical conception a production economy, just as the market economy is. The difference is that the system was managed by an incompetent state bureaucracy rather than by a wildly speculating capital market. The planned economy was not a genuine and balanced social economy either.

Here we must pause for a moment. A glance at the situation will cause us to realize that the concept of a *social economy*, can be given a broad or a narrow interpretation. In its broad sense it includes the totality of economic undertakings in which a specific society engages (in the actual case a regional comunity). The most important components of the economic totality are - according to the socio-ecological model of the Umeå Group - the natural resource economy, the institutional economy and the productive economy. Most varieties of economic activity in society can be placed into one of these three categories. In its narrow sense the term refers more definitely to the *institutional economy* as opposed to the other two levels. It is rather in this latter sense that we have hitherto used the term in discussion of the contemporary economic crisis.

Both aspects of the word are of course equally important and equally legitimate in the economic discussion which is taking place here. The schemes are in reality quite closely related and in practice often merge, just as do the decision-making and administrative organs in the public sector.

What then do we mean by the concept of *sustainable social economy*, to which we referred a few moments ago? In the socio-ecological vision which we are interpreting it means first and foremost economy in the broad sense, with equilibrium between the three main components which a functioning economy has to possess in order to maintain *external* ecological balance with the surrounding human space and the biospherere.

But upholding the *internal* systemic balance is an equally important challenge. Good internal balance assumes that the subordinate levels of the social economy have been so organized that dynamic interaction between them can take place, to the benefit of the community and its citizens.

We shall go on to describe in turn the three levels which - in our model - form part of a functioning social economy: *the institutional economy, the natural resource economy* and *the productive economy.*

THE INSTITUTIONAL ECONOMY

The institutional level of the socio-ecological economy embodies a welfare policy which safeguards society's long-term development goals. The interest of the institutional economy is focused on the structural and evolutionary base of society. This social base is seen as a framework not only for the present generation but for the human species, as it rises from its history and continues along its path into the future. It encompasses not only the material needs but the whole fabric of social and cultural, productive and spiritual manifestations in the shared life of the community.

All these aspects capture the interest of the institutional economy, since they imply major public contributions for the preservation and development of the community. To handle such a complex task requires a common assumption of responsibility, within clearly defined outer boundaries. The field of responsibility of a well-functioning institutional economy consists of a social territory, which has a definite, judicially established extent.

The municipality and the nation-state represent classical social territories which have emerged for different reasons. In the socio-ecological model we have nominated the micro-region as an optimum space for the fundamental needs of life, for reasons which have been stated earlier.

The main economic capital of the ecological society is represented by the total regional infrastructure for a functioning community life, together with the permanent institutions by which public activities are run. This is an essentially inert material, intended for long-term use, and the transformation process through which it goes is normally a slow one. Management of this social infrastructure is the main task of the institutional economy in practice. It follows a forward-looking development plan which is reported on regularly and which leaves room for spontaneous features, if the popular will and the social budget allow this.

A cornerstone of the institutional economy is formed by the individual or common home, which encompasses the stationary dwelling, and the basic social functions housed therein. In contemporary economic thought, this has been described as the informal sector of the market economy. In a social ecological system, however, the latter should be entitled a formal part of the

THE SOCIO-ECONOMIC CYCLE
according to the neoclassical model

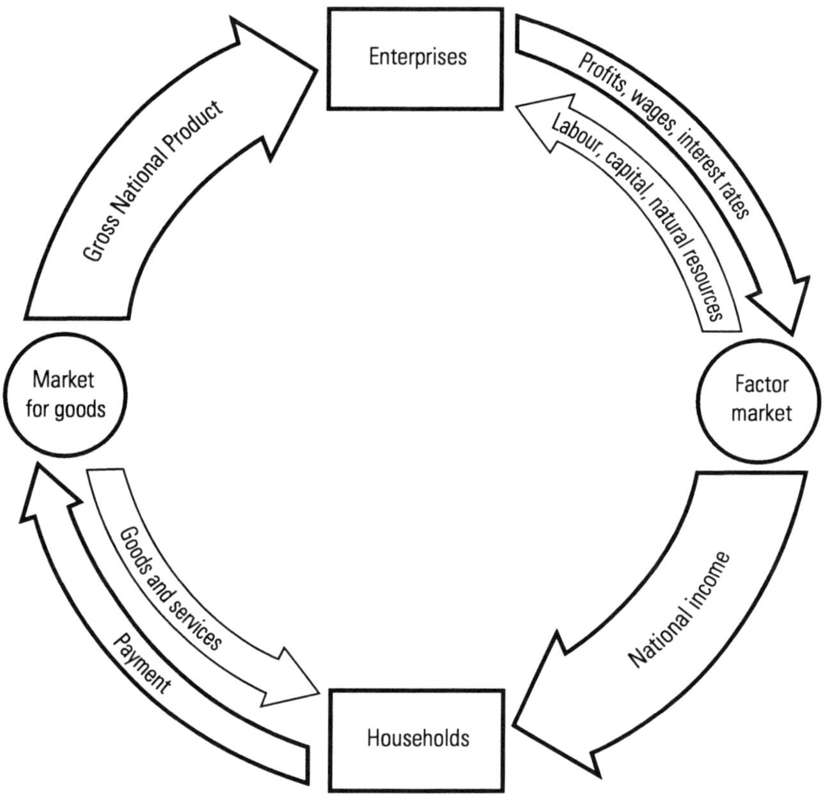

THE SOCIO-ECONOMIC CYCLE
according to the socio-ecological model

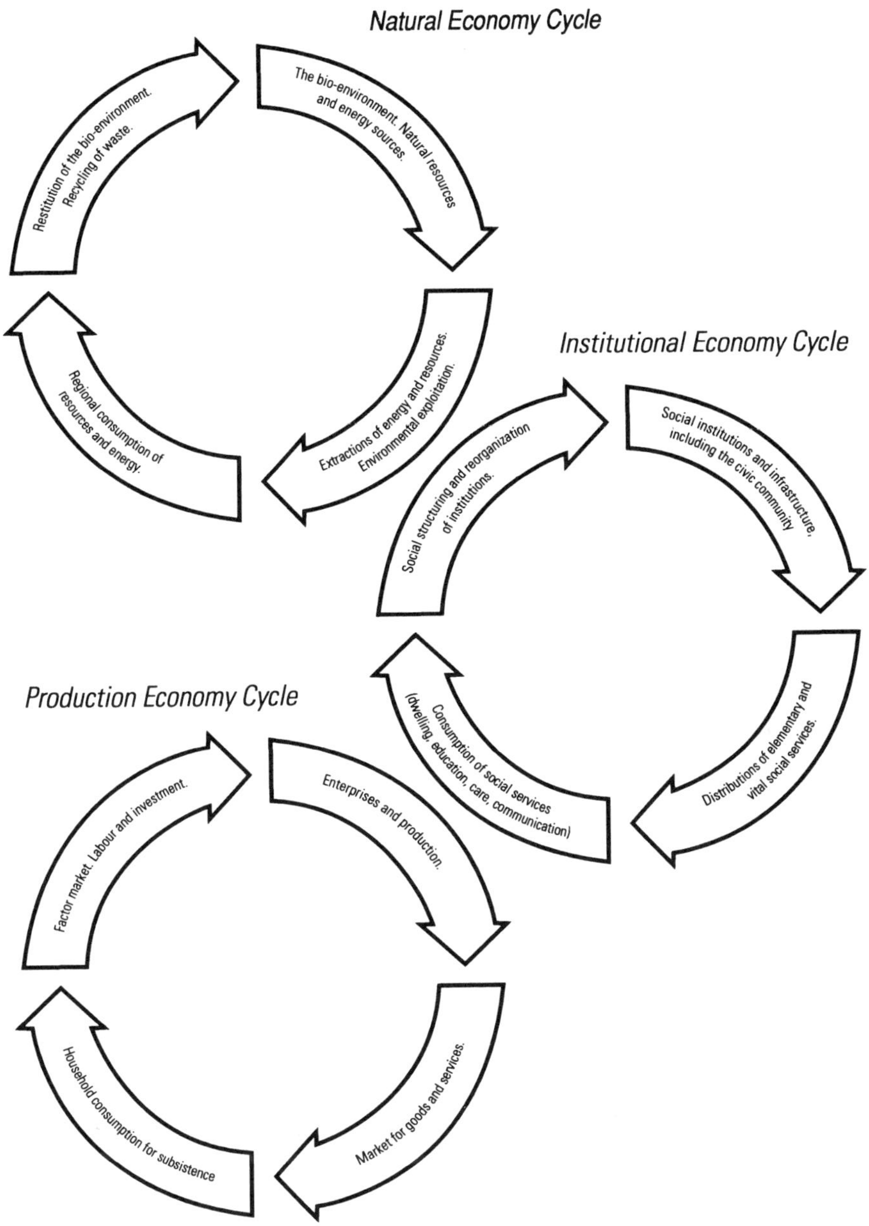

institutional economy. Access to a comfortable dwelling of one's own and to the security of a home, are fundamental social rights which cannot be compromised. When today's political forces allow the value of the home to be subjected to the fluctuations of the market economy and then let society itself tax it accordingly, this entails a violation of the individual which in many cases would be regarded as criminal if we had a legal system based on human ecological thought.

The dynamic force in the institutional economy consists primarily of the creative work of the population. "Creative work" may take place over the whole spectrum of social, cultural and economic activities which form part of the established social model, and at all relevant organizational levels. As all activities form necessary components of the whole, no order of priority can be laid down between sectors. The whole society benefits from all citizens being given the opportunity to make a contribution to the common good in the areas they best command.

The maintaining of an efficiently functioning institutional structure demands a carefully prepared social budget. In the joint budget the sectors are allocated the resources necessary to allow them to perform their continuous duties and finance desirable changes. At the same time there is an assessment of the social cost of the sectors by means of different criteria and from different time perspectives. Each sector ties up labour and capital and claims social space. Each new investment at the wrong time entails a social cost in the form of the depreciation in value of the existing infrastructure that it implies. In the institutional budget each investment is assigned a numerical value which reflects its social cost to the region. This numerical value enables different investments and sectors in the economy to be compared.

Another important task is the establishing of a clearing system between the institutional and the productive economy to facilitate the necessary transfers between the two interacting levels of the regional economy. But still more important is to take measures in order to safeguard the basic sectors of the regional economy from disturbing influences from the complementary sectors. Separate but exchangeable currencies is only one of several techniques which can be used to achieve this objective.

A further important factor is the calculation and distribution of the socio-economic profit. The general view is that this consists of the institutional good which is produced by the joint exertions. Almost by definition, this is a common asset, which is made up of the cultural facilities, technical and communicative systems and social services which can be maintained in a society with the life region's demographic characteristics. The effect of the efforts must of course be monitored, but there is no real need to make a distribution policy problem out of these self-evident matters.

Civic access to the common facilities satisfies an elementary aspect of *economic democracy,* which is at the centre of a socio-ecological vision. But economic democracy also means society's control of its own institutional and socio-cultural infrastructure, as this is defined by territorial boundaries. By virtue of their citizenship the citizens share the public responsibility for the regional economy. Each citizen possesses an audible voice in the regional household, as in the regional culture as a whole.

However this solution is not based on an entrepreneurial philosophy but on an ecological view. In bioecological reality, organisms' control of their own social life system forms a precondition for systemic balance and survival of the species. No species population will therefore voluntarily surrender power over its elementary conditions of life to others. There is no serious reason to believe that different principles apply in the case of the human ecosystem.

THE NATURAL RESOURCE ECONOMY

The subject with which we are dealing under this title invites even deeper consideration of the economic relationship between man, the institutions and the factors of production. If we adopt a serious ecological approach we realize that the socioeconomic contract must be supplemented with an even more important contract, which regulates the economic relationship between society and its natural surroundings.

The most elementary economic cycle is acted out in utilizing the biotic natural resources which enclose our habitat. A balanced relationship in the interplay of economics and nature, i.e. between the utilization and the conservation of natural resources, is the opening paragraph of a declaration of ethics for local and global survival. Lack of attention to the natural ecological balance sheet leads to ruthless exploitation which devastates not only the local conditions for life but also, in the last resort, the global ecosystem.

The contemporary economy shows clearly pathological features in that it quite consciously threatens or transgresses the boundaries of the biotic systems' viability. The counterfire which environmental activists have long been igniting all over the world therefore contains the idea of environmental balance sheets, in which environmental destruction and pollution of the social environment are regarded as a cost and are given a price. The initiative has led to widespread activity among economists in recent years, and the internalization of what were once considered irrelevant externalities is now even taking place in the economic textbooks.

The basic attitudes of environmental ecology and the scheme of environmental balance sheets form an idea which socio-ecological theory embraces and incorporates as one of its foundation stones, underlying even the institutional economy. On the other hand, a precondition of such a step is that the question of ecological pricing is not allowed to dominate the debate.

The environment must not be transformed into a department store in which those with sufficient resources can buy polluting rights as they desire and society may perhaps make a nice profit. Environmental destruction ought in principle never to be negotiable. Much larger questions must head the agenda of the natural-ecology economy.

Creating a social system that is in all respects environment-friendly is the only *realistic* way of implementing in the long term the vision of an ecologically sustainable planet. This is the main vision that has to guide work at environmental-ecology level in the regions, too. The Umeå University Group's model of an alternative regional society is a concrete proposal for a society practising balanced ecological resource management.

THE PRODUCTIVE ECONOMY

An economy based on socio-ecological principles has three levels: at the bottom, a natural resource economy which regulates the biotic interactions, then an institutional economy for the permanent social structures and at the top a productive economy for short-term provision of basics. This division follows the same pattern as we have previously applied to the social and cultural levels of ecoexistence, and is expected to give a comprehensive coverage of needs.

Above the stable level of the institutional economy, therefore, comes the mobile level of the everyday economy. Its specific task is to handle the productive activities which ensure the survival and material welfare of the citizens in the short term. At the same time it is closely related to the institutional and the natural resource levels and interacts with both.

The central role here is played by regional industry, farming and commerce, with their production units, distribution systems and organized trade in goods. The most important target groups are private households, the public sector and the inter-regional market. The circulation systems which have been devised to link production and demand in the regional community differ radically from the routines of the institutional economy. Light units, rapid transport, flexibility and dynamic change are their hallmark. Responsiveness to new and varying needs, offensive strategies, temporary alliances and competitiveness are natural characteristics of the productive sector - and distinguish it both in an ethical and practical sense from the two basic economic levels.

Ecologically the market economy - defined as a production economy, be it noted - forms an appropriate instrument for the satisfaction of individual and common everyday needs in the eco-society, and for creating a certain surplus which permits new investment and a rise in standards. The most important difference by comparison with today's market economy is that the anarchic features have been succeeded by respect for the socio-economic and resource-economic rules of the game. The protection of the resource base and the socio-sphere is a postulate of the economic systems of the eco-society, as together they form the basis of a flourishing commercial life.

Another significant difference is the division into a local and a global market sector or, more exactly, between an intra-regional and an inter-regional economic sphere. In the *intra-regional market sector,* local companies compete in the markets of the home region. They are relatively small-scale companies and co-operatives, producing a selection of basic necessities for the households. A comparatively permanent and foreseeable demand gives a guaranteed basic sustenance on the elementary level and at the same time a stable labour market. Economic production as a way of life is a common phenomenon here.

A complementary task for the intra-regional market is to reduce the vulnerability of the home region to external threats - climatic, political or cyclical. In the event of a failure of the cosmopolitan economy, the whole sector should serve as an economic safety net and a buffer against regional unemployment.

In the *inter-regional market sector* the region acts instead in a cosmopolitan market by exporting and importing specialized products on a principle of comparative advantages and market segmentation. Relatively large-scale production units use domestic resources and/or skills, and form a labour market for large sections of the population. By participating in the inter-regional sector, it is hoped that a significant surplus will be generated; this permits the hyperstandard that we associate with the modern technological society.

At the same time there is an accumulation of capital which can be ploughed back to stimulate the growth of inter-regional production. Thanks to the intra-regional safety net which has been described above, the participants can operate more offensively on the cosmopolitan arena, and allow themselves to take greater risks.

Much overlapping of the endogenous and the exogenous market sectors, to some extent of an improvised character, is to be expected. At the same time different types of co-ordination are needed between the market sectors and between the market and the other levels of the social economy.

In the ecological life region the parties in the market are given a mandate to formulate a commercial policy which is administered by an independent trade and industry department. This body and corresponding responsible agencies in the institutional and the natural-resource economy sector collaborate on long-term economic matters. The organizational structure promotes a short-term creativity, and also a long-term stability, in the economic development of the regional society.

The Science of Social Ecology

THE BERMUDA TRIANGLE OF THE SCIENCES

The alternative economic model described above is only a preliminary outline. That despite a limited expertise we have ventured into this field is explained by our conviction of the legitimacy - even the necessity - of also approaching the field of economics from a socio-cultural perspective and incorporating the economy in an all-embracing vision of society. And in this specific context we are in fact all amateurs.

When it comes to filling in the details and achieving balance between the different levels, however, we must ask for the assistance of the professional economists and preferably also for their creative curiosity. But can we really expect an interest on the part of the economists?

I am thinking of the branches of economics that are taught in our universities and colleges: subjects such as political economy or national economics, regional economics, business administration etc. One may surmise that the intellectual climate within these disciplines will not be particularly receptive to the socio-economic experiments, and that they will quickly be rejected with arguments of varying substance.

This is not particularly surprising in view of the symbiotic relationship existing between the science of economics and the market, or the mass education for the market which is given by the departments. Holistic social thinking, ecological responsibility and political welfare theory are hardly the main focus of interest even of academic economists. They presumably suspect that serious involvement in such matters would undermine the academic status of the discipline. Even internal deviants within the science of economics have always had difficulty in asserting themselves against the prevailing historical consensus. All this belongs to the well-known "trivia of academia".

In all probability the same dilemma applies to the other aspects of the socio-ecological vision which we have discussed in this book. It would be naive to imagine that radical social and humanist propositions are especially welcome in the leading human and social sciences. I have in mind subjects such as sociology, political science, intellectual history and also the aesthetic disciplines.

I have personal experience of this, having once made an attempt to launch the concept of "ecological aesthetics" in my original field, the history and theory of art. The intention was to try to start up humanistic basic research in the subject, but the only result of the initiative was to bring ignominy on its author. Socio-ecological ideas imply too much questioning of established subject traditions and trespass over too many inter-disciplinary boundaries to find much favour.

Nor is cross-disciplinary fraternization a realistic alternative in the paradigmatic discussion which arises here. Upstarts, particularly in the humanities, are doomed to fall between the stools of an academic environment in which the idea of humanistic futurology is regarded almost as a contradiction in terms.

Still, there are many people today who argue that the modern university is not satisfactorily fulfilling its deepest duty: to maintain the frontline of scientific knowledge in society (whether this is represented by research or by education). The university is described as a basically conservative institution, perhaps the most conservative of all society's prestigious institutions.

The reason for this contradictory state of affairs is considered to be the university's position of dependency on the authorities and, more recently, on the market, both of which influence the universities by their orders and their need for self-affirmation. This situation poses a real threat to the intellectual freedom and creativity of the university, a threat which is aggravated by the present dominance of the task of mass education and by the bureaucratization of university life.

Together these factors have a stifling effect on new thinking in scientific research, particularly in the humanities and social sciences. Although today's society is positively crying out for new humanist and social knowledge with which to repel the growing threat of the technocratic order, humanist social research and qualitative futurological studies remain the universities' most neglected areas.

Historical factors also play a part here. The socio-ecological approach actually operates in a blind zone of university research, in a veritable Bermuda Triangle between the established humanities, the social sciences and the natural sciences. Not even modern cross-disciplinary science seems to have discovered it, and anyone who enters the field takes the risk - in a figurative sense - of disappearing. It is the artificial separation of the intellectual traditions of aethetic humanism, social management and natural analysis that has given raise to this wide desert at the center of human thinking and understanding.

In reality this triangle of death to avant-garde projects constitutes a gold mine for a radical search for human knowledge, thanks to the creative fusions which tend continually to occur at the intersection of the three traditions. It ought to be called the "innovative zone" of scientific futurology on account of its merits as a catalyst of a knowledge-intensive creativity, particularly in the socio-cultural sphere.

The Umeå Futurology Group has long been working in this sphere, and our statement is based on our own experiences. It is our decided opinion that the "innovative zone" between the traditional disciplines is far too important an area to be neglected by the scientific community in the long run. One or perhaps more disciplines ought to be established in the sector.

The mere act of making the zone visible would be very significant. The scientific legitimation of the socio-ecological approach of the Umeå Group at the university might serve as an experiment and a touchstone.

ARGUMENTS FOR A SCIENCE OF SOCIAL ECOLOGY

We here propose that the socio-ecological perspective be established as an independent field of science at the university. (As an alternative the term "cultural ecology" can be used.) In presenting the idea, we have two main motives.

The first is intra-disciplinary and relates to the needs of the modern human and social sciences for a firmer theoretical foundation for their work. Whereas the natural sciences can point to a long series of epoch-making advances in basic research and formulation of theory, the human and social sciences have advanced slowly. Nevertheless, a flourishing empirical-descriptive research with a historical, socio-geographical, anthropological and artistic profile has emerged and established a rich tradition. A special methodology - hermeneutics - has also been designed to safeguard the social, historical and spiritual values which the field contains.

But the substantial corpus of knowledge contains few attempts at generalization from experience, and even fewer attempts at systematization in the form of consistent theories about man, society and civilization. Consolidated human and social knowledge which can be used in political and technical social planning is virtually non-existent. Whereas technology and the natural sciences swamp us with innovations which reshape society at a hectic pace, the cultural and social sciences more often than not stand as spectators, silently recording the changes.

When asked the reason for their defensive attitude, human sciences refer first to a scientist's theory of science which banishes the general study of mentality from the brotherhood of sciences, and then to a humanist's theory which readmits it as the study of private and unique phenomena. This sleight of hand neutralizes the role of the human and social sciences in society, while retaining it in the world of learning.

Hereby an illusion of the comprehensiveness of the search for knowledge is created, but in reality the whole scientific community becomes an instrument for the legitimation of a one-sided technological culture. It is only the destructive social effects of social and cultural contempt (and self-contempt) in the sciences that have opened our eyes to these dubious internal connections.

The socio-ecological project is a serious attempt to penetrate the scientific myths which surround contemporary human and social research. Its primary aim is to investigate human existential needs - including their qualitative aspects - by means of empirical studies of mentality. A second stage goes on to describe the psycho-social and ecological prerequisites of the formation of human society, with its social, cultural and productive facets. In a third stage there follows an outline of societal models which permit an existential satisfaction of needs for larger groups of the population, still with qualitative criteria as a guideline.

The practical value of the models may finally be tested in full-scale experiments. The knowledge gained from such research can be utilized in many ways to uphold basic social and cultural ecological values in social policy and planning.

The socio-ecological school begins to be well equipped - not least by its broad range of applications - to fulfil the functions of a basic science in the social science and humanities faculties. There is no reason to see it as a rival to the established, specialized disciplines. Instead it may function both as a theoretical resource base and as a discussion partner to them.

The second crucial motive for establishing the socio-ecological perspective as a science is the global community's acute need of consistent models for social welfare development. Two centuries of revolutionizing industrial progress have created an advanced technological society whose imbalance today threatens its own socio-cultural and ecological foundations.

More and more local and international bodies (particularly within the UN family) are now taking part in the work of planning a sustainable global future. But to restore a balanced relationship between man and the planetary ecological environment the disturbed homeostasis in man's social ecosystem must - as has repeatedly been stated in this book - also be restored. And to achieve this, applicable knowledge of man's socio-cultural basic requirements is needed, knowledge which none of today's established sciences can offer.

The socio-ecological initiative is to a high degree intended to fill this unfortunate gap in the self-understanding of contemporary society. In a stimulating environment and with the support of the necessary resources, it can serve as one of the generators of knowledge which the global society strongly needs in order to realize the vision of a socially and ecologically enduring planetary system.

The scientization of social ecology and its installation as an academic discipline require scrupulous consideration and organization of the material. The boundaries with competing and parallel disciplines must be drawn with care. The socio-ecological approach differs from, for example, academic human ecology in its distinct orientation towards the socio-cultural sphere of problems, its sincere scientific ambitions and its clear emphasis on social and cultural problem-solving.

In order somewhat to clarify the vision of the alternative discipline, a conceptual curriculum is presented below. This must only be seen as an outline closely associated with the Umeå University Group's version. Nevertheless, the rough draft may provide a basis for more detailed discussion of the arrangement of the material for scientific, educational and social development purposes.

SOCIAL OR CULTURAL ECOLOGY AS A DISCIPLINE

1. ***SYSTEMS ECOLOGY***

Theoretical ecology
Scientific basis of ecology and philosophical assumptions. Organization of the ecosystem at different levels of evolution.

Man's ecology
Human society and settlement as an ecosystem. Also the ecological constitution and socio-ecological evolution of the human species.

Ecological pathology
Evolutionary faults in the ecosystem, their causes and consequences. General view of eco-pathological symptoms in the human social system.

2. ***HUMAN ECOLOGY***

Ecology of natural interaction
The basis of the human ecoidentity and its interaction with the natural surroundings.

Ecology of community life
Social behaviour and socializing characteristics of the human species.

Ecology of everyday action
The local dynamics of individual action in their aesthetic, communicative and productive functions.

Ecology of cognitive reflection
Functional and cross-boundary dynamics on a level with the cognitive logoidentity.

3. *SOCIAL ECOLOGY*

Ecological organization of society
Organization of man's habitat and its socio-cultural functions on anthropo-ecological principles.

Life space of the eco-society
The ecological life-region as the logical life environment of the eco-identity. Urban alternatives and sectoral variants.

Inter-regional structure of the eco-society
Organizational networks of the macro-society. Interaction between eco-regions in varying functional combinations.

Eco-politics and eco-institutions
Self-reliance and self-management as generators of greater social and ecological sustainability in society. A socio-ecological perspective in politics and planning.

Society and ecological ethics
Interaction within the human ecosystem and with other species in the global ecosphere, considered from an ethical and judicial perspective. Human rights, ecological rights etc.

4. *CULTURAL ECOLOGY*

Cultural crisis analysis
The erosion of cultural values in the techno-society. Disintegration of spiritual, ethical and aesthetic normative systems. Their replacement by pseudo-doctrines and substitute pleasures.

Ecological base of aesthetic culture
The roots of the aesthetic and cultural value systems in man's psycho-ecological nature. Their spontaneous expression and normal functions in the interactive life system.

Cultural life in the eco-society
Cultural conventions and institutions in the eco-society. The perception of nature, the cultural landscape, popular culture, architecture and the city, the metaphysical arts etc., as aspects of an aesthetic life culture.

Philosophy and cultural ecology
Eco-philosophical and eco-cosmological perspectives. The extent and boundaries of the ecological world view. Its forms of materialization and spiritual content.

5. *ENVIRONMENTAL ECOLOGY*

The local and global environmental threats
The ecological environmental threats on a local and global level. Their relationship to society's enterprises and welfare ideologies.

Global strategies and environmental policy
From doomsday scenario to public eco-politics. Agents at local and global level. The message of the environmental conferences.

Local strategies and environmental work
Environmental activists and municipal environmental planners. National

strategies for the local level. The need for competence and knowledge for long-term projects.

Ecological environmental research
Examples of surveys and research projects from different environmental areas.

6. *COMMUNITARIAN ECOLOGY*

Social crisis analysis
Social disintegration in the community. Connections between the social crisis and the global environmental crisis.

Socio-pathological symptoms
The psycho-somatic and compensatory symptoms of social disintegration. Their destructive effects on society and the ecosystem.

Social crisis and public health strategies
Introduction to the methodology of public health work and social policy. Social strategies in international development bodies and alternative movements.

A socio-ecological health and welfare policy
The outlines of an intersectoral social policy based on socio-ecological knowledge and civil competence. Its tasks in general social policy.

7. *SOCIO-ECOLOGICAL PLANNING*

The civic eco-cycle
Initiatives for strengthening social identity at the civic level. Socio-ecological renewal of neighbourhoods and local areas.

The local eco-cycle
The local community as a theatre for socio-ecological mobilization. The public sector as an agent of new visions. Contributions to renewal which reinforce the local quality of life.

The regional eco-cycle
The region as a framework for the cosmopolitan aspirations of the eco-society. Initiatives which give the region and its social network a cosmopolitan identity.

The global eco-cycle
The need of global bodies which represent a humanitarian and socio-ecological vision and link this with the environmental movement. The role and responsibilities of the UN and the nation-state.

INSTITUTIONAL LOCATION

The institutional location of a science of social ecology is an important issue if this discussion is to be pursued to a logical conclusion. Certainly it is desirable for the new discipline to be introduced at the national universities, in the faculty of either social sciences or humanities. In the latter case the designation *cultural ecology* may possibly be used as a synonym. The realistic aim and theoretical rigour of the discipline may, it is hoped, justify these suggestions.

From this position the newcomer may provide some of its well-established neighbours with new basic perspectives and analytical instruments. A discussion of socio-ecological (or cultural ecological) competence in community work can begin and regular instruction can start.

Once this has been said, it is necessary to mention an ideological difference in nature between the new approach and the modern university. The modern university is a child of eighteenth-century rational enlightenment and nineteenth-century learning tradition. The natural horizon of the university's activities is international. In every aspect of its attitude the classical university addresses the universal sphere of knowledge which has made the advances of rational consciousness possible in the specialized disciplines. It is a university of an elitist kind, both in its scientific claim to the highest levels of the intellect and in providing training for the most responsible positions in contemporary society.

This also implies that the universities and their sciences have long been regarded as elevated above the everyday existential conditions of the surrounding society; they have assumed a distant - *exogenous* - attitude to the life of the community. That is why the big universities loom like spaceships from distant worlds in the local societies in which they happen to be located. The modern university is in itself evidence of the communicative abyss which exists between the search for knowledge and community life in our epoch.

Unfortunately this abyss is incapable of being bridged by research information or distance courses alone. This is the backside of the heavy coin which scientific development in the modern era represents.

The socio-ecological initiative starts from the opposite perspective. Its insistence on scientific rationality is no less demanding, but both its vantage point and the object of its study are different. The socio-ecological scientist stands in the centre of the reality he intends to describe. We say that he adopts an *endogenous* approach to his study. From this vantage point he also observes a different kind of reality from that of the exogenous observer, and chooses other themes for his analyses. The natural subject of socio-ecological research is the everyday community which surrounds the observer here and now and the dynamic human being who is this community's average member.

It is the totality of the immediate social surroundings and the interaction between their components which forms the focus of interest. This also means that scientific analysis does not proceed from a doctored hard-data version of the social environment, but is based on a holistic picture in which the "soft" humanist qualities of life are integrated.

The endogenous approach thus ensures a cultural analysis of much greater realism, and in many cases this guarantees more fruitful results, i.e. results which can be used in practical planning and social policy. This should not be understood in a narrow-minded sense as a recommendation of biased local chauvinism in the social sciences; what is decribed is, on the contrary, also a universal perspective.

Many representatives of the classical university take the view - tacitly or openly - that such a methodology lacks scientific legitimacy and has no place in the university. Undeniably this traditional attitude constitutes an obstacle to the introduction of the new approach. For this reason an interesting alternative placing ought to be discussed. This is *the regional universities.*

These are spread around the provinces and constitute an educational and cultural resource which looks directly to the regional community. The regional universities are limited in size, but many wish to maintain a research capacity. Sometimes the research profile is a poor imitation of that of the elite university, but there is often a specialization in more local and regional problem areas which is promising. The creation of an autonomous regional skill and knowledge base, which also includes a good familiarity with theoretical aspects and intraregional conditions, is regarded as highly desirable in many regions.

In the circle of regional universities the socio-ecological initiative could play an inspiring part. Its immediate goal - the generation of knowledge for social and economic welfare planning purposes in the provinces - coincides with the visions of the future entertained by many regions of county size. Under the designation of *Regional Ecology* the socio-ecological perspective could be established as a basic science at the regional university and serve as an earmarked resource for local development.

For far too long the regions have had to wait for a science which works uncompromisingly for their legitimate interests. Such an initiative would give the regional universities an independent scientific profile, and possibly also a higher status as centres of learning. That the creation of viable and ecologically conscious regions is at the same time a constructive contribution to the vision of a sustainable global future should hardly make the initiative less interesting.

THE NEW PARADIGM

Before we conclude this chapter it is necessary to admit that an even deeper intention underlies the socio-ecological project. There have already been many glimpses of it in this book. The project is very much in sympathy with the paradigmatic criticism which many people today level against the prevailing Cartesian world view of our society.

This is a criticism which emanates from the innermost circle of scientific culture and includes such well-known personalities as the philosopher Arne Naess, the historian Theodore Roszak, the peace researcher Johan Galtung and the physicists Ilya Prigogine and David Bohm. Among the predecessors of these scientists are a number of representatives of the Frankfurt School, such as Herbert Marcuse, Theodor Adorno and Jürgen Habermas. The name best known to the wider public is that of the American Fritjof Capra, whose best-seller "The Turning Point" (1982) sketches the contours of the alternative world view which today attracts a growing number of radicals.

According to these critics, the Cartesian paradigm, whose mechanistic view of knowledge has long dominated Western thought, has now reached the end of the road. Its theoretical and practical limitations have been revealed not only in the laboratories of natural science but also in the blunders and disasters that a runaway technological culture has perpetrated in our living environment.

When it was born in the seventeenth century, Cartesianism represented a breakthrough for rationalism and faith in reason. But its conquest of society led in practice to an exclusive instrumentalism and a cult of technology, which has proved unfortunate in the long term. The result is a rationalism which, by dismissing the social and cultural dimensions of society, has come to imply such strong anti-human tendencies that it can no longer be called reasonable in the normal sense of the word.

Fritjof Capra and several others hold out, in marked contrast to the mechanistic model, an alternative world view based on the organic rationality of the life sciences and a holistic systems approach to human thinking. They introduce a concept of dynamic, cyclical processes, which embraces not only nature and the multiplicity of species but also man and his whole culture. The scheme also includes a new view of humanity, in which the daily interaction of the individual with the surrounding world takes place not only on a local but also on a global and, ultimately, a cosmic level.

The new vision is supported by a series of epoch-making observations in the field of modern physics. But its ideological essence is a deepened respect for the elementary life processes in both society and nature which is founded on true humanitarian empathy. This respect also extends to the ethical and aesthetic overtones of global existence. From this it is clear that the new rationality implies a dramatic expansion of the field of action and responsibility of human reason by comparison with the Cartesian definition.

The core of the holistic vision offered by the new paradigm consists of an anthropic - or more precisely an organismic - perspective. At the centre of the creative cosmos there is a materialized subjective consciousness, possibly as a part of a universal consciousness. In the evolution of the universe, man - like the other organisms and species on the planet - is a co-creator, and his frame of reference consists of the micro-universe which is comprehended by his ego. In the course of interaction with his environment man forms his habitat, his community, and also his consciousness of the world and of the cosmos. The holistic paradigm is not fully developed until it contains a vision of man, a vision which also makes the principle of self-understanding part of the interpretation of reality.

Such an anthropo-organismic approach is to be found within the deep-ecology movement, which has been a rivulet trickling alongside the current of environmental ecology for at least twenty years (see Erich Jantsch: "The Self-Organizing Universe" (1980) and Henryk Skolimowski: "Living Philosophy. Eco-Philosophy as a Tree of Life" (1992)). The anthropic angle, understood as a general principle of the vital forces, also forms the epistemological foundation of the socio-ecological study which the University Group for Humanist Futurology tries to accomplish in this book.

AN ALTERNATIVE SCIENTIFIC PARADIGM

This new ecodynamic paradigm is still only a dream on the horizon of radical consciousness. But the dream is sufficiently attractive to inspire the persistent methodological work which is required before we can claim that a new world picture has been created. Some of the radical dreamers also seem to be quite well established already. Empirical and conceptual research is, in fact, being carried out in scientific niches at many universities around the world.

One of the hot areas is *Epistemology* and the closely related *Philosophy of Science.* This value judgment is based on the conviction that the theory of knowledge represents the ideological frontline of rational thinking. If ideas and theories still play a part in the process of civilization, this is the very locking mechanism in the door which may lead to an intellectual breakthrough. With a knowledge of the epistemological code, we can in our minds cut keys which lead to success in the paradigmatic enterprise.

Professional philosophers at well-known universities are continuing to settle accounts (a process begun by the Frankfurt School) with the epistemology of reductionist positivism sanctioned by the Vienna Circle and logical empiricism early in the 20th century. Alongside the recently popular relativism of post-modernism this school still - often implicitly - dominates the view of knowledge in large sections of the scientific community (see John O'Neill: "Worlds Without Content. Against Formalism" (1991), and Tom Sorell: "Scientism. Philosophy and the Infatuation with Science" (1991)).

A pioneer of philosophical criticism in the field is Edmund Husserl. His long overlooked "The Crisis of European Sciences" (English translation published in 1970) has inspired a radical philosophical criticism, which repudiates positivist epistemology's monopoly of truth and dominion over the faculties. The book gives striking examples of how positivist rationalism has led to a technification of the pursuit of knowledge, whose narrow-minded zeal threatens to reduce the scientific description of society to trivialities - and the theory of science to pure methodology.

Positivism is an approach which has been inspired by classical natural science, whence it has spread covertly to the social and human sciences. There it has in particular affected the academic milieu by declaring taboo a serious scientific research into the advanced mental processes and by its reservations against an active social commitment from the sciences.

Within large and growing areas the modern theory of knowledge represents a narrow scientism, whose basis in reality is often in inverse proportion to its totalitarian claims. As a phenomenon of the times and a cleansing process scientism has played an important part. From both intellectual and social standpoints, however, it lacks real depth and explanatory value.

What is the alternative? The new epistemological criticism does not succumb to the temptation to accept an anarchic relativism of post-modernist or any other type. One reason is that relativism in these context is scientism's closest ally, in that it questions the possibility of a cultural-scientific basic research. Instead the new critics look for the precepts of a more *realistic rationalism* than that offered by the logical empiricists.

New paradigmatic foundations are erected by introducing the observer - the rational subject - as a perceptible participant in the search for objective knowledge. (See Roger Trigg: "Rationality & Science. Can Science Explain Everything?" (1993).) Reality cannot be faithfully described from any higher non-human standpoint as logical empiricism implies, they argue. The anthropic perspective represents a precondition for the scientific observer as for all other people, and must be clearly stated as the starting point of scientific description.

But the observer's position has to be specified even more precisely: he is himself irrevocably involved in the reality conception that he is describing, being simultaneously an observer and a part of the universe of knowledge.

Is "objective" knowledge possible at all under such conditions? Roger Trigg's answer is that the clarification of the total receptive situation at least creates the conditions for a deeper form of objectivity than that of pure scientism, though this does not, for him, conclude the epistemological analysis. On the contrary, it is here that the truly serious business begins: defining subjective rationality and revealing its inner composition is the big challenge facing radical epistemology. There are several paths to take and Trigg himself tries that of metaphysics.

EPISTEMOLOGY OF SOCIAL ECOLOGY

The socio-ecological initiative tested in this volume also offers an approach to this fascinating field, although in a rather different style. Its basic ideas are quite close to epistemological concepts which have recently been put forward by representatives of naturalist philosophy (see W. Quine: "The Pursuit of Truth" (1992) and R. Rorty: "Philosophy and the Mirror of Nature" (1980)).

The empirical foundation of the naturalist theory of knowledge is the neural processes of the human brain, as they are described by modern neurophysiology. The biochemical processes in the brain serve indirectly the purpose of maintaining the individual's contact with the world around, and have arisen from an evolution in which no definite break can be discerned between the animal and the genuinely human stage. The knowledge of the world which man obtains derives ultimately from the receptive conditions created by functional bio-evolution.

The naturalist approach may appear at first glance to be scientism of the purest kind, and can easily be used to explain away the advanced functions of consciousness at the level of civilized man. On the other hand it may also - used more correctly - serve exactly the opposite purpose: that of giving scientific realism to complex cultural processes, which have been undermined in our civilization by being constantly defined as private.

What is promising in naturalist epistemology is that it draws attention strongly towards the subjective pole in the observation of reality, and at the same time points out that "subjectivity" also contains general features which are identical from person to person.

How, then, do we describe the position of the "Umeå School"? The whole socio-ecological discussion is, to be sure, based on an analysis of the effects of environmental interaction on subjective consciousness, including man's perception of reality (this is a discussion which has been carried on internally within the Umeå Futurology Group for over ten years). What is exclusive to the project is the uncompromising ambition to also allocate a place in the system to the higher, even the very highest, categories of consciousness.

Admittedly, the overall aim of the project has been pragmatic, that of marking out the foundations of a humane social planning strategy. But our observations also offer the basis for a discussion of an anthropic theory of knowledge, and for a specification of the components which ought to be included in the scheme of *subjective rationality*. The epistemological approach that I have in mind here ought, in accordance with the main message of the book, to be designated *ecological* or *socio-ecological*. Let me, purely as a declaration of intent, indicate the main points of such a proposal.

The starting point of a socio-ecological epistemology is the idea of a bipolar construction of the conception of reality. Our consciousness of reality is composed of an external - *exogenous* -and an inner - *endogenous* - sphere, where the latter is involved in the former, but at the same time its only verification. This may be called a logical interactionist cycle, controlled by crosswise conditioning. The definition also indicates the ontological boundaries which apply to the reality we are capable of surveying, with the rationality or sensibility we possess.

Whether this epistemologically closed world is open to observation by forces outside the system - possibly via the neural system of the organism - is something about which we can only speculate.

However, we can make subjectively conceivable reality - within the framework of our conceptions - the subject of rational analysis. We then apply a "strictly objective" (empirical) assessment of the structures which surround us. We describe the constituents of their objects and study the causal and casual processes which connect or distinguish them. In the overall model of *the ecological identity* which we decoded earlier in the book, this represents the view of reality comprised by *logo-identity*, being the uppermost level of the hierarchy of eco-identity.

It is man's talent for mastering long temporal processes in his use of the surrounding world which forms the eco-psychological basis of the rational perspective of the logo-identity. In this capacity man differs radically from the other species and represents an advance in the evolution of consciousness on the planet.

But the older, animal levels of consciousness also survive in man. They signify more fundamental attitudes to the world around, and are therefore indispensable. Whether we recognize them or not, they form the foundations of rational consciousness and its functioning. Just below the level of the logo-identity in the Umeå model comes *the ego-identity*, which we have described as the level of sensory consciousness, or the phenomenalist level.

The phenomenalist level contains several complementary outlooks and can be roughly sorted into *the sensory-aesthetic, socio-communicative* and *instrumental-productive* archetypes. All the archetypes in sensory consciousness are associated with a dynamic attitude to the world around. Together, and in continuous interaction, they form the vivid picture of reality which constitutes the empirical basis of sovereign logo-identity and its rational analysis.

The eco-identity contains even deeper attitudes to the world. They derive from still more archaic stages, but have been transformed, "mutatis mutandis", into parity of value with the advanced logoidentity. At the level which in the model is designated as that of *species identity,* we find a mentality which may be described as structure-oriented. The striving for structural organization is expressed in clear patterns of action, whose aim is to arrange the content of societal life as integrated wholes; the home, the workplace and the local community are examples of such entities. We suppose that the evolutionary origin of this behaviour lies on a bio-ecological level, and that our own body is in that case an example of the organic significance of the principle.

Among existing theories, systems theory (e.g. in Capra's interpretation) and, to some extent, cybernetics describe the type of processes that by analogy take place in our consciousness and in rational thinking. But ontologically the striving for structural organization of reality comes before both the phenomenal and the logical level.

At the very bottom of the ecological identity model we find *the natural identity*. The identification of the absolutely primary categories of experienced reality are imagined to occur at this level. Subjective consciousness is determined almost automatically as the starting point of anthropic reality, and the borderline is sharply drawn between its inner (subjective) and outer (environmental) spheres. At the same time the constant flow of impulses in both directions is regulated in the border zone between the two spheres.

As a fundamental tone of the dynamic process runs the intuition that the world around - i.e. nature - is the very source of life and the ego (merely) an expression of life, still closely connected to its origin. This definite bottom level in our eco-consciousness remains the basis of the identity levels described above, and ultimately of human rational thinking. The evolutionary roots of the natural identity may lie on the cellular level of organismic development.

AN ENDOGENOUS THEORY OF KNOWLEDGE

The conclusion drawn by ecological epistemology from the observations described is that rational thinking, despite its role as evolution's most advanced level of consciousness, can never free itself from its foundations in our minds. Shifts of emphasis may occur, different levels and sectors may be brought into focus with special acuity, but the other aspects of consciousness remain in the background as a corrective and, after the act of will, organic balance returns to the system.

A common shift of emphasis in man is of course the sharpening of the cognitive (rational) perspective, from which science in particular has benefited. It is by a systematic accentuation of this perspective that the modern sciences and elite universities have actually come into being. Using the terminology employed in the book this designates the attitude of exogenous rationality. It cultivates a specialized top-down view of reality and is inclined to excel in abstract methods and models.

Ecological epistemology regards this attitude as quite legitimate, just as long as it retains contact with the epistemological base levels and is prepared to use them as an instrument for verification. And, of course, for giving a fuller and more realistic description of "reality".

However, what has alarmingly happened within the rational tradition of knowledge and science is that exogenous rationality has suffered an attack of hubris and is trying to distance itself from its epistemological roots. In our time it has claimed to be able to carry out alone the rational analysis of reality and also attempted to monopolize the right to explain it.

The end result of this experiment is a veritable falsification of reality. The empirical world of phenomena has been replaced by an abstract mathematical myth, which hovers in anonymous space without much contact with observable reality - except in those cases where it possess technological significance. In ecological epistemology this well-known phenomenon cannot be designated anything else than an overstrung form of rationalism.

Ecological epistemology offers another possibility. We may call this alternative the attitude of *endogenous rationality*. The alternative relates directly to the new philosophical discussion of subjective rationality.

It implies that the rational thought, during its scientific work, keeps all senses and receptors wide open to the archaic and phenomenal levels of consciousness. Without losing sight of the rational requirement for truth, the logo-identity allows the impressions of these levels to wash over it. With the immeasurably stronger illumination of reality which is thus attained, our reason receives a more multi-dimensional picture of reality than by logical analysis alone. And with the broader calibration of the intellect an interpretive apparatus with a considerably greater differentiation is obtained.

The result will be a more vivid, and presumably also more credible, description of external reality. It is a *bottom-up perspective* on reality which is maintained in ecological epistemology.

The ecological theory of knowledge also adds to a comprehensive understanding of reality in another way. With the interactive basic view the boundaries between *subjective* and *objective* world become less dramatic. The inner universe of the subjective consciousness may be the object of as much scientific curiosity as the external universe. The enquiring eye shuttles between the external panorama and the internal landscape and is stimulated by the duelling observations to reach entirely new and unexpected conclusions.

Utopian holistic rhetoric suddenly finds an anchorage in deeper human strivings, social communication can be understood as an evolutionary force, aesthetic perception and the spectrum of emotions appear all at once to be filled with rational logic. The cosmos is revealed as a natural ecosystem, as our self-evident habitat. Everywhere in the life world the new paradigm reveals its unexpected pictures and simultaneously explains them.

Furthermore, endogenous rationality also creates new prospects for the genuine scientific search for knowledge. They seem to lie in a number of specific areas, and at the same time imply a chance of organic co-ordination of the type long sought by cross-disciplinary science.

The really great potential of the endogenous perspective, however, lies within the human and social sciences. Here the preconditions for a rational search for knowledge are at last created within a sphere which, due alternately to declarations of sanctity and condemnation, is today a scientific no-go area.

This makes it possible at last to generate the knowledge of humanist goals and social values of life that society inevitably needs in order to uphold human values against the forces of prejudice which today dominate the social stage. In the concrete role of basic science at the regional universities, endogenous rationality might perhaps be able to realize these ideals best of all. An entirely new field of study for human and social science and its educational theory is on the way to open up.

Finally: the greatest of all the limitations of contemporary human and social science is the denial of creativity, the banishment of creative intuition from science. Perhaps we are now on the way to a stage where historical innovation and evolutionary creativity can, at last, become an essential part of the scientific effort. A stage where empirical science is at the same time dethroned from its omnipotence but acknowledged as an indispensable instrument for giving technical, social and cultural creativity its critical and historical basis.

This is a view which undeniably opens fascinating new perspectives.

REFERENCES:

Bergström, Sören, 1994: Naturekonomi (Natural Economy). Carlssons, Stockholm.

Brett, E.A., 1986: The World Economy Since the War: The Politics of Uneven Development. MacMillan Education Ltd., London.

Capra, Fritjof, 1982: The Turning Point. Simon and Schuster, New York.

Costanza, Robert, 1991: Ecological Economics. The Science and Management of Sustainability. Columbia University Press, New York.

Daly, H., Cobb, J.B., 1989: For the Common Good. Beacon Press, Boston.

Dicken, Peter, 1992: Global Shift. The Internationalization of Economic Activity (2. ed.). Paul Chapman Publishing Ltd., London.

Dillard, Dudley, 1986: Västeuropas och Förenta staternas ekonomiska historia (Economic Development of the North Atlantic Community). Liber Förlag, Malmö.

Drivkrafter för produktivitet och välstånd [Driving Forces For Productivity and Prosperity]. Produktivitetsdelegationens betänkande. SOU 1991:82. Allmänna Förlaget, Stockholm.

Ekins, P., Hillman, M., Hutchison, R., 1992: Wealth Beyond Measure. An atlas of new economics. Gaia Books Lt., London.

Eklund, Klas, 1993: Vår ekonomi. En introduktion till samhällsekonomin [Our Economy.

An Introduction to National Economics] (4. uppl.). Tidens förlag, Stockholm.
Hewitt de Alcántara, C., 1993: Real Markets: Social and Political Issues of Food Policy Reform. Frank Cass, London.
Husserl, Edmund, 1970: The Crisis of European Sciences. Northwestern U.P., Evanston.
Jantsch, Erich, 1980: The Self-Organizing Universe. Pergamon, Oxford.
Jespersen, J., Brendstrup, S., 1994: Grøn økonomi (Green Economy). Jurist-og Økonomforbundets Forlag, København.
Larsson, Mats, 1991: En svensk ekonomisk historia [An Economic History of Sweden]. SNS Förlag, Stockholm.
Mishan, E., 1967: The Cost of Economic Growth. Penguin Books, Harmondsworth.
Mishan, E., 1986: Economic Myths and the Mythology of Economics. Humanities Press Int., Atlantic Highlands.
Myrdal, G., 1975: Against the Stream. Critical Essays on Economics. Vintage (Random House), New York.
Nilsson, Ingvar, Wadeskog, Anders, 1985: Ekonomi för 90-talet [Economics for the 90s]. Akademilitteratur, Stockholm.
North, Douglas C., 1990: Institutions. Institutional Change and Economic Performance. Cambridge University Press, Cambridge.
O'Neill, John, 1991: Worlds Without Content. Against Formalism. Routledge, London.
Quine, W. van Orman, 1992: Pursuit of Truth. Harvard U.P., Cambridge, Mass.
Robertson, James, 1978: The Sane Alternative. Signposts of a Self-Fulfilling Future. Villiers Publications Ltd., London.
Rorty, Richard, 1980: Philosophy and the Mirror of Nature. Blackwell, Oxford.
Samuels, W.J., 1988: Institutional Economics, vol 1-3. Edward Elgar, Aldershot.
Sandkühler, Hans Jörg, 1991: Die Wirklichkeit des Wissens. Suhrkamp, Frankfurt am Main.
Schumacher, E.F., 1973: Litet är vackert: ekonomi som om människor betydde något (Small is Beautiful. A Study of Economics as if People Mattered). Prisma, Stockholm.
Skolimowski, Henryk, 1992: Living Philosophy. Eco-Philosophy as a Tree of Life. Arkana, London.
Sorell, Tom, 1991: Scientism. Philosophy and the Infatuation with Science. Routledge, London.
Söderbaum, Peter, 1993: Ekologisk ekonomi. Miljö och utveckling i ny belysning [Ecological Economics. Environment and Development in a New Light]. Studentlitteratur, Lund.
Thorbek, Ole, 1992: "Frie Faelleskaber - I stedet for markedsøkonomi" [Free Communities - Instead of Market Economics], Politisk Revy.
Trigg, Roger, 1993: Rationality & Science. Can Science Explain Everything? Blackwell, Oxford.
Törnqvist, Gunnar, ed., 1986: Svenskt näringsliv i geografiskt perspektiv [Swedish Industry and Commerce in a Geographical Perspective). Liber Förlag, Stockholm.
Törnqvist, Gunnar, 1993: Sverige i nätverkens Europa [Sweden in the Europe of the Networks]. Liber-Hermods AB, Malmö.
van der Wee, Herman, 1987: Prosperity and Upheaval. The World Economy 1945-1980. Penguin Books Ltd., Harmondsworth.

VI APPENDIX

"Angelica Silvestris". Photo by Margareta Klingberg

Conference Programme I

THE REGION AS LIVING-SPACE
A R&D Conference in Härnösand, Friday 24th April 1992

10.00 Opening of the Conference by City Architect *Hans Thunell.*

MORNING:
European Regionalism: Doctrines and Alternatives

10.05 Regional Development in the Pan-European Community. *Walter B. Stöhr,* Professor, Vienna.
10.50 Swedish Regions and the Regions of Europe. Panel Debate. *Johan Carlström,* Swedish Association of Local Authorities.
Kurt Jansson, Municipal Commissioner, Kramfors.
Lars Nilsson, Deputy County Architect, Västernorrland.
Per Råberg, Assoc. Prof. Umeå University.
Walter B. Stöhr, Wirtschaftsuniversität, Vienna.

12.30 LUNCH.

AFTERNOON:
The Case for an Alternative Regionalism

13.45 Region, Territory and Identity. *Torsten Malmberg,* Professor, Lund University.
14.30 The Forces of the Social Living Space. *Gunilla Lundahl,* Critics of Architecture, Stockholm.
15.30 The Idea of the Ecological Region. *Per Råberg,* Umeå University.

16.30 DINNER.

EVENING:
Alternative Regionalism in Planning Practice

17.45 The Settlement Heritage of Västernorrland.
Hans Thunell, Municipality of Härnösand.
18.15 Life-Region Västernorrland. Outline of a Socio-Ecological Plan of Renewal.
Jan Henriksson, Professor, Royal Institute of Technology, Stockholm.
19.45 Concluding Debate.

FOCUS ON THE REGION

In the social debate of the early nineties, growing attention is coming to focus on questions of regional development. With European integration an imminent reality, the field of regional policy is dramatically revalued by politicians and regional researchers. Our society, however, is already undergoing a restructuring which many actors are describing as if it were foreordained. Regions which are already strong are mobilizing to increase their competitive power in the new Europe. They will attract national competence and economic resources more powerfully than ever. In the name of national interest they are requesting State participation for the accomplishment of their plans.

With these new signals, the inter-metropolitan and peripheral regions are being left to their fate. Perhaps they make up as much as 70% of the country's territory and 60% of its infrastructure. The infrastructure consists of industrial plants, a landscape cultivated by succeeding generations and a historical culture of architectural heritage and urban centres. This is a social capital which, even today, is in use and in good condition. The regions have a population which, for decades, has loyally contributed towards national concentration, on the assumption that a policy of this kind would pay the best dividends for them too. What is now going to happen to these regions? Will they continue to decline until they collapse and a dramatic destruction of capital sets in? Or is it possible to form new strategies which, on new premises, will create viable regions? Strategies which will guarantee, not just continuing survival, but regional prosperity?

A conference will be held in Härnösand on 24th April 1992 to discuss these outstandingly vital issues. It is plainly but eloquently entitled: THE REGION AS LIVING SPACE. The conference is being organized by the Municipality of Härnösand in association with the Umeå University Group for Humanist Futurology. It is intended for planning practitioners and architects in the County of Västernorrland and for other groups actively concerned with questions of local and regional development. Its purpose is to bring about a general debate on ideas concerning new paths for physical planning in regions threatened by current development. Interesting ideas for future R&D and planning work can result when practitioners from a disadvantaged county are confronted with committed researchers. The conference is planned to include a panel debate in which regional policy experts from the central European stage will be taking part.

But the conference in Härnösand also includes a creative approach. Researchers at the Umeå University Group have collaborated with architects from the Royal Institute of Technology in Stockholm, to devise a model of alternative regional development. Their proposal differs from the main present-day trends in their endeavour to put the socio-cultural policy objectives at the centre. The model also represents an attempt to introduce an indepth ecological approach to regional development. To focus planning measures more directly on the primary social and ecological objectives is a strategy which the Umeå project is intended to try out. Faced with the human and social poverty often entailed in purely market-related planning, a strategy of this kind appears to be economically as well as socially valid.

Perhaps it can be shown that the socially, culturally and ecologically optimal society is a regional society, with a geopolitical structure and way of life fairly similar to those of the County of Västernorrland. What implications would such a finding hold for Swedish and Nordic regional policy?

Conference Programme II

CULTURE AND THE LIVING REGION
A R&D Conference in Örnsköldsvik, Friday 26th June 1992

10.00 Opening of the Conference by *Elvy Söderström,* Ö-vik.

MORNING:
Culture and Region: Doctrines and Alternatives

10.05 Culture and Regional Identity
Michel Bassand, Professor, École Polytechnique Fédérale, Lausanne.
10.50 The Region as a Cultural Arena. Panel Debate.
Karin Backman, Swedish Association of County Councils
Michel Bassand, École Polytechnique Fédérale, Lausanne
Jan Henriksson, County Council of Norrbotten.
Per Råberg, University of Umeå.
Harry Ördell, County Council of Västernorrland.

12.30 LUNCH.

AFTERNOON:
Concepts for a Regional Culture

13.45 The Avant-Garde as a Door to the Cosmos.
Jean Fischer, Cultural Sociologist, Copenhagen.
14.30 Mass Culture Meets Man.
Sidsel Mörck, Author, Oslo.
15.30 Nature as Culture
Per Råberg, Assoc. Prof. University of Umeå.

16.30 DINNER.

EVENING:
Alternative Culture in Regional Practice

17.45 The Culture of Västernorrland in Retrospect.
Harry Ördell, Cultural Director, County Council of Västernorrland.
18.15 Regional Cultural Mobilization
Three Sketches for Cultural Renewal.
Reports from Local Working Groups.
19.45 Final Debate.

A CULTURE FOR THE LIVING REGION

Europe is being remodelled. The economic powers of development are bursting societal frames and building new networks on higher and wider arenas than we, a short time ago, thought possible. Structures and schemes, people and borders are in a state of flux driven by integrative powers which, from the countries of Europe, want to create a new functional, and maybe, political unit. At the same time the process of integration triggers a search for the ethnic and historical origin, after a locality-bound identity, among different national groups. These tendencies appear most evidently in the strong endeavors to the regional accumulation of power we have experienced during the last decade. The emerging new Pan-Europe is, according to a current mode of expression, a Europe of the regions.

The extensive discussions about the future European regions has also initiated a debate about the cultural dimension. Which role has culture traditionally possessed on the European scene and what significance may culture have in the future as a factor for development? Defining the implications and shape of a regional cultural concept has quickly become a marketable skill among cultural scientists and social debaters on the continent. In the Nordic countries it is principally the Norwegians and Danes who have profiled themselves in the debate while the Swedes have preferred to bide their time.

Different messages waft through the air. At the side of the cultural economists, who try to fit the world's art into their economic calculations, there are the anthropologists who are hoping for a revival for folk culture. Through the haze of debate sweep the chilly winds of practical politics with a reminder that the deeper undercurrents of development must also be included. In the general turbulence it is easy to forget that which constitutes the quiet prerequisite for the discussion: the creative culture and the circle of working artists. The silence from the center of the debate arouses a feeling of vague anxiety that culture itself is being crushed in the mill of debate and that the cultural assets go up in smoke in the hands of the developmental powers.

A conference will be arranged in Örnsköldsvik on June 26, 1992, with the aim of illuminating the many issues which surround the regional cultural concept. Its caption, which appears on the flyleaf, will be "Culture and the Living Region". The Cultural Affairs Committee of Örnsköldsvik is arranging the event in collaboration with the University Group for Humanist Futurology at Umeå. The conference will primarily be of interest to cultural workers in the district of Västernorrland and to cultural administrators on the municipal and regional level. A number of radical debaters from Sweden, Norway and Denmark have been invited as lecturers. We look forward with particular expectations to the contribution of Michel Bassand, a Professor of Sociology from Lausanne, who has recently concluded a project for the Council of Europe with the name "Culture and Regions".

The purpose of the conference in Örnsköldsvik is, however, not only to give an account of the current regional culture debate. Hope is that it will also be able to leave a creative contribution. Particular attention will be devoted to the question of a regional model of aesthetic culture, a theme often treated stepmotherly by the debaters. To what extent are the classical models relevant in a regional culture of the twenty-first century and to what degree are radical new formulations called for? A new theoretical projection, introduced by the Group for Humanist Futurology has inspired an alternative model for regional cultural policy. Working teams from the district will give an account of proposals for a practical application of the radical outline.

Conference Programme III

SOCIAL IDENTITY AND LIFE-REGION

A R&D Conference in Örnsköldsvik, Friday 11th June 1993

10.00 Opening of the conference by municipal representative.

MORNING:

Social environment and global sustainability

10.05 The Vision of an Ecological Health.
Ilona Kickbusch, Dr. Dir. WHO, Copenhagen.

10.50 Ecological Health in Social Practice.
Stig Ernestål, Dir. of Social Services, Örnsköldsvik.

11.20 Strategies for a Radical Public Health Policy. Panel Debate.
Roland Andersson, Head of Department Västernorrland County Council.
Ilona Kickbusch, WHO Copenhagen.
Birgit Mollberg, Dir. of Social Services,
Västernorrland County Administration.
Per Råberg, Umeå University.
Stig Ernestål, Social Services Örnsköldsvik.

12.30 LUNCH

AFTERNOON:

Parameters of social development

13.45 Social Mobilization and Local Development.
Alf Ronnby, Ass.Prof. Östersund College.

14.30 Global Communication - the Social Basis.
Lars Qvortrup, Lecturer Odense University.

15.30 The Regional Dimension of Social Life.
Per Råberg, Assoc. Prof. Umeå University.

16.30 DINNER

EVENING:

Social alternatives in regional practice

17.45 The Social History of Västernorrland County.
Birger Stark, Former Social Dir. Sundsvall.

18.15 Social Mobilization. Reports from Regional Groups.

19.45 Concluding Debate.

THE REGION AS A SOCIAL LIFE SPACE

In times of profound social crisis there always used to be an innermost line of defence to rely on: the social networks. When the imperial edifice of power collapsed, it was their invisible meshes that caught the individual and held the collective together. Society was spared disintegration and could be reconstructed through the patient efforts of local forces. The social system of the civil community was the backbone of the regions and nations.

The situation today is different. As a result of the transformatory process of modernization the social networks have been torn into pieces in most parts of society. This loss has coincided with structural changes which most people have long regarded as a welfare gain. The transmission of individual experience has been superseded by pedagogics, communication by information, personal care by professional support, mutual assistance by social services. But social rationalization has exacted a heavy price: helplessness and desperation on the part of many individuals, and the alienation of large groups from society. We are all closely familiar with the symptoms in the form of social rejection, psychosomatic disturbances, violence and crime.

During decades of material growth we have dismissed social dissolution as a problem relating to the individual. But in today's profound crisis we are inexorably confronted with the structural effects of the dismantling process. When the public security system beats a retreat, there are no longer any informal networks to rescue us. Equally alarming is the growth of social anarchy. All over the world we are experiencing a growth of acts of warfare between ethnic groups and regions. This is a socio-pathological process which no longer appears to be politically controllable.

To overcome the deep social crisis, there will have to be a radical change of attitude on the part of politicians, professional groups and those most immediately affected: the citizens. New social visions will have to be outlined which can repair the broken networks and at the same time successfully inscribe them in the new world order which is dawning. Above all, it is important that we identify the arenas of sustainable social living.

In a vision of sustainable development, a major focus is on the provincial regions. This is a circumstance which the regions can turn to their advantage. Many regions are seriously threatened by the convulsions which plague ageing modern society. In the future, presumably, social sustainability will, even more than at present, be a precondition of regional development and economic prosperity.

This is why the Municipality of Örnsköldsvik and futurology researchers at Umeå University have decided to join forces in the R&D conference SOCIAL IDENTITY AND LIFE-REGION. In a dialogue between social researchers and regional players we will scrutinize strategies for reinforcing the social networks of the region. Experts on social mobilization, communication theory, social medicine and human ecology have been invited to illuminate the theme from their vantage points.

A panel debate is being arranged in order to discuss a scheme which is likely to be a foundation of radical social debate for a long time to come: The WHO agenda for ecological public health. In addition, ideas for socially upgrading the County of Västernorrland, formulated by local study groups, will be presented at a special session. The entire spectrum of social issues is taken into account, ranging from neighbourhood care to leisure culture and the women's movement.

Hopefully this project will generate ideas which can be applied in future social reform work in the county's municipalities. But we make no bones about the fact that the conference also constitutes a contribution to the discussion of a sustainable global development, an issue recently highlighted by the 1992 World Conference in Rio de Janeiro.

ACTA UNIVERSITATIS UMENSIS

1 **Bertil Malmbergs skolpojksbrev.** Utgivna av Sven Vallmark. 1975.
2 **Saga-Marianne Norlén & Teut Wallner.** Skriftatlas för datering av Birger Sjöbergs manuskript. 1974.
3 **Karin Eriksson.** Studier i Umeå stads byggnadshistoria. Från 1621 till omkring 1895. Diss. 1975.
4 **Sylvia Danielson.** Läroboksspråk. En undersökning av språket i vissa läroböcker för högstadium och gymnasium. Diss. 1975.
5 **Sigurd Fries.** Svenska växtnamn i riksspråk och dialekt. 1975.
6 **Folke Nordström.** Mediaeval baptismal fonts. An iconographical study. 1984.
7 **Sven-Bertil Jansson.** Folket, de bildade och landsmålslitteraturen. Svensk dialektdikt 1875–1900. 1975.
8 **Jan Nilsson.** Plurala ortnamn på Island. Morfologiska iakttagelser. Diss. 1975.
9 **Sylvia Danielson.** Samuel Columbus' språkprogram i En Swensk Ordeskötsel. 1976.
10 **Birger Liljestrand.** Strindbergs Mäster Olof-dramer. En studie i 1800-talets dramaspråk. I. Diss. 1976.
11 **Gudrun Utterström.** Tillnamn i den karolinska tidens Stockholm. 1976.
12 **Dialectology and Sociolinguistics.** Essays in honor of Karl-Hampus Dahlstedt 19 April 1977. Edited by C.-C. Elert, S. Eliasson, S. Fries and S. Ureland. 1977.
13 **Ann-Maria Ullsten.** Umeå-pressen 1917–1922. En studie i lokalpress. Diss. 1977.
14 **Gustaf Utterström.** Fattig och föräldralös i Stockholm på 1600- och 1700-talen. 1978.
15 **Gunnar Eriksson.** Kartläggarna. Naturvetenskapens tillväxt och tillämpningar i det industriella genombrottets Sverige 1870–1914. 1978.
16 **Iréne Johansson.** Funktionella aspekter på satsintonationen i svenska. Diss. 1978.
17 **Inger Nilsson.** Grönlandsfrågan 1929–1933. En studie i småstatsimperialism. Diss. 1978.
18 **Tom Ericsson.** I fosterlandets tjänst. En studie i den svenska arméunderbefälskårens fackliga strävanden 1901–1922. Diss. 1978.
19 **Brit Uppman.** Samhället och samerna 1870–1925. Diss. 1978.
20 **Teater i Göteborg 1910–1975.** I–III. 1978.
21 **Dag Nordmark.** Samhället och scenen. En studie i Rudolf Värnlunds drama *Den heliga familjen*, dess litterära och sociala förutsättningar. Diss. 1978.
22 **Björn Jernudd.** The Language Survey of Sudan. The First Phase: A Questionnaire Survey in Schools. Diss. 1979.
23 **Ulf Bäcklund.** Restricted Adjective-Noun Collocations in English. 1981.
24 **Nord-Skandinaviens historia i tvärvetenskaplig belysning.** Utg. av Evert Baudou och Karl-Hampus Dahlstedt. 1980.
25 **Rolf Hedquist.** Emotivt språk: en studie i dagstidningars ledare. Diss. 1978.
26 **Monica Loeb.** Vonnegut's Duty-Dance with Death-Theme and Structure in Slaughterhouse-Five. Diss. 1979.

27 **Algot Hellbom.** Ordbok över Njurundamålet. 1980.

28 **Kerstin Elert.** Portraits of Women in Selected Novels by Virginia Woolf and E. M. Forster. Diss. 1979.

29 **Veronica Bonebrake.** Historical Labial-Velar Changes in Germanic. Diss. 1979.

30 **Birger Liljestrand.** Strindbergs Mäster Olof-dramer. En studie i 1800-talets dramaspråk. II. 1980.

31 **Maria Bolander.** Predikativens funktion i svenskan. Diss. 1980.

32 **Konstkultur i Norrland.** En informationsskrift om forskning vid Institutionen för konstvetenskap. 1980.

33 **Internordisk språkförståelse.** Utg. av C.-C. Elert. 1981.

34 **Magnus von Platen.** Privatinformation i skolan. En undervisningshistorisk studie. 1981.

35 **Marja Sjöberg.** Brott och straff i Västernorrland 1861–90. Diss. 1981.

36 **Tvåspråkighet.** Utg. av E. Ejerhed och I. Henrysson. 1981.

37 **Svenskans beskrivning 12.** Utg. av C.-C. Elert och S. Fries. 1981.

38 **Ingeborg Nordin Hennel.** Dömd och Glömd. En studie i Alfhild Agrells liv och dikt. 1981.

39 **Kjell Söderberg.** Den första massutvandringen. Diss. 1981.

40 **Claes-Christian Elert.** Ljud och ord i svenskan. 2. 1981.

41 **Rolf Hedquist.** Språklig kommunikation med förbehåll. En studie i det politiska språkets pragmatik. 1981.

42 **Bo Sundin.** Ingenjörsvetenskapens tidevarv. Ingenjörsvetenskapsakademin, Pappersmassekontoret, Metallografiska institutet och den teknologiska forskningen i början av 1900-talet. Diss. 1981.

43 **Readings on Unbounded Dependencies in Scandinavian Languages.** Edited by Elisabet Engdahl and Eva Ejerhed. 1982.

44 **Inger Hultgren.** Kvinnors organisation och samhällets beslutsprocess. Fredrika-Bremerförbundet och Husmodersförbundet Hem och Samhälle på riksplanet och lokalt i Jämtlands län 1925–1975. Diss. 1982.

45 **Teater i Stockholm 1910–1970.** I–III. 1982.

46 **Roger Qvarsell.** Ordning och behandling. Psykiatri och sinnessjukvård i Sverige under 1800-talets första hälft. Diss. 1982.

47 **Suppression, Struggle and Success.** Studies on Three Representatives of Cultural Life in Sweden: Fredrika Bremer, Andreas Kempe and Linnaeus. Edited by Claes-Christian Elert and Gunnar Eriksson. 1982.

48 **Åke Jonsson.** Den omsorgsfulle ordmålaren. Studier i Sven Jerrings radiospråk mot bakgrund av radions allmänna syn på språket under de första decennierna. Diss. 1982.

49 **Nordsvenska.** Språkdrag i övre Norrlands tätorter. Utg. av C.-C. Elert och S. Fries. 1982.

50 **Per Frånberg.** Umeåsystemet. En studie i alternativ nykterhetspolitik 1915–1945. Diss. 1983.

51 **Maria Haar.** The Phenomenon of the Grotesque in Modern Southern Fiction. Some Aspects of Its Form and Function. Diss. 1983.

52 **Tage Boström.** Okänd soldat och kända soldater. Beteenden, attityder och struktur i Väinö Linnas krigsroman. 1983.

53 **Birgit Petersson.** "Den farliga underklassen." Studier i fattigdom och brottslighet i 1800-talets Sverige. Diss. 1983.

54 **Birger Liljestrand.** Tal i prosa. Om svenska författares anföringsteknik. 1983.

55 **Taisto Määttä.** Hur finskspråkiga uppfattar svenskans vokaler. En studie i kontrastiv fonetik med användning av naturligt och syntetiskt tal. 1983.

56 **Sverker R. Ek & Lennart Forslund.** Marionetten som avantgardist. Kring Marionetteatern i Stockholm. 1984.

57 **Från barnatid till åldrig dag.** Minnesanteckningar av Sophie Björkman. Utgivna av Christina Nordström. 1984.

58 **Ingeborg Nordin Hennel.** "Ämnar kanske fröken publicera något?" Kvinnligt och manligt i 1880-talets novellistik. 1984.

59 **Nordic Prosody III.** Papers from a Symposium. Ed. by C.-C. Elert, I. Johansson & E. Strangert. 1984.

60 **From Sounds to Words.** Essays in honor of Claes-Christian Elert 23 december 1983. 1983.

61 **Florilegium Nordicum.** En bukett nordiska språk- och namnstudier tillägnade Sigurd Fries den 22 april 1984. 1984.

62 **Anders Brändström.** "De kärlekslösa mödrarna." Spädbarnsdödligheten i Sverige under 1800-talet med särskild hänsyn till Nedertorneå. Diss. 1984.

63 **Ulf Bäcklund.** Realize and Recognize. A Study in Immediate Contextual Dependency. 1984.

64 **Birgitta Hene.** "Den dyrkade Lasse och stackars lilla Lotta." En syntaktisk-semantisk studie av personbeskrivande adjektiv och adverb i populära ungdomsböcker. Diss. 1984.

65 **Gull-Mari Rosén.** Konst- och konsthantverkscirklar under 1970-talet. De fyra största studieförbundens syn på denna verksamhet med anledning av den nya kulturpolitiken. Diss. 1984.

66 **Claes Rosenqvist.** Hem till historien. August Strindberg, sekelskiftet och "Gustaf Adolf". Diss. 1984.

67 **Elina Helander.** Om trespråkighet – en undersökning av språkvalet hos samerna i Övre Soppero. Diss. 1984.

68 **Karl Johan Danell.** Om franska munkars nöjesliv på medeltiden, om George Brassens' gorilla, om hertig d'Harcourt och grönkål samt andra kåserier kring franska fenomen och frågetecken. 1985.

69 **Eva Strangert.** Swedish speech rhythm in a cross-language perspective. Diss. 1985.

70 **Texten och läsaren i ett historiskt perspektiv.** Föredrag vid nordiskt symposium i idé- och litteraturhistoria. Utg. Raoul Granqvist. 1985.

71 **Hugo Olsson.** La concurrence entre *il, ce* et *cela (ça)* comme sujet d'expressions impersonnelles en français contemporain. 1985.

72 **Inga-Britt Lindblad.** Lokalradiospråk. En studie av tre lokalradiostationers sändningar. Diss. 1985.

73 **Inger Fredriksson.** Studier i Mikael Agricolas bibliska företal. Diss. 1985.

74 **Agneta Guillemot.** Rask Resolut Trogen. De indelta soldaterna i det svenska agrarsamhället. Diss. 1986.

75 **Ulf Magnusson & Gunnar Persson.** Facets, Phases and Foci. Studies in Lexical Relations in English. 1986.

76 **Kerstin Laitinen.** Begärets irrvägar. Existentiell tematik i Stig Dagermans texter. Diss. 1986.

77 **Gunilla Domellöf.** I oss är en mångfald levande. Karin Boye som kritiker och prosamodernist. Diss. 1986.

78 **Jan Nilsson.** Relation och namn. Två kapitel om nordiska prepositionsnamn och liknande ortnamnskonstruktioner. 1986.

79 **Ingrid Melander.** The Transcendent Flame. Thematik/Structural Complexity in Francis Warner's Experimental Sonnets. 1987.

80 **Lennart Pettersson.** Verbböjningar i jukkasjärvifinskan. Diss. 1987.

81 **Inger Bergström.** Grammatical Correctness and Communicative Ability. A Performance Analysis of the Written and Spoken English of Swedish Learners. Diss. 1987.

82 **Christina Florin.** Kampen om katedern. Feminiserings- och professionaliseringsprocessen inom den svenska folkskolans lärarkår 1860–1906. Diss. 1987.

83 **Per Råberg.** The Space of Man. New Concepts for Social and Humanist Planning. 1987.

84 **Per Ringby.** Författarens dröm på scenen. Harald Molanders regi och författarskap. Diss. 1987.

85 **Jörg Lindner.** Den svenska Tysklands-hjälpen 1945–1954. Diss. 1988.

86 **Tom Ericsson.** Mellan kapital och arbete. Småborgerligheten i Sverige 1850–1914. 1988.

87 **Axel Groundstroem.** Finnische Kasusstudien. 1988.

88 **Eva Vikström.** Sverigespel. Kent Anderssons och Bengt Bratts dramatik 1967–71 mot bakgrund av samhälls- och teaterdebatten. Diss. 1989.

89 **Carl Eber Olivestam.** Idé och politik III. De politiska partierna – skolan och ideologin. 1989.

90 **Instead of Flowers.** Papers in Honour of Mats Rydén on the Occasion of his Sixtieth Birthday, August 17, 1989. 1989.

91 **Älvdal i norr.** Människor och resurser i Luledalen 1300–1800. Sune Åkerman och Kjell Lundholm (red). 1990.

92 **Gunnar Persson.** Meanings, Models and Metaphors. A Study in Lexical Semantics in English. 1990.

93 **Anders Öhman.** Äventyrets tid. Den sociala äventyrsromanen i Sverige 1841–1859. Diss. 1990.

94 **Brigitte Kusche.** Frauenaufklärung im Spätmittelalter. Eine philologisch-medizinhistorische Untersuchung des gynäkologisch-obstetrischen Gks 1657, Kopenhagen. Diss. 1990.

95 **Lena Kjersén Edman.** I ungdomsrevoltens tid. Svensk ungdomsbok och dess mottagande åren kring 1968. Diss. 1990.

96 **Börje Harnesk.** Legofolk. Drängar, pigor och bönder i 1700- och 1800-talens Sverige. Diss. 1990.

97 **Barbro Nilsson.** Le mot chien n'aboie pas. *Quand tu va chez les femmes.* Un roman autobiographique de Christiane Rochefort analysé à partir de ses œuvres précédentes. Diss. 1990.

98 **Ulla Lidén.** Der grammatische Tigersprung. Studien zu Hermito von Doderers Sprachterminologie. Diss. 1990.

99 **Signs and Signals.** Popular Culture in Africa. Edited by Raoul Granqvist. 1990.

100 **Språket som kulturspegel.** Umeforskare berättar. Red. Karl Johan Danell, Gunnar Persson & Astrid Stedje. 1991.

101 **Alf Arvidsson.** Sågarnas sång. Folkligt musicerande i sågverkssamhället Holmsund 1850–1980. Diss. 1991.

102 **Kent Larsson.** Svenska brev. Några bidag till studiet av en texttyps historia. 1991.
103 **Marianne Liliequist.** Nybyggarbarn. Barnuppfostran bland nybyggare i Frostvikens, Vilhelmina och Tärna socknar 1850–1920. Diss. 1991.
104 **Karl-Hampus Dahlstedt.** Gosshuvudet. Skolpojksminnen från Härnösand, Ådalen och Vilhelmina. 1992.
105 **Gerald Porter.** The English Occupational Song. Diss. 1992.
106 **Tuija Määttä.** Kvantitativa studier av syntaxen i finsk ungdomslitteratur. Diss. 1992.
107 **Lars-Erik Edlund & Gunnar Persson.** Language – The Time Machine. 1992
108 **Papers in Museology I.** 1992
109 **Tryggve Sköld.** Wortstudien. 1992
110 **Lars Johansson.** In the Shadow of Neocolonialism. Diss. 1992
111 **Österreich. Beiträge über Sprache und Literatur.** C. Pankow (Hg.) 1992
112 **La culture dans la langue.** I Söhrman (Red.) 1993
113 **Mikael Svonni.** Samiska skolbarns samiska. Diss. 1993
114 **Per Sörlin.** Trolldoms- och vidskepelseprocesserna i Göta hovrätt 1635–1754. Diss. 1993
115 **Eva Lambertsson Björk.** Campus Clowns and the Canon. Diss. 1993
116 **Alexander Pankow.** Prinzipien der literarischen Übersetzungsanalyse. Zu Übersetzungen russischer Lyrik ins Deutsche. Diss. 1993.
117 **Svante Norrhem.** Uppkomlingarna. Kanslitjänstemännen i 1600-talets Sverige och Europa. Diss. 1993.
118 **Sigurd Fries.** Växtnamn då och nu. Artiklar sammanställda till Sigurd Fries sjuttioårsdag den 22 april 1994.
119 **Sprache als lebendiger Kulturspiegel.** M. Todtenhaupt & I. Valfridsson (Hg.). 1994.
120 **Magnus Nordén.** Logische Beziehungskonzepte und Inferenzprozeduren. Diss. 1994.
121 **Seyyed Mohammad Fazlhashemi.** Förändring och kontinuitet, Al-Ghazâlîs politiska omsvängning. Diss. 1994
122 **Per Nordahl.** Weaving the Ethnic Fabric, Social Networks Among Swedish-American Radicals in Chicago 1890–1940 Diss. 1994
123 **Karl Johan Danell.** Le phénomène de concurrence en français moderne – Réflexions à partir de an-année, jour-journée, paraître-apparaître, et d'autres.
124 **Sune Åkerman and Jack L. Granatstein.** Welfare States in Trouble. 1995
125 **David Bell.** Ardent Propaganda: Miners' Novels and Class Conflict 1929–1939. Diss. 1995
126 **Raija Kangassalo.** Mastering the Question. The Acquisition of Interrogative Clauses by Finnish-Speaking Children. Diss. 1995
127 **Henryk Kardela & Gunnar Persson.** New Trends in Semantics and Lexicography. 1995
128 **John A Stotesbury.** Apartheid, Liberalism and Romance. A Critical Investigation of the Writing of Joy Packer. Diss. 1996
130 **Elisabeth Mårald.** In Transit: Aspects of Transculturalism in Janice Kulyk Keefer´s Travels. Diss.1996
131 **Per Råberg.** The Life Region. The Social and Cultural Ecology of Sustainable Development

Distribution:
Swedish Science Press, Box 118, S-751 04 Uppsala Sweden.